Superconductivity

Physics and Applications

Superconductivity
Physics and Applications

Kristian Fossheim and **Asle Sudbø**

The Norwegian University of Science and Technology
Trondheim, Norway

John Wiley & Sons, Ltd

This publication is designed to provide accurate and authoritative information in regard to the subject
matter covered. It is sold on the understanding that the Publisher is not engaged in rendering professional
services. If professional advice or other expert assistance is required, the services of a competent
professional should be sought.

Other Wiley Editorial Offices

John Wiley & Sons Inc., 111 River Street, Hoboken, NJ 07030, USA

Jossey-Bass, 989 Market Street, San Francisco, CA 94103-1741, USA

Wiley-VCH Verlag GmbH, Boschstr. 12, D-69469 Weinheim, Germany

John Wiley & Sons Australia Ltd, 33 Park Road, Milton, Queensland 4064, Australia

John Wiley & Sons (Asia) Pte Ltd, 2 Clementi Loop #02-01, Jin Xing Distripark, Singapore 129809

John Wiley & Sons Canada Ltd, 22 Worcester Road, Etobicoke, Ontario, Canada M9W 1L1

Wiley also publishes its books in a variety of electronic formats. Some content that appears
in print may not be available in electronic books.

Library of Congress Cataloging-in-Publication Data

Fossheim, K. (Kristian)
    Superconductivity : physics and applications / Kristian Fossheim and Asle Sudbo.
       p. cm.
    Includes bibliographical references and index.
    ISBN 0-470-84452-3 (alk. paper)
  1.   Superconductivity.   I. Sudbo, Asle. II. Title.
    QC611.92.F67 2004
    537.6′23 – dc22

                                              2004002271

British Library Cataloguing in Publication Data

A catalogue record for this book is available from the British Library

ISBN 0-470-84452-3

Typeset in 10.5/13pt Times by Laserwords Private Limited, Chennai, India
Printed and bound in Great Britain by Biddles Ltd, King's Lynn
This book is printed on acid-free paper responsibly manufactured from sustainable forestry
in which at least two trees are planted for each one used for paper production.

Contents

Preface

Writing this textbook was motivated by the opinion of the authors that the time has come for an updated look at the basics of superconductivity in the aftermath of progress during the last couple of decades, both through the discoveries of new superconductors, and the ensuing theoretical development. High-T_c superconductor research since 1986 represents an almost unlimited source of information about superconductivity. This is an advantage in the sense that there is ample material with which to fill new books, but a disadvantage in the sense that only a very small fraction of all the efforts that were made, and the results that came out, can be discussed here. In this sense the situation is entirely new: The older texts, like those of de Gennes and Tinkham could discuss or refer to almost all aspects of superconductivity of importance in the 1960s and 1970s. With tens of thousands of papers published after 1986, there is no possibility to take such an approach any more. We apologize to the numerous researchers in the field whose work we could not mention. This situation leaves it even more to the taste of the authors to choose. First and foremost we have wanted to review the basics of superconductivity to new students in the field. Secondly, we wanted to allow those who take a serious interest in the subject at the PhD level, to follow the ideas to old heights like in the BCS theory, or to new heights like in the theory of the vortex system in high-T_c cuprates. Superconductivity is now a far richer subject thanks to the discovery of high-T_c cuprates by Bednorz and Müller. Suddenly, superconductivity became an arena for the study of critical behaviour in three-dimensional superconductors, an unthinkable situation in the low-T_c era. Our book seeks, among other things, to clarify this new aspect of superconductivity. In addition, we wanted to respect the wish of students to learn where physics meets the real life of applications. We have concentrated the material here to the central topics, basically how to describe and exploit the properties of Josephson junctions on the small scale, and on the large scale to give some insight into the makings of wires and cables. A special feature of this book is the inclusion of a chapter containing Topical Contributions from distinguished scientists in various areas of superconductivity research and development, from the smallest to the largest scale. Each of these scientists were invited to contribute their leading edge knowledge to give a clear idea of the state of the art in several important sub-fields as of September 2003 when the writing of this book came to a conclusion.

Kristian Fossheim
National High Magnetic Field Laboratory, Tallahassee,
Florida and The Norwegian University of Science
and Technology Trondheim, Norway
Asle Sudbø
The Norwegian University of Science and Technology
Trondheim, Norway

Acknowledgements

From *Kristian Fossheim*: My part of the writing was done in about equal parts during a sabbatical at the National High Magnetic Field Laboratory (NHMFL) at Florida State University, in Tallahassee, Florida during the academic year 2000/2001, and upon my return to Trondheim, Norway, at the Physics Department of the Norwegian University of Science and Technology (NTNU). I greatfully acknowledge the warm hospitality of the NHMFL, and in particular Jack Crow, Director, and Hans Schneider-Muntau, Deputy Director of the NHMFL. I also wish to express my thanks to my wife Elsa for her patience and her willingness to type, from day to day, my handwritten notes during our stay in Tallahassee. I also thank the Physics Department of NTNU, Trondheim for relieving me of some of my teaching duties during one semester of writing. In this book we also wanted to make Nobel laureates in the field more visible to the students, as human beings, and not just as scientific references. I therefore visited ten laureates, videotaping interviews with them. I want to thank them for the friendliness with which they all met me. It gave me some of the best moments of my scientific life to meet these great scientists and fine individuals in their own settings, and hear their own story told. In the Historical Notes, Chapter 14 a few and brief references to these encounters are mentioned. The rest of the material will have to be presented elsewhere. I also want to thank Ulrik Thisted, my PhD student who undertook the task of weeding out the mistakes in my drafts, and Torbjørn Hergum for various assistance, in particular for drawing the major part of the figures.

I dedicate this book to the memory of my parents, Nikolina and Øyvind Fossheim.

From *Asle Sudbø*: I thank the Department of Physics at the California Institute of Technology, and in particular Professor Nai-Chang Yeh, for warm hospitality during the initial stages of this project in the autumn of 2000, which I had the privilege of spending at CalTech on sabbatical leave. It gave me an opportunity to try out part of the material in this textbook on some of the brightest students one can possibly have. I also thank Professor Hagen Kleinert and the Department of Theoretical Physics at the Freie Universität Berlin for their hospitality during 4 months of 2001, where part of the writing was completed. I also want to thank Dr Anh Kiet Nguyen, Dr Joakim Hove, and Dr Sjur Mo for several helpful suggestions and for making some of the material in their PhD theses available to me for this book. Having them as graduate students was truly a pleasure and a privilege. Finally, I want to express my sincere thanks to Professor Zlatko Tesanovic, for the innumerable and inspiring discussions I have had with him over the years on the physics of strongly correlated systems.

I dedicate this book to the memory of my good friend Anthony Houghton.

The authors would like to extend their special thanks to scientific colleagues in the US, Europe and Japan for their topical contributions in Chapter 13: D. K. Christen, Ø. Fischer, T.H. Johansen, K. Kadowaki, Y. Maeno, J. Mannhart, M. Murakami, M. Muralidhar, N. P. Ong, H. Schneider-Muntau and J. R. Thompson.

PART I

Basic Topics

1

What is Superconductivity? A Brief Overview

1.1 Some introductory, historical remarks

Progress in science in general, and in physics in particular, is characterized by many great discoveries both in theory and experiment. On superficial observation these sudden advances may look like a series of accidental occurrences. But in most cases such an impression would be wrong, in many cases even totally misleading. Very often discoveries occur when the time is ripe, i.e. when certain preconditions have been met, such as adequate technical capability, or attainment of a necessary intellectual and theoretical level. The discovery of superconductivity – the property of certain conductors to display zero DC electrical resistance – by Heike Kamerlingh Onnes and co-workers in 1911 illustrates the point [1]. It was a true discovery, and indeed a remarkable one, because there were no valid arguments around to predict such a phenomenon. Yet, to draw the conclusion that it was accidental would be unjustified. The necessary technical basis and opportunity for the discovery had been solidly established in the same group by the liquefaction of the inert gas helium in 1908. And research, both experimental and theoretical, on the electrical conductivity of metals at temperatures approaching the absolute zero, was ongoing and considered an important issue by leading physicists. They were even puzzled by an apparent similarity between the temperature dependences of electrical conductivity and heat capacity in metals. On travelling down the infinite road towards zero degrees Kelvin, new discoveries could and should be expected, since temperature is the most universal variable by which the (equilibrium) state of matter is defined. In fact, the absence of discoveries along this route should have been the real surprise. Even the cryogen itself, liquid helium, was later to offer a number of opportunities for important discovery. And many more were to come in other condensed matter systems below ambient temperatures.

Superconductivity: Physics and Applications   Kristian Fossheim and Asle Sudbø
© 2004 John Wiley & Sons, Ltd   ISBN 0-470-84452-3

Even the choice by Onnes of the metal to study, mercury, while appearing perhaps surprising today, was a judicial and natural one at that time: It was the metal that offered the most ideal conditions for the study of intrinsic properties of metals at low temperatures due to the purity at which it could be obtained. Purity was indeed a valid consideration, and would be even today, when studying electrical conductivity versus temperature. Research in the Leiden laboratory was carried out by coworkers and students of Onnes in a strictly systematic manner. Metal upon metal was measured, and the results contemplated. Gold was found to have almost immeasurably low resistance in the liquid helium range; but mercury became the first substance found to be superconducting, at a temperature near 4 K. The distinctive feature was a sudden drop of resistance by several orders of magnitude on lowering the temperature below what appeared to be a sharply defined temperature. Lead and tin were soon added to the list. Onnes dismissed the prevailing idea that electrons would 'freeze to the atom', and surmised instead that 'the free electrons would remain free' while the 'vibrators' (atoms) would become 'practically immovable'[1]. The timeliness of their research is also illustrated by the fact that thermal properties were much in focus, thanks to the blackbody problem, the early quantum theory of Planck, and the attempts by Einstein and others to explain low-temperature heat capacity, as well as by ongoing research on thermal and electrical conductivity of solids at low-temperatures. The electron, a relatively recent discovery, had brought electrical phenomena to the forefront of research. The fact that the atomic nucleus was discovered that same year, 1911, makes this a truly remarkable year in the history of scientific discovery.

The Leiden group attached considerable hope to the technological potential of superconducting coils for generation of magnetic field, foreseeing fields as high as 10 T. But they soon found an unexpected obstacle: An upper limit for current that would flow at zero resistance in lead or tin superconductors in the self field of the coil, what we today call the critical current I_c. This difficulty could not be circumvented until decades later, when it was realized, first through theoretical work, that a different kind of superconductor was needed. This new class of superconductors was to be labeled 'type II' as opposed to type I for Sn and Pb and similar conductors which had been studied in Leiden. Only upon discovering, understanding, and further developing type II materials could critical current density be raised to much higher, practical values. From the 1960s on a gradual development of superconductor technology has taken place, until today when superconducting magnets are commonplace in laboratories and hospitals all over the industrialized world. And the remarkable SQUID technology developed for measurement of small magnetic fields, has taken on a wide range of applications, and is showing promise in ever new fields by use of both old and new superconductors.

---

[1]For an account of early work on superconductivity, see Dahl [2].

Other great discoveries were to follow. The magnetic properties of super-conductors attracted considerable attention through the 1920s and 1930s. A breakthrough came in 1933 when Meissner and Ochsenfeld [3] showed that in magnetic fields below a certain threshold value the flux inside the superconductor was expelled, and that this defined a new thermodynamic state and was not a consequence of infinite conductivity. The phenomenon became known as the Meissner effect, and laid the foundation for a thermodynamic treatment of superconductivity, later to be expanded when Abrikosov [4, 5] was finally permitted to report his theory of the magnetic properties of superconductors at a meeting in Moscow in 1957. That same year Schubnikow's [6] important experimental work from the 1930s could also finally be recognized publicly and posthumously in the Soviet Union. This work had already established important magnetic properties of type II superconductors two decades earlier, and had shown promise for superconductors to carry larger current densities than materials previously studied by Onnes, and by Meissner and others.

1957 was also the year when the beautiful quantum theory, known as the BCS theory, was published by Bardeen, Cooper and Schrieffer [7], finally explaining the fascinating properties of superconductors from first principles. It had taken 46 years from the time of the discovery. A few years later Josephson's [8] astonishing predictions regarding the physical properties of inhomogeneous superconductors were announced, soon followed by experimental verifications, and by a variety of applications as well as further theoretical development.

It would take another 25 years before superconductivity made the transition from being mainly an exotic laboratory phenomenon, known mostly to physicists, to become almost a household word. That development was to take place by the discovery of a new class of superconductors, known as high-T_c cuprates, by Bednorz and Müller [9] at the IBM laboratory in Rüschlikon near Zürich in 1986. This is an example of a great discovery made on the basis of systematic, goal oriented research. From then on the world outside of physics became fully aware of the almost magical properties of superconductors. The discovery started a race towards applications that is still ongoing, with great economic prospects. Equally intense was the desire to understand the mechanism. At the time of writing this book, that goal has not yet been reached. In the meantime, research has benefited greatly from the famous semi-phenomenological theories of the London brothers from the 1930s, and of Ginzburg and Landau [10] published in 1950, in particular.

It has been estimated that well over 50 000 scientific papers have been published on superconductivity since the high-T_c discovery in 1986. A single book can only give a modest insight into this vast amount of knowledge and results. The present one aims to shed light on what the authors regard as the most central issues, with special emphasis on developments taken place during the last 15–20 years.

1.2 Resistivity

Two fundamentally important and intuitively startling properties are associated with superconductivity:

- The transition from finite resistivity, ρ_n in the normal state above a super-conducting transition temperature T_c, to $\rho = 0$, i.e. perfect DC conductivity, $\sigma = \infty$, below T_c.

- The simultaneous change of magnetic susceptibility χ from a small positive paramagnetic value above T_c to $\chi = -1$, i.e. perfect diamagnetism below T_c.

These aspects are qualitatively illustrated in Figure 1.1. Actual measurements of resistivity as a function of temperature in a high-temperature superconducting material are shown in Figure 1.2. We shall return to the second of the above statements in Section 1.3. Let us now discuss the implications of the first statement.

The traditional way to measure resistance is by balancing a Wheatstone bridge, using the sample as the fourth and unknown resistance. When balanced, the precision of the measurement is as good as that of the resistors used in the circuit, or possibly limited by the zero-readout accuracy of the galvanometer used in balancing the bridge. There is a limit to how low resistance one can claim to measure: Zero resistance is not accessible to direct measurement. By standard methods we can measure a reduction in resistance by many orders of magnitude, *but we can never measure zero resistance in a strict sense*. If we try to measure the inverse quantity, i.e. conductivity $\sigma = 1/\rho$, the impossibility

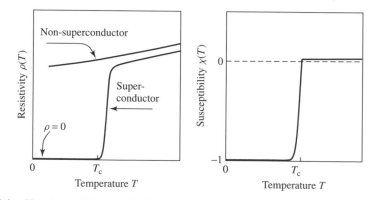

Figure 1.1 Sketches of the two basic characteristics of a superconductor: The left figure shows the drop to zero resistivity, $\rho = 0$, at a temperature T_c, compared to a non-superconducting behavior. The right hand figure shows the corresponding drop in susceptibility to the ideal diamagnetic value of $\chi = -1$ below T_c. The onset of the diamagnetic response corresponds quite closely to the point where $\rho \to 0$ on the temperature axis. The figure also indicates that χ is positive but quite small above T_c.

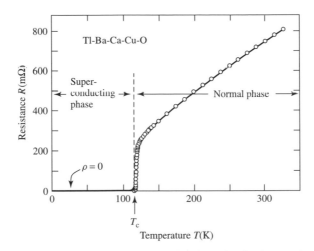

Figure 1.2 Experimental data on the resistance of a Tl-Ba-Ca-Cu-O ceramic sample from the early days (1988) of high-T_c discovery (unpublished data from the Trondheim group). As was often the case during the early days the material was not of precisely determined composition. The value of T_c indicates that the material was close to $Tl_2Ba_2Ca_2Cu_3O_{10}$ composition.

remains. No method is available to measure an infinite quantity. For the discoverers of superconductivity this was a real dilemma. Kamerlingh Onnes held on to the belief that a 'micro-resistance' remained below T_c.

It is, however, still possible to arrive at the reasonable conclusion $\rho \to 0$, or equivalently $\sigma \to \infty$, by inference from real measurements. Experiments have been devised for this purpose. In such experiments the magnetic field associated with an induced current has been found to remain constant during a time span as long as 1 year. This allows an estimate to be made of the lower bound of the decay constant, and of the upper bound of the resistance. The resulting analysis of such a measurement leads to the conclusion that the lower bound on the decay constant τ for the current in the superconductor is of the order of 100 000 years, implying that the total time for the current to die out completely would be millions of years.

Let us see how such an estimate can be made. To be specific, we assume that a current I_0 is set up by induction in a superconducting closed loop at time $t = 0$, at a temperature below T_c. Next, the associated magnetic field is monitored over a very long period of time over some area in the loop, as illustrated in Figure 1.3.

As long as a resistance R exists in the superconducting loop, with inductance L in the presence of a current $I = I(t)$, conservation of energy requires

$$\frac{d}{dt}\left(\frac{1}{2}LI^2\right) + RI^2 = 0 \tag{1.1}$$

with the well-known solution

$$I(t) = I_0 e^{-(R/L)t} \tag{1.2}$$

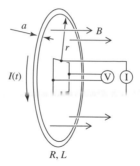

Figure 1.3 Sketch of a simple setup to monitor the possible decay of current $I(t)$ via its associated magnetic induction $B(t)$ in a closed loop of inductance L and resistance R. The wire has a diameter a and a loop radius r. The field B is normal to the loop area. The magnetic sensor is a Hall probe.

where I_0 was the initial current at time $t = 0$. The decay time of the current, as well as of the corresponding magnetic induction B, is $\tau = L/R$. Now, if the B-field surrounding the loop is measured to be the same after a time t_1, say 1 year later, it appears to mean that $I(t_1 = 1 \text{ year}) = I(t = 0) = I_0$, and similarly for the corresponding B-field. However, this statement is not an exact one. The B-field we measure, resulting from the current $I(t)$, can only be measured with a certain accuracy, determined by the instruments used. If we assume that the change of B-field can be detected with a relative instrumental resolution $\delta B/B_0 = 10^{-5}$, an observation of 'no change' in reality only sets an *upper bound* on the possible decay δI of current and of field δB that may have occurred. What the measurement can determine is therefore only the maximum amounts by which the current I and the field B may have decayed. In reality the decay may have been much lower. Hence we can estimate

$$\delta I > I_0 - I(t_1) = I_0(1 - e^{-(R/L)t_1}) \tag{1.3}$$

and

$$\delta B > B_0 - B(t_1) = B_0(1 - e^{-(R/L)t_1}). \tag{1.4}$$

From this it follows that

$$\frac{\delta I}{I_0} = \frac{\delta B}{B_0} > 1 - e^{-(R/L)t_1} \tag{1.5}$$

Solving for R, we find its upper bound

$$R < -\frac{L}{t_1} \ln\left(1 - \frac{\delta B}{B_0}\right) = -\frac{L}{t_1} \ln\left(1 - \frac{\delta I}{I_0}\right) \tag{1.6}$$

where all quantities on the right can be determined from the experiment.

If we assume that we have used a single-loop inductor with a diameter a and loop radius r, and use the inductance formula for a single loop which applies for $r \gg a$: $L \approx \mu_0 r \ln(r/a)$, we find the expression for the upper bound on R as

$$R < -\frac{\mu_0 r \ln(r/a)}{t_1} \ln\left(1 - \frac{\delta B}{B_0}\right) \tag{1.7}$$

Assume that we have set up the experiment with a single 5-cm radius loop of Al wire, whose radius is 0.5 mm. If the observations went on for a year, we have $t_1 \approx 3 \cdot 10^7$ s. We assume as above that the instrumental resolution allows us to measure $\delta B/B_0 = 10^{-5}$. Using Eq. 1.7 we obtain $R < 10^{-19}\,\Omega$. What resistivity ρ can we expect to observe below the superconducting transition of Al?

Resistance and resistivity are related by $R = \rho\frac{l}{A}$, where l and A are, respectively, the length and the cross section of the wire in the loop. With $R < 10^{-19}\,\Omega$ found before, we obtain:

$$\rho < 10^{-19}\Omega\left(\frac{A}{l}\right) \approx 2.5 \times 10^{-25}\,\Omega\,\text{m} \tag{1.8}$$

We should therefore feel reasonably justified in treating the superconducting state as one of zero resistivity. We have found by our estimate that the resistivity at T_c drops by a factor of about 10^{13} if we take ρ_n in the normal state of high purity Al to be $2.5 \times 10^{-12}\,\Omega$ m at helium temperatures. This also tells us that the resistivity in the superconducting state of aluminium is at least 17 orders of magnitude lower than that of good copper at room temperature, since for copper $\rho_{Cu}^{273K} = 1.56 \times 10^{-8}\,\Omega$ m.

What is the lower bound on the decay time $\tau = L/R$ in the case discussed? Using the above numbers we find $\tau = L/R > 5.8 \times 10^{13}$ s. Because 1 year is close to 3×10^7 s, the lower bound on the decay time is about 2 million years. This is a factor of 10 longer than the observations referred to above, so we were a bit generous. But we have to remember that the real observation was of *no change*. The instruments are the limitation, and we may not have overestimated the actual decay time at all. On the contrary, it might have been found to be much longer had instruments with higher resolution been used. The quest for an even better experimental verification of $\rho = 0$ in the superconducting state could be continued. If we had used a SQUID superconducting detector of the magnetic field (see Chapters 5 and 11), the detection limit for the small changes in magnetic field would be lowered by several orders of magnitude more, and the upper bound on resistance and resistivity might have been lowered correspondingly. The lower bound on the decay time $\tau = L/R$ would then be raised by many orders of magnitude above 10^{13} s that we have found so far. τ might even approach the lifetime of the Universe. This would be in accordance with theoretical arguments in ideal situations. What better measure could we ask for to conclude that zero resistivity is physically possible?

Having satisfied ourselves that $\rho = 0$ is an appropriate statement about the electrical resistivity in a superconductor, we must immediately caution against one particular misinterpretation: We have to keep in mind that this state of affairs only applies to DC resistivity. An applied AC electric field accelerates charge. We cannot avoid continually transferring energy to the electron system as long as we keep accelerating it, i.e. as long as we expose it to an AC electric field. We shall return to this question in detail in Chapter 11, and find that the resistance increases with the square of the frequency, but with a very small prefactor. Consequently, there is no abrupt change from DC to the AC resistivity in the superconducting state as we start increasing the frequency from zero. This loss becomes important only at very high frequencies. Still, because of the fact that in the actual technical construction of superconducting cables, normal metals unavoidably have to be used as part of the whole structure, there are losses, so-called coupling losses in the normal metal surrounding the superconductor even at standard electric grid frequencies, i.e. at 50 Hz or 60 Hz. In addition, losses caused by displacement of supercurrent vortices in a magnetic field are important in such cases, even in DC situations. These questions are treated in Chapters 8 and 12.

That supercurrents in ideal situations can run without observable loss for years, indeed perhaps much more than 10^6 years as we found previously, is one of Nature's most fascinating phenomena. It contradicts all experiences we have from physical systems in the macroscopic world, because it implies frictionless or lossless motion of matter, in this case charged matter, which is displaced relative to the ions. A classic analogue to electronic supercurrent transport in an electrical circuit could be that of sliding a pebble along a flat, icy surface. The pebble soon comes to rest because it spends all its energy working against friction. However, in the superconducting circuit the charged particles (current) continue to run, hence there *is* no friction. It is precisely because this situation is so contrary to all our experience with other real systems that superconductivity is such a startling and counterintuitive phenomenon.

In fact, as we shall see, superconductivity is a pure quantum physics phenomenon on a macroscopic scale. Superconductivity has an additional property compared to the lossless motion of charged particles in atoms, namely phase coherence of the manybody wavefunction. Indeed one of the most remarkable feats of science was the development of a manybody quantum theory that finally explained superconductivity, and with predictive power.

1.3 The Meissner effect: perfect diamagnetism

From the discovery of zero electrical resistivity by Onnes in 1911, 22 years would pass until the next important discovery by Meissner and Ochsenfeld.

They found that the superconducting state possesses a second characteristic, defining property – *perfect diamagnetism*. Prior to this discovery there had been considerable debate regarding the magnetic properties, and the discussion had been based on an interpretation of the superconducting state with perfect conductivity as the (only) basic characteristic. Meissner and Ochsenfeld managed to escape this conceptual trap by observing that a magnetic field, which was applied above T_c would be expelled from the body of the superconductor on cooling below T_c, leading to $B = 0$ inside. Using the constitutive equation for a magnetic body we therefore have to write, in the superconducting state:

$$B = \mu_0(H + M) = 0 \tag{1.9}$$

which means that *in* the superconductor

$$M = -H. \tag{1.10}$$

Therefore, the susceptibility takes on the ideal value for a perfect diamagnet:

$$\chi = \frac{\mathrm{d}M}{\mathrm{d}H} = -1. \tag{1.11}$$

Because permeability is $\kappa = 1 + \chi$ we also have $\kappa = 0$ below T_c. These two ways of expressing the Meissner effect are of course equivalent, but the diamagnetic statement $\chi = -1$ is perhaps physically more 'descriptive' and appealing.

What happens is that shielding currents arise in the superconductor surface. These currents create a field both inside and outside the superconductor such that on the inside the applied and the induced fields exactly cancel, while outside they add. The result is that we observe expulsion of the B-field: zero B-field inside, and an increased field near the sample, on the outside. This is illustrated in Figure 1.4. While in the normal state Faraday lines would pass straight through the metal, practically as if there was only vacuum, in the superconducting Meissner state all lines are forced to pass on the outside. The highest

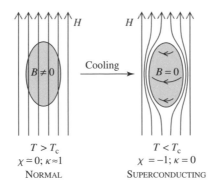

Figure 1.4   The Meissner effect in a superconductor.

density of flux lines is found near the equator, while near the poles there is a lower flux density than before the flux expulsion.

That this situation represents a thermodynamic state is demonstrated by the experimental observation that the state is uniquely defined by the values of the thermodynamic variables, temperature T and applied field H, independent of how that state was reached. It is important in this context to observe that this is not possible if perfect conductivity was the only special property that the superconductor attained on passing below T_c. This can be confirmed by a simple thought experiment, as in the two sequences illustrated in Figure 1.5.

Imagine (i) that a specimen which is originally at room temperature, is cooled in zero applied field to below a temperature T_c where it acquires zero resistance, $\rho = 0$. Now turn on an external magnetic field. Lenz' law requires currents to arise in the specimen so as to screen out the applied field. As $\rho = 0$ this situation will persist forever, and no field will penetrate. Let us call this state I. Next, remove the field. Because the prevailing requirement from Lenz' law is to allow no change of flux in the specimen, the currents in the sample will diminish to zero as the field is turned off. At the end of the sequence the sample is still free of magnetic flux inside. Call this state II. In the next sequence, (ii), start the experiment at room temperature, this time in an applied external field. This field penetrates the sample completely and almost totally undistorted due to the low paramagnetism of non-magnetic metals. Now cool the specimen in the magnetic field to below the same temperature T_c as before. Although the sample, by assumption, acquired $\rho = 0$ on

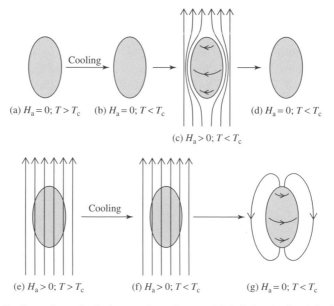

Figure 1.5   Cooling a hypothetical normal conductor with infinite conductivity in zero external field, and in a finite field, and comparing the resulting states of the material. The figure is explained in the text, and important conclusions are drawn from that discussion.

passing through T_c, Lenz' law requires the flux inside to remain perfectly constant. Call this state III. Next, remove the external field. Again Lenz' law requires the flux to stay constant. This can indeed be implemented by a spontaneous current arising on the surface. Since by assumption $\rho = 0$, the situation will persist forever. Call this state IV. Now compare, in particular, state II and state IV. The thermodynamic variables T and H are the same, but the states are quite different as just described. The difference was unavoidable due to the assumption that the only new property that the specimen acquired on passing below T_c was perfect conductivity. However, on exposing real superconductors to precisely the same sequences the superconductor ends up in exactly the same state, namely one of total flux expulsion in both cases, and again the situation is persistent. We have two important conclusions to draw from these observations:

- Superconductivity is more than just $\rho = 0$.

- Superconductivity is a thermodynamic state, contrary to a state characterized by just $\rho = 0$.

What happens in the Meissner effect is that spontaneous currents arise in the surface to exactly cancel the B-field inside. Diamagnetism is well known to exist in all matter, but usually the corresponding susceptibility is very small, typically $\chi = -10^{-6}$ in atomic and molecular diamagnetism. Perfect diamagnetism clearly is a totally different phenomenon. The reason for the large negative value is precisely the fact that currents encompass the entire body, and are not broken up into tiny currents circulating individual atoms. We might add that diamagnetism exists in metals in the normal state too, due to surface currents in a magnetic field. But this effect is more than overcome by the paramagnetism of the conduction electrons.

In the discussions of the superconducting state we have so far assumed that currents and fields employed were below the threshold where superconductivity breaks down. In Chapter 8 we discuss the limits as to how large currents and fields a superconductor can tolerate before being forced to allow magnetic flux to penetrate. As long as the superconductor is at a field below such values, and $\chi = -1$ is maintained, we say that the superconductor is in the Meissner state, or Meissner phase, as shown in Figure 1.6.

These remarks point to the need of defining more precisely the thermodynamic range of superconductivity in the (H, T)-plane. This has been well established for a great number of metals and alloys, as well as high-T_c and other classes of superconductors. We return to this problem in the next section.

1.4   Type I and type II superconductors

Superconducting materials have the ability, as we have seen, to exist either in the normal state or the superconducting state, depending on the external

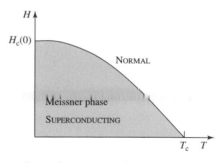

Figure 1.6   The Meissner phase of a superconductor. Under the curve, in the shaded area, the material is perfectly screened against an external magnetic field.

magnetic field they experience. It turns out that if we increase the magnetic field beyond a certain critical value H_c or H_{c1}, which is different for different materials, the Meissner effect breaks down. That is to say, flux penetrates into the material. A sketch is shown in Figure 1.7. This figure shows the negative of the magnetization along the positive vertical axis. The slope of the line from the origin is then exactly 1 as long as the sample is in the Meissner state, corresponding to $\chi = -1$. Let us here first discuss the ideal case which best brings out the underlying physics: Imagine we study a long, thin needle-like specimen, with the applied field parallel to its axis, in which case the demagnetization factor $n = 0$.[2] A pickup coil can be used to measure the amount

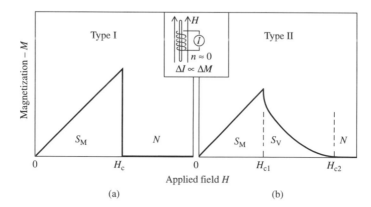

Figure 1.7   (a) shows the typical behaviour of a type I superconductor which switches abruptly from the Meissner state S_M to the normal state N at H_c. (b) shows a different behaviour in a type II superconductor: At a lower critical field H_{c1} the superconductor changes state from the Meissner state with complete screening, S_M, to a state of vortex line penetration S_V, and finally, at H_{c2}, to the normal state N. The insert indicates that the magnetization development can be monitored by a coil around the rod-like sample (demagnetization factor $n \approx 0$) such that the change of current ΔI is proportional to the change of magnetization ΔM.

---

[2]We refer to Section 1.7 for a discussion of demagnetization factors.

of flux which penetrates the sample. Now turn on a magnetic field. At low fields the Meissner state, corresponding to complete screening, is found. But as the field increases further we will discover two distinctly different outcomes, depending on the materials we investigate. This will allow us to sort all known superconductors into two categories as follows:

- Type I: The superconductor switches abruptly over from the Meissner state to one of *full penetration of magnetic flux,* the normal state, at a well-defined critical field, H_c. Examples of such materials are Hg, Al, Sn, In. Figure 1.7a shows a sketch of this behaviour. The ideal behaviour is only observed when $n = 0$.

- Type II: The superconductor switches from the Meissner state to a state of *partial penetration of magnetic flux,* the *mixed state,* at a critical field H_{c1}. Thereafter it *crosses over continuously* to *full flux penetration,* the normal state, at an upper field H_{c2}. Examples: Nb_3Sn, NbTi, and all high-T_c cuprates. Figures 1.7b and Figure 1.8b illustrate this behaviour.

The distinction between type I and type II turns out to be so important that hereafter we shall think of every superconductor we encounter as belonging to one class or the other. Several chapters of this book are occupied with aspects

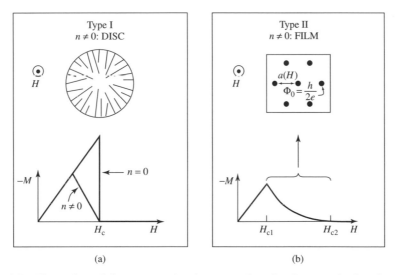

(a)   (b)

Figure 1.8   Illustration of flux penetration in cases when the demagnetization factor $n \neq 0$. In the (a) upper part, normal lamina penetrate the disc from the periphery when $n \neq 0$. In (b) vortices with flux lines carrying an elementary amount of flux $\Phi_0 = \frac{h}{2e}$ arrange themselves in a hexagonal pattern in a field normal to a type II film. In the case of type I the laminar penetration is seen only when $n \neq 0$. When $n = 0$ the transition from the Meissner state to the normal state is abrupt, no lamina appear. In the case of type II superconductor the observed mixed state behaviour is inherent to the material, irrespective of the value of n.

related to their distinction, which turns out to be essential from a practical as well as a theoretical point of view.

At this point we will limit the discussion to the descriptive level, and refer the more advanced discussion to later chapters. In order to see another difference between the two types of superconductors, let us relax the requirement on the demagnetization factor, and allow n to be non-zero. This forces flux to penetrate inhomogeneously even into a type I superconductor. Using various techniques to decorate and picture the flux distribution the following facts emerge, as illustrated in Figure 1.8a and b.

In type I materials the flux penetrates in the form of continuous lamina, flat or meandering. The lamina are alternating normal and superconducting layers, parallel to the field, whose relative thicknesses depend on both temperature and applied field. The following constraints have been found to apply to these structures: Superconducting lamina are flux-free in a Meissner-like state, while the normal ones contain a magnetic flux density corresponding to the critical field, i.e. $B_c = \mu_0 H_c$. The relative thickness of normal and superconducting lamina is fixed by a combination of this fact, and by flux conservation. We refer to Section 7.3 for a further discussion.

In type II materials, on the contrary, flux penetrates in tiny, precisely quantized units of flux $\Phi_0 = \frac{h}{2e}$ where h is Planck's constant, and e is the magnitude of electronic charge. We call these objects flux lines or vortex lines, a distinction in terminology which will be made clear in the next Section. The density of such flux lines increases with increasing applied magnetic field, as is reflected in the M versus H diagram (Figure 1.8b).

The magnetic energy density $H_c^2/2\mu_0$ required to break down the Meissner state completely at H_c in the type I case, with $n = 0$, measures directly the stability of the superconducting phase against that of the normal one. Hence this crucial information is obtainable by relatively simple measurements. H_c, it turns out, differs from one material to another, not a surprising fact considering that all metals have different electronic properties, like Fermi surfaces and band structures. If we study flux penetration into specimens of other shapes by allowing $n \neq 0$, we will find that when the Meissner state breaks down, the details of how flux penetrates and the precise fields at which it first occurs, and where it is completed, depend on the shape of the specimen due to the demagnetization effects alluded to above. An important point regarding these shape dependent effects is that whenever the demagnetization factor for a chosen combination of geometry and field direction is nonzero, the external field varies along the surface of the specimen. This causes flux to penetrate inhomogeneously whenever $n \neq 0$, unlike the case when $n = 0$. In the former case the transition from the Meissner state to the fully penetrated state is gradual when M versus H is measured even in type I. We underscore again the importance of the fact that partial flux penetration in type I superconductors, often referred to as the *intermediate state*, is very different from that of the mixed

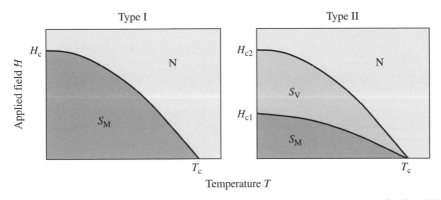

Figure 1.9 Phase diagram of type I (left) and type II (right). S = superconducting, M = Meissner phase, V = vortex phase, N = normal phase.

phase of type II. It is precisely this difference that makes type II superconductors useful on a large scale, while type I are not. Figure 1.9 illustrates the complete phase diagram for type I and type II superconductors in the $H, T-$ plane.

1.5 Vortex lines and flux lines

We want to stress a very important conceptual difference between what we call *vortex lines* and *flux lines*. Often, these terms are used interchangeably. This is permissible in most conventional type II superconductors, where the ratio between the upper and lower critical fields H_{c2} and H_{c1}, is not very large. In the cuprate high-T_c superconductors, however, this ratio is enormous. It then turns out to become important to distinguish between the concepts of flux lines and vortex lines.

As already mentioned, Abrikosov predicted the existence of the so-called mixed phase, in what has become known as type II superconductors. The distinguishing characteristic of these, as we have seen, is their ability to allow magnetic field penetration in the form of magnetic vortices of quantized circulation. This cannot happen in type I superconductors. In the centre of each magnetic vortex superconductivity is destroyed, i.e. the density of superconducting electrons is zero, while outside the centre, it is non-zero. Hence, far away from the centre of the vortex the associated magnetic field goes to zero. As we will see in later chapters, the ratio of the penetration depth λ to the length over which a magnetic field destroys superconductivity, ξ, determines the extent to which a superconductor is type I or type II. When the ratio λ/ξ is small it is type I, and when the ratio increases roughly beyond one, the superconductor becomes progressively more type II. This situation is illustrated in Figure 1.10.

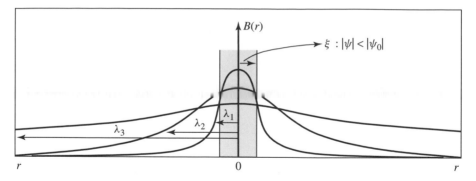

Figure 1.10   The parameters ξ of the vortex line and λ of the flux line. As explained in the text, the distinction between the concepts of vortex lines and flux lines becomes increasingly important as λ becomes greater and greater compared to ξ. λ measures the radius of circulation of supercurrent around the vortex core. The wavefunction is severely depressed for $r < \xi$, and goes to zero on the axis.

Flux lines are tubes of confined magnetic flux. They have a diameter given by the distance with which a magnetic field can penetrate into the superconductor.

On the other hand, a vortex line is a line with a diameter given by the distance over which a magnetic field suppresses superconductivity. This length is conceptually different from the magnetic penetration depth. It is then easy to imagine that a thick line with a large diameter has a typical bending length which is much longer than a very thin line. Moderate (conventional) type II superconductors are characterized by the fact that the two length scales described above are roughly equal. In extreme type II superconductors, such as the high-T_c cuprates, this is not at all the case. The magnetic penetration length, i.e. the diameter of the flux line, is typically 100 times larger than the diameter of the vortex line. Hence, the vortex line resides deep inside the flux line and moreover fluctuates on a vastly different length scale than the flux-line. In essence, they therefore represent different degrees of freedom of the system. This will turn out to be of crucial importance in the theoretical treatment presented in Chapters 9 and 10. In these contexts, therefore, the conceptual distinction between flux lines and vortex lines is essential.

In the extreme type II case, where flux lines no longer can be defined, basically being treated as infinitely thick objects, a vortex line is still perfectly well defined.

1.6   Thermodynamics of the superconducting state

We have already pointed out that the superconducting state has a limited stability against application of a magnetic field. This became apparent in the discussion of the magnetization diagrams, $M(H)$, in Section 1.4. The $M(H)$ diagrams can be read as phase diagrams, and hence may be subjected to thermodynamic analysis. A very useful aspect of thermodynamics is as a tool to derive relationships between measurable quantities by analysis of free energies. We will now look into this possibility for superconductors. Let us first recall the so-called thermodynamic square, Figure 1.11, a useful mnemonic device from which one can quickly read out the basic relationships, in this case with respect to magnetic properties, of completely general validity. The starting point is the following square:

Figure 1.11 Thermodynamic square.

Here M is magnetization, T is absolute temperature, A is Helmholtz free energy, U is internal energy, G is Gibbs' energy, S is entropy, E is enthalpy, and H is the applied magnetic field. The idea here is that each free energy should be considered as depending on the two variables next to it. Following the suggestion of the arrows it may be used to read out thermodynamic derivatives. Going in the direction of the arrows generates a positive sign, going in the opposite direction generates a negative sign. Some examples:

$$\left(\frac{\partial G}{\partial H}\right)_T = -\mu_0 M$$

$$\left(\frac{\partial G}{\partial T}\right)_H = -S \tag{1.12}$$

$$\left(\frac{\partial U}{\partial S}\right)_M = T$$

When discussing the superconducting state in an external magnetic field H, and the resulting magnetization, we should use the Gibbs' energy G, and consider it dependent on T and H. We rewrite the first relation in Eq. 1.12 as

$$\mathrm{d}G = -\mu_0 M \mathrm{d}H \tag{1.13}$$

This relation applies both in the normal and superconducting state. Next, integrate from zero to H_c at constant T in the superconducting state:

$$\int_0^{H_c} \mathrm{d}G_s = -\mu_0 \int_0^{H_c} M_s(T, H)\, \mathrm{d}H \tag{1.14}$$

Here $M = M(T, H)$, but varies only with H as T is constant during the integration.

We now consider a type I superconductor. The Meissner state, in which we now carry out the integration, was found in Section 1.3 to be characterized by $M = -H$. Using this, Eq. 1.14 yields

$$\int_0^{H_c(T)} dG_s = -\mu_0 \int_0^{H_c(T)} (-H)dH = \frac{\mu_0}{2} H_c^2(T) \qquad (1.15)$$

On the left-hand side we have

$$\int_0^{H_c} dG_s = G_s(T, H_c) - G_s(T, 0) \qquad (1.16)$$

The result is:

$$G_s(T, H_c) - G_s(T, 0) = \frac{\mu_0}{2} H_c^2(T) \qquad (1.17)$$

We want to calculate the energy difference between the normal and the superconducting states, which addresses the question of stability of the superconducting state. To do that we recall first that χ for a normal metal is small, $\chi \ll 1$. This implies that a normal metal in a field H_c has practically the same Gibbs energy as in zero field. Hence we write

$$G_n(T, H_c) = G_n(T, 0) \qquad (1.18)$$

This corresponds to saying that the integral

$$\int_0^{H_c} dG_n = -\mu_0 \int_0^{H_c} M_n dH \qquad (1.19)$$

taken in the normal state can be treated as zero.

From the fact that the two phases coexist in the intermediate state of type I, with a field H_c in the normal lamina and zero in the superconducting ones, we conclude that

$$G_n(T, H_c) = G_s(T, H_c) \qquad (1.20)$$

Comparing Eqs 1.18 and 1.20 we see that

$$G_n(T, 0) = G_s(T, H_c) \qquad (1.21)$$

Putting this back into Eq. 1.17 gives:

$$G_n(T, 0) - G_s(T, 0) = \frac{\mu_0}{2} H_c^2(T) \tag{1.22}$$

The interesting aspect of this relation is that the difference in energy density between the (Gibbs') energy in normal and superconducting states *without a field present* can be expressed by the characteristic energy $\frac{\mu_0}{2} H_c^2$ in the simple fashion of Eq. 1.22. From this relationship it becomes clear that H_c^2 is in a deeper sense a measure of the 'condensation energy' of the Meissner state, or the stability of the Meissner state with respect to the normal one. The energy density is *lower* in the superconducting state by the amount $\frac{\mu_0}{2} H_c^2(T)$. Alternatively we can write it as $\frac{1}{2\mu_0} B_c^2(T)$. We will later discover (Chapter 8) that the condensation energy can be found on a microscopic basis, leading to a relationship between H_c and the superconducting energy gap.

It is useful at this point to calculate other thermodynamic aspects of the Meissner phase. Analytically this can be done if we can find an expression for $H_c(T)$. It turns out that experiments, to a good approximation, give a unanimous answer, which we can use in the further analysis: $H_c(T) = H_c(0) \left(1 - \frac{T}{T_c}\right)^2$. This is a simple parabolic shape with the top of the parabola at $H_c(0)$ on the vertical H-axis, and an approximately linear shape where it meets the T-axis. With the important result for the condensation energy and the functional expression for H_c in hand, we choose a relationship from the thermodynamic square, that for entropy:

$$S_s = -\left(\frac{\partial G_s}{\partial T}\right)_H \tag{1.23}$$

We can now find the entropy change at the phase boundary by taking the temperature derivative of Eq. 1.22 directly and obtain:

$$S_n(T, 0) - S_s(T, 0) = -\mu_0 H_c \frac{dH_c}{dT} \tag{1.24}$$

We note here that the quantity $-\mu_0 H_c dH_c/dT$ is always positive, meaning that the entropy in the superconducting state is *always* lower than in the normal state. Hence, *the superconducting state is characterized by greater order than the normal state.* An entropy drop at (T_c, H_c) accompanies the superconducting transition. The exception is when $T = T_c$, where $H_c = 0$. We see that in this case $S_n = S_s$, and there is no entropy drop. At this particular point the entropy is therefore continuous, characteristic of a 2nd order transition. There are two aspects of the superconducting state that show the increased order below T_c indicated by the thermodynamic analysis: Formation of Cooper pairs represent a partial ordering of the electron gas, and the superconducting wavefunction shows an ordered *phase of the wavefunction*, or *phase coherence*. These facts are of course related.

Another quantity closely related to entropy is the latent heat. Here we find

$$L = T(S_n - S_s) = -T\mu_0 H_c \frac{\partial H_c}{\partial T} \geq 0 \tag{1.25}$$

Again, this is a positive quantity as long as we are considering finite H_c. At T_c, where $H_c = 0$ we get $L = 0$, again characteristic of a second order or continuous transition.

Finally, let us look at the heat capacity, which from thermodynamics is found by differentiation of the entropy

$$C_H \equiv T \left(\frac{\partial S}{\partial T} \right)_H \tag{1.26}$$

Taking the 2nd derivative of $\frac{\mu_0}{2} H_c^2$ we find:

$$C_s(T) - C_n(T) = \mu_0 T \left[\left(\frac{dH_c}{dT} \right)^2 + H_c \frac{d^2 H_c}{dT^2} \right] \tag{1.27}$$

which, at T_c becomes

$$C_s(T_c) - C_n(T_c) = \mu_0 T_c \left(\frac{dH_c}{dT} \right)_{T=T_c}^2 > 0 \tag{1.28}$$

We find a discontinuity at T_c, an upward jump on lowering the temperature through T_c as all quantities in the right-hand side are positive. Experiments have verified this aspect in every case it was measured in the low-T_c superconductors. In high-T_c materials the story is more complicated. There, fluctuations play a dominant role near T_c, and a different description altogether is necessary, as will be discussed in several chapters of this book. The discontinuity found in Eq. 1.28 is characteristic of mean-field theories and mean-field systems. The BCS theory, which eventually explained superconductivity in low-T_c materials, belongs to that class, and reproduces the experimental results nearly perfectly.

Figure 1.12 illustrates examples of two different phase transitions. The left figure illustrates the BCS mean-field like behaviour, measured on a low-T_c material. The right figure illustrates the situation in a high-T_c material and shows a completely different form, with no jump, and with fluctuation contributions on both sides of T_c. We shall return to these questions which are of major importance in high-T_c materials, with a ratio $\lambda/\xi \gg 1$, the so-called high-κ materials.

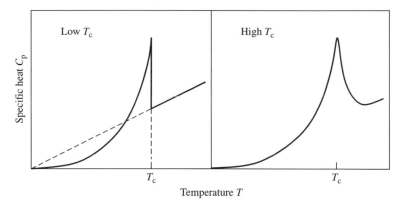

Figure 1.12   Sketch of typical forms of specific heat curves measured from above the super-conducting transition in a mean-field low-T_c metallic superconductor like Al (left), and in a substance with strong superconductivity phase fluctuations like the high-T_c compound YBCO (right).

1.7   Demagnetization factors and screening

When a specimen of ellipsoidal shape is subject to a homogeneous external field along one of its principal axes, it is well-known from magnetostatics (see for instance Jackson, 1975 [11]) that the internal fields $\boldsymbol{B}_i$ and $\boldsymbol{H}_i$ are both parallel to the applied field $\boldsymbol{B}_a$. The constitutive relation is

$$\boldsymbol{B}_i = \mu \boldsymbol{H}_i = \mu_0(\boldsymbol{H}_i + \boldsymbol{M})$$
$$= (1 + \chi)\mu_0 \boldsymbol{H}_i \tag{1.29}$$

As already discussed, in the Meissner state $\chi = -1$, which leads to $\boldsymbol{B}_i = 0$. These two statements are equivalent, and consistent. For the applied fields we have similarly

$$B_a = \mu_0 H_a \tag{1.30}$$

provided no magnetized matter is surrounding the magnetized body under study. These relations apply separately inside, and outside the measured body, and do not answer the question: what is the relationship between the applied field $\boldsymbol{H}_a$ and the internal field $\boldsymbol{H}_i$? This is not a trivial question. What complicates the relationship between the applied fields and the internal fields B_i, H_i, B_a, H_a is the so-called demagnetizing field. We refer to standard books on magnetostatics for a discussion of the theory of demagnetization effects. Here we will use the results without derivation. The word 'demagnetization' was originally coined for use in ferromagnets. It applies with opposite consequences in superconductors as compared to ferromagnets, but is equally important to consider here due to the fact that magnetization effects and demagnetization are very strong in superconductors. The reason for this is that in superconductors the susceptibility $\chi = -1$, the ideal and complete diamagnetism.

Briefly stated, when a magnetic field H_a is applied to a superconductor, the screening currents which appear near the surface, create an additional field in the specimen. This is called the demagnetizing field H_D. We now have:

$$H_i = H_a - H_D \qquad (1.31)$$

We will only address situations where the internal field is homogeneous, i.e. cases like those we mentioned in the introductory statement above. With this restriction we can still treat all ellipsoidal shapes. These all have a uniform field inside the body when exposed to a uniform external field. We may then write

$$H_D = nM \qquad (1.32)$$

where n is the (scalar) demagnetizing factor, and M is the uniform magnetization of the body. We now have the relationship

$$H_i = H_a - nM \qquad (1.33)$$

In a superconductor $M = -H_i$; which, upon insertion in Eq. 1.33 leads to the important relationship

$$H_i = \frac{1}{1-n} H_a \qquad (1.34)$$

precisely the relationship which we pointed out was missing previously.

To take a concrete example, let us examine the situation when a compact ellipsoidal body of superconductor is placed in an external magnetic field applied along one of the principal axes. At the surface of the superconductor the tangential component of H is continuous. The internal field is now parallel to the applied field which exists everywhere on the circumference of the extremal cross-section of the sample as viewed along the field direction. In case of a spherical body this circumference corresponds to the equatorial circle. This makes the field H_a just outside the circumference equal to the field H_i just inside. Eq. 1.34 now allows us to calculate the field just inside, and because this equals the field just outside, the field just outside is equal to $H_a/(1-n)$. We learn from this that *a demagnetization factor $n < 1$ makes the field at the maximum circumference larger than the applied field H_a*.

Some important examples are the following, illustrated in Figure 1.13:

- For a sphere, $n = \frac{1}{3}$, always giving a tangential field of $\frac{3}{2} H_a$ at the equator.

- In case the field is applied normal to a long rod of circular cross-section, $n = \frac{1}{2}$, and the tangential field is $2H_a$.

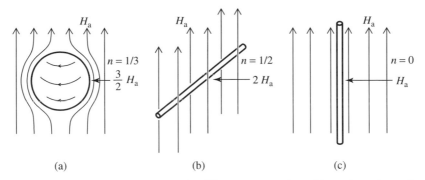

Figure 1.13 Demagnetization factors for different geometries with n given for each case. Horizontal arrows indicate the value of applied field at the surface in each case.

- With the field applied along the axis of the long rod, $n = 0$; no demagnetization occurs. The field along the entire surface of the rod is tangential to the (extremal) crossection, and equal to the applied field H_a.

Demagnetization effects are very important in superconductors, where $\chi = -1$ in contrast to normal metals where $\chi \approx 10^{-4}$ or less. In practical cases, when the body under study is not ellipsoidal, one approximates the real geometry with an inscribed ellipsoidal surface which gives a best possible description of the real body, and uses the demagnetizing factor which applies to that ellipsoidal shape. This gives quite good results.

A paradox: Having just proven that the field just outside the superconductor (usually) exceeds the applied field, one needs to consider what will happen when the applied field approaches H_c, the critical field of a type I superconductor. If we let the applied field H_a take the value $(1 - n)H_c$, then the internal field H_i becomes everywhere precisely equal to the critical field H_c. This looks like we now should expect the entire sphere to be driven into the normal state.

However, this would cause M to become zero, since in the normal state this is the case with good approximation. This gives $H_i = H_a < H_c$, and we should conclude that we have a body driven into the normal state by a field $H < H_c$. This situation is an impossible paradox.

Solution: What happens, is that the specimen allows some field to penetrate parts of the superconductor with normal lamina (type I), or vortices (type II). We have a coexistence of alternating normal and superconducting lamina in one case and vortices in the other. Such a thermodynamic coexistence of two phases is commonly found in nature. The situation is therefore normal, in agreement with the thermodynamic laws. The existence of such mixed normal and superconducting volumes is well documented in the scientific literature.

Because of the effects discussed here, one finds in real situations that the transition from superconducting to normal state does not occur abruptly and completely at the thermodynamic field, H_c, but rather over a certain range of fields, the width of which is determined by the demagnetization factor. We refer again to Figure 1.8 for an illustration of these points. With $n = 0$, as it is with the field along the axis of a long thin rod, the transition in a type I superconductor occurs abruptly at H_c, as we are led to expect from the foregoing, otherwise not.

2

Superconducting Materials

2.1 Introductory remarks

After the initial discovery of superconductivity in an element of the periodic table, Hg, almost 20 years went by before research was undertaken in alloys. Another 40 years would pass before organic superconductors were synthesized in the 1970s. Then, another decade would pass before superconducting cuprates were discovered in 1986, followed by fullerenes shortly after. There is a line of progression of structures from the very simple to the quite complex. At the same time, T_c has increased by a factor of 40 from the beginning in mercury to the record T_c in cuprate perovskite; and there is a time span of 90 years between their discoveries. This seemingly slow development is governed by the general development of physics in a broader sense. Theoretical physics could not handle the manybody quantum theory necessary until the mid-1950s when the BCS theory was worked out. And the mechanism developed there was the only one to be relied upon until the superconducting cuprates were found 30 years later. Even today, 90 years after the initial discovery, there is a fierce debate about the mechanism for the high T_c cuprate superconductors. Superconductivity is still a subtle and very complex phenomenon.

2.2 Low-T_c superconductors

2.2.1 Superconducting elements

After superconductivity in Hg had been found, Sn and Pb followed suit. This brought T_c up from 4 K to 7 K. By the time the Meissner effect was discovered, several more elements of the periodic table had been added to the list. Meissner, among others, undertook studies of the transition elements with high melting point, called 'hard' metals. Discovery of superconductivity was announced in

Superconductivity: Physics and Applications   Kristian Fossheim and Asle Sudbø
© 2004 John Wiley & Sons, Ltd   ISBN 0-470-84452-3

Table 2.1 Superconductors in the periodic table, including thin films. Adapted from Ref. [12]

Legend:
T_c = Superconducting transition
F = Film
P = Pressure applied

1	2	3	4	5	6	7	8	9	10	11	12	3	4	5	6	7	8
Li F	Be 0.03																
												Al 1.2	Si FP	P P			
		Sc 0.01	Ti 0.4	V 5.4	Cr F						Zn 0.9	Ga 1.1	Ge FP	As P	Se P		
		Y P	Zr 0.6	Nb 9.3	Mo 0.9	Tc 7.8	Ru 0.5				Cd 0.5	In 3.4	Sn 3.7	Sb P	Te P		
Cs FP	Ba P	La 4.9α 6.3β	Hi 0.1	Ta 4.4	W 0.02	Re 1.7	Os 0.7	Ir 0.1			Hg 4.2	Tl 2.4	Pb 7.2	Bi FP			

Ce P				Eu F					Lu 0.1
Th 1.4	Pa 1.4	U P		Am 1.0					

tantalum in 1928 with $T_c = 4.4\,\text{K}$, thorium in 1929, with $T_c = 1.4\,\text{K}$, and Nb in 1930 with $T_c = 9.2\,\text{K}$. The latter remains the highest T_c found in any element. Table 2.1 shows the elements with known superconducting T_c, a total of over 40 when observations of superconductivity under high pressure and in thin films are included. We notice in particular that superconductivity is neither found in the magnetic compounds, nor in the noble metals or copper.

This indicates that superconductivity is incompatible with magnetism, and absent in metals with the highest electrical conductivity. Both of these rules will turn out to be understandable in light of the BCS theory; magnetism breaks up the Cooper-pairs, and is therefore a destructive influence; and excellent electrical conductivity is a signature of weak electron–phonon interaction, a property that reduces the effect of the electron–phonon mechanism for superconductivity in the BCS theory.

In pure form the elements of the periodic table have provided excellent materials for scientific research in superconductivity. None of these pure elements have, however, contributed to applications of superconductivity on a large scale, like wires and cables for magnets. However, on a small scale Pb and Nb have been used for advanced development of Josephson technology. For SQUIDs (superconducting quantum interference device) niobium has been the best material overall, and is widely preferred in those particular low-T_c applications, to be discussed in Chapters 5 and 11.

2.2.2 Binary alloys and stoichiometric compounds

The earliest work on binary alloys has been largely forgotten, partly due to the overwhelming number of compounds that have been made and tested in later years. One person alone, Bernd T. Matthias, is said to have made some 3000 different alloys in his heroic attempt to achieve high-T_c superconductivity during the 1950s and 1960s. These efforts never succeeded, a fact which in itself gave strong impetus to go in totally new directions in more recent years. The history of the development of T_c is shown in Figure 2.1.

The record shows [2] that research on binary alloys was started already in 1928 in Leiden by de Haas and Voogd. They found superconductivity in

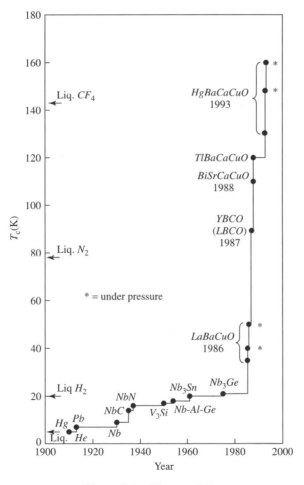

Figure 2.1 History of T_c.

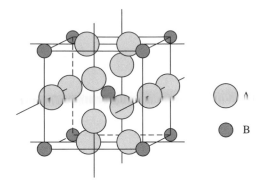

Figure 2.2   The A15-structure, A$_3$B.

SbSn, Sb$_2$Sn, Cu$_3$Sn, and Bi$_5$Tl$_3$. They noticed that the combination of a superconducting element with a non-superconducting one was successful. In Bi$_5$Tl$_3$, T_c was raised by a factor of 2–3 compared to pure Tl. What was even more interesting, was that the magnetic threshold for destruction of superconductivity in this material was much higher than in any of the elements known to be superconducting at the time. The material would remain superconducting up to 0.5 T at 3.4 K, and by extrapolation it was predicted to tolerate 0.9 T at 1.3 K. But soon they found an even more promising material; a Pb–Bi eutectic alloy with a critical field $B_c = 2.3$ T at 1.9 K. Unfortunately, the substance was so difficult to make and to handle that it never was to fulfill its promise as a material for wires, which might otherwise have made it an important material for superconducting electromagnets.

After an intense period of research on binary alloys around 1930, not much happened in the materials area until Bernd T. Mattias and John K. Hulm started a new programme in the US in the early 1950s. Their 'materials approach' to superconductivity would bear rich fruit. A number of new compounds were made, with impressively high T_c and high critical fields. Throughout the 1950s the materials that were developed for use as superconductors included: solid solutions of NbN and NbC with $T_c = 17.8$ K; V$_3$Si with $T_c = 17$ K; Nb$_3$Sn with $T_c = 18$ K; NbTi with $T_c = 9$ K. Later (1973) Nb$_3$Ge was added to this list with the highest T_c of all, at 23.2 K, a record that lasted until 1986. We refer to Table 2.2 for some additional facts.

The intermetallic compounds, just mentioned, belonging to the A$_3$B type of materials are classified as A15 compounds (Figure 2.2). They share a special feature, namely a softening of an elastic mode above T_c, a property which suggests relating their high transition temperatures to enhanced electron–phonon interaction caused by the softening. The two materials that have led to successful industrial production of low-temperature superconducting magnets are Nb$_3$Sn and NbTi. The first one is a stoichiometric, intermetallic compound, the second an alloy. These have paved the way for superconducting technology to

Table 2.2 Some binary alloys and stoichiometric compounds. Values of T_c and $B_{c2}(0)$ may vary somewhat depending on precise composition

Compound	T_c [K]	$B_{c2}(0)$ [T]
V_3Si	17	25
Nb_3Sn	18	24
Nb_3Ge	23.2	38
V_3Ga	14	21
NbTi	9	15
VTi	7	11

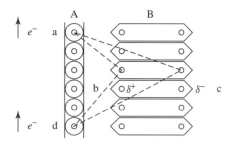

Figure 2.3 Predicted model substance for organic superconductor (Adapted from Little, 1964 [14]).

conquer the market for large laboratory magnets and magnetic resonance imaging magnets over the entire industrialized world. Behind this development lay great efforts in processing and materials science, as well as daring ventures into science-driven technological developments. Among the technical achievements were processing methods to make wires consisting of a large number of very thin filaments, a necessary procedure in order to stabilize the superconductor in a normal matrix.

2.3 Organic superconductors

2.3.1 Polymer and stacked molecular type

Research on electronically conductive organic materials dates back to the 1940s. High electrical conductivity was first discovered in 1954, in perylene bromine complex. Much later the discovery of a pronounced conductivity peak in (TTF)(TCNQ) near 60 K in 1973 [13] stimulated a lot of effort in the direction of low-dimensional systems, so-called charge transfer salts.

Superconductivity in a polymer material was first found in $(Sn)_x$ in 1975. This was followed by the discovery in 1979 of superconductivity in a molecular salt,

Table 2.3 Some selected organic superconductors

Material	Symmetry of counter molecule	T_c [K]
$(TMTSF)_2PF_6$	Octahedral	0.9
$(TMTSF)_2ClO_4$	Tetrahedral	1.4
$\beta_L - (ET)_2I_3$	Linear	1.5
$\kappa - (ET)_2Cu(NCS)_2$	Polymeric	10.1
$\kappa - (ET)_2Cu[N(CN)_2]Br$	Polymeric	11.8
$\alpha - (ET)_2RbHg(SCN)_4$	Polymeric	0.5
$\kappa_H - (ET)_2Ag(CF_3)_4 \cdot TCE$	Planar	11.1
$\kappa_L - (ET)_2Ag(CF_3)_4 \cdot 112DCBE$	Planar	4.1
$\kappa_H - (ET)_2Au(CF_3)_4 \cdot TCE$	Planar	10.5
$\lambda - (BETS)_2GaCl_4$	Tetrahedral	8

TCE:	1,1,2-trichloroethane
112DCBE:	1,1-dichloro-2-bromoethane

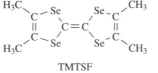

TMTSF

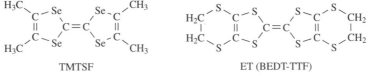

ET (BEDT-TTF)

BETS (BEDT-TSF)

$(TMTSF)_2$ FF_6 under 1.2 Gpa pressure, and with a T_c of 0.9 K [15]. Since then, a long list of organic superconductors have been synthesized, some of which are seen in Table 2.3. T_c remains low, although it has increased by a factor of more than 10 since the first discovery. In this sense progress has been remarkable. In another sense it has been disappointing, since predictions had been made for room-temperature superconductivity in stacked organic structures. This prediction was set forth in 1964 by Little [14] in a paper where he suggested the possible exis-tence of superconductivity in an organic substance consisting of a long unsaturated polyene chain, called the 'spine', with an array of side chain molecules attached at regular intervals (see Figure 2.3). He showed that even if the spine was ini-tially an insulator because the valence bond was full and the conduction bond was empty, the addition of side chains could increase the effective electron–electron attraction to the point where it became energetically favourable to enter the super-conducting state by mixing in states of the conduction band. It was concluded that superconductivity at room temperature could result.

This paper in particular stimulated a worldwide effort to discover supercon-ductivity in substances of appropriately stacked organic structures. The mecha-nism proposed by Little has later become known as the polaron mechanism. It is an alternative to the phonon mechanism, which was the one worked out in detail in the BCS-paper. Due to the higher energies involved in this new mechanism, T_c was expected to be much higher in the polaronic superconductors than in those with phonon-mediated pairing.

Some of the organic superconductors are best described as a stack of two-dimensional superconducting sheets with Josephson coupling between them. A good example is the so-called κ-type BEDT-TTF salts. They have different zero-degree coherence length $\xi_{0\parallel}$ and $\xi_{0\perp}$ in the stacking plane and normal to it, respectively. Table 2.4 shows T_c and coherence lengths for four different ET-compounds, as determined by analysis of magnetic measurements. One finds the remarkable result that the interplanar coherence length $\xi_{0\perp}$ is shorter than the interplanar distance by a factor of about 5. Supercurrent in the direction normal to the stacking planes must therefore be carried by Cooper pair tunnelling, so-called Josephson tunnelling. We refer to Chapter 5 for a discussion of the physics of Josephson tunnelling.

What determines the important coherence length? Disregarding anisotropy for the moment, let us call this length ξ_0. A simple argument by means of the uncertainty principle leads to the following expression:

$$\xi_0 \approx \frac{\hbar v_F}{k T_c} \tag{2.1}$$

where $\hbar$ is Planck's constant divided by 2π, v_F is the Fermi velocity, and k is the Boltzmann constant. This indicates an inverse relationship between T_c and ξ_0, and proportionality between ξ_0 and v_F. Ideally we would want both ξ_0 and T_c to be high, T_c for obvious reasons, ξ_0 because a higher value provides better stability of the superconducting state. But clearly, a compromise is the best one can hope for. We see from the values in Table 2.4 that $\xi_{0\parallel}$ in the 2D-plane is much longer than in the direction normal to the plains. This gives an indication for a 'design-criterion' for stacked superconductors. We shall encounter a similar situation in the inorganic high-T_c compounds in Section 2.5.

Table 2.4 T_c, and coherence lengths normal to the stacking plane $\xi_{0\perp}$, and parallel, $\xi_{0\parallel}$, for some 'ET'-compounds

Material	$T_c[K]$	$\xi_{0\perp}[nm]$	$\xi_{0\parallel}[nm]$
$\kappa - (ET)_2Cu(NCS)_2$	8.7 ± 0.2	0.31 ± 0.05	2.9 ± 0.5
$\kappa - (d_8 - ET)_2Cu(NCS)_2$	9.0 ± 0.2	0.32 ± 0.05	2.9 ± 0.5
$\kappa - (ET)_2Cu[N(CN)_2]Br$	10.9 ± 0.2	0.58 ± 0.1	2.3 ± 0.4
$\kappa - (d_8 - ET)_2Cu[N(CN)_2]Br$	10.6 ± 0.2	0.57 ± 0.1	2.3 ± 0.4

After Ishiguro *et al.* [13].

2.3.2 Fullerene superconductors

The Buckminster fullerene C_{60} is a carbon molecule containing 60 atoms located at the vertices formed by the intersections of 12 pentagonal and 20 hexagonal faces, altogether forming a closed cage, as shown in Figure 2.4. We note that all atoms are in equivalent positions, where each is surrounded by one pentagon and two hexagons. One π electron and three σ electrons form a sp^2 hybrid orbital.

C_{60} condenses into an fcc structure, leaving room for introduction of small interstitial dopants like K and Rb. Other structures are bct K_4C_{60} and bcc K_6C_{60}. A significant transfer of electrons from K $4s$ orbitals to the C_{60} conducting band takes place. The electrical conductivity increases with dopant concentration, with a maximum for K_3C_{60}. The resulting substances have a well-defined Fermi surface.

Superconductivity has been observed in a number of fullerene-based materials. The initial discovery was made in K_3C_{60} [16]. As shown in Table 2.5, transition temperatures go as high as 40 K. In Table 2.8 in the last section of this chapter a T_c as high as 45 K is indicated. Critical fields H_{c2} are quite high, 28 T in K_3C_{60} and 38 T in Rb_3C_{60}.

The fullerenes have been carefully studied with respect to isotope effects in T_c-values. The presence of significant T_c-dependence on atomic masses indicates electron–phonon mediated superconductivity. The presence of the Hebel–Slichter peak in nuclear magnetic resonance (NMR) (see Chapter 4) is considered definite proof that the BCS mechanism discussed in Chapter 3 is active.

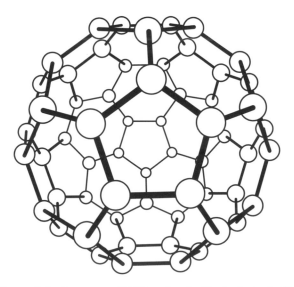

Figure 2.4   Structure of C60-molecule (Buckminster ball).

Table 2.5 Structure and T_c's of some fullerene type superconductors

Material	Symmetry of the salts	$T_c[K]$
K_3C_{60}	fcc	19.3
Cs_2RbC_{60}	fcc	33
$(NH_3)_4Na_2CsC_{60}$	fcc	29.6
Cs_3C_{60}	bct/bcc	40
$NH_3K_3C_{60}$	Orthorhombic	28
$Rb_x(OMTTF)C_{60}$ (benzene)		26

fcc = face-centered cubic, bct = body-centered tetragonal, bcc = body-centered cubic, OMTTF = octamethylenetetrathiafulvalene.

2.4 Chevrel phase materials

Superconductors referred to as Chevrel phase materials were discovered in 1971 (Figure 2.5) [17]. These are ternary molybdenum chalcogenides of composition MMo_6X_8, where X is one of the chalcogenes S, Se, or Te. M can be one of many different metals or rare earths. A most remarkable property of these compounds is the high critical magnetic field of some of them, as listed in Table 2.6. Figure 2.6 shows the temperature dependence of $B_{c2}(T)$ for a number of low-T_c superconductors.

2.5 Oxide superconductors before the cuprates

Superconductivity was found in 1964 in the perovskite oxide $SrTiO_3$ [18], whose structure is shown in Figure 2.7. The following oxide compounds were found to exhibit superconductivity: doped $SrTiO_{3-\delta}$, NbO, and TiO. They had T_c's

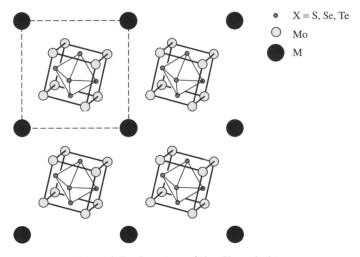

X = S, Se, Te
Mo
M

Figure 2.5 Structure of the Chevrel phase.

Table 2.6 Critical temperatures T_c, and critical fields B_{c2} in some Chevrel phase materials

Compound	$T_c(K)$	$B_{c2}(T)$
$SnMo_6S_8$	12	34
$PbMo_6S_8$	15	60
$LaMo_6S_8$	7	45

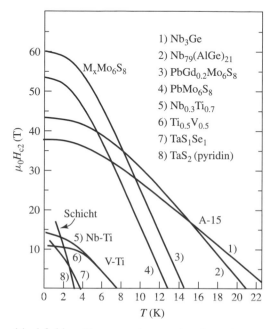

Figure 2.6 Upper critical field $\mu_0 H_{c2}$ versus temperature in various superconductors. After Fischer [19].

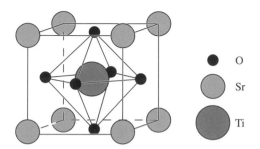

Figure 2.7 The cubic ABX_3 structure. A prominent example is $SrTiO_3$.

in the range 0.7 K to 2 K. In the years after 1964, bronzes were found to be superconductors: K_xWO_3, K_xMoO_3, K_xReO_3. These had T_c's in the 4–6 K range. The 1970s brought discoveries of $LiTi_2O_4$ and $Ba(PbBi)O_3$, both with $T_c = 13$ K. Naturally, they all had low carrier density. By now the results were encouraging, and the development pointed to oxide perovskites as a possibly promising class of materials. Some groups therefore pursued such a course.

2.6   High-T_c cuprate superconductors

2.6.1   The discovery of cuprate superconductors

The breakthrough to a new era in higher superconducting transition temperatures came in 1986 by the discovery of superconductivity in the $La_{2-x}(Ba,Sr)_xCuO$ compounds by two scientists at the IBM Zurich laboratory, J. George Bednorz and K. Alex Müller [9]. Figure 2.8 shows the corresponding undoped mother perovskite La_2CuO_4. In their article in *Zeitschrift für Physik* they cautiously announced: 'Possible high T_c superconductivity in the Ba-La-Cu-O system.' Their material showed onset of superconductivity at about 30 K, well above previous records (see Figure 2.9). Initially the reaction from the scientific community was somewhat reluctant, but this changed to intense interest and competition, as soon as their results were confirmed by other groups, in Japan, in the US and China.

The next important development occurred when the Houston group lead by Chu showed that external pressure could raise T_c substantially, going above 40 K under 13 kbar pressure [20]. An effect which is equivalent to application of

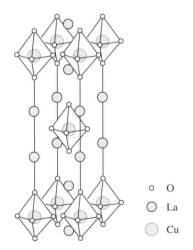

O   O
O   La
○   Cu

Figure 2.8 La_2CuO_4 tetragonal crystallographic unit cell.

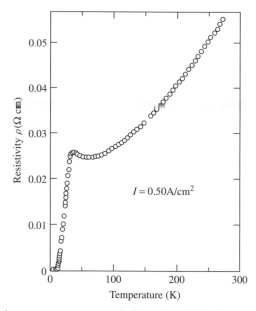

Figure 2.9 Resistivity versus temperature in $La_{2-x}Ba_xCuO$. After the original publication by Bednorz and Müller, 1986 [9].

high pressure is achieved by replacing some of the ions with smaller ones having the same chemical properties. Thus, replacing Ba with the smaller ion Sr led to a T_c of 38 K. Naturally, one would think that a similar procedure ought to be tried at the La-site. The Huntsville group led by Wu proceeded in collaboration with the Houston group to replace both La and Ba, with Y and Sr, respectively. This suddenly brought T_c above 90 K [21].The news spread quickly all over the world. Soon the successful new compound turned out to be $YBa_2Cu_3O_7$ (Figure 2.10). The discovery was made more or less simultaneously in several laboratories, in Tokyo, Beijing and at Bell Labs in the US. This breakthrough was of historic proportions. T_c had now moved well above the boiling point of liquid N_2 at 77 K. The best values of T_c turned out to be in the range 91–93 K in $YBa_2Cu_3O_{7-\delta}$, depending on the value of δ (see Figure 2.11).

The ensuing response from the scientific community was without parallel in the history of science. Suddenly, the efforts were joined by thousands of scientists and students around the world, trying to understand $YBa_2Cu_3O_{7-\delta}$ and the other newly discovered compounds, and to push T_c even higher.

Several new compounds with higher T_c were soon synthesized. With one exception, $Ba_{1-x}K_xBiO_3$, all had one common feature: a quasi-2D network of CuO_2 – i.e. they were all cuprate perovskites. The new materials are very different from traditional metals, being doped oxides. Their normal state properties are different from metals, and they are so strongly anisotropic that they could be shown in some cases to possess metallic-like conductivity in directions parallel to the CuO_2 planes – although not free-electron like – while behaving

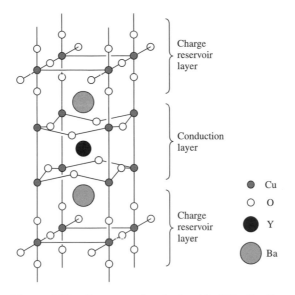

Figure 2.10 Structure of orthorhombic $YBa_2Cu_3O_7$.

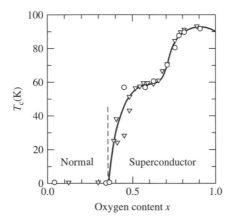

Figure 2.11 T_c versus x in $YBa_2Cu_3O_{6+x}$. Data from Cava *et al.* (circles) [21] and from Jorgensen *et al.* (triangles) [22].

like semiconductors along the c-axis, normal to CuO_2 planes. This profound anisotropy affects essentially all physical properties [23].

In the ensuing years huge efforts were spent on achieving superconductivity at still higher temperatures, even with the hope of reaching room temperature. This did not happen, but the efforts resulted in a long list of new superconducting compounds with complex structures and intriguing properties. A superconducting transition temperature as high as 163 K was eventually reported in the Hg-based $HgBa_2Ca_2Cu_3O_8$ compound at a high external pressure. We refer again to Figure 2.1 which summarizes the main historical development of T_c

versus time. For further details about the historical record we refer the reader to Dahl [2].

2.6.2   Composition and structure

Referring now to Figure 2.10 showing the YBCO structure, one clearly sees the connection to the simple cubic perovskites. Essentially, YBCO is a stack of three perovskite unit cells where, compared to the basic model ABX_3 (Figure 2.7), with atoms removed in some places and substituted in other places. The typical oxygen octahedra, so familiar in the ABX_3 compounds like $SrTiO_3$ have been split in two by a middle layer, and Ba replaces Y as the central atom in two out of three layers. We notice also that $YBa_2Cu_3O_7$ is a modification of the structurally simpler compound $YBa_2Cu_3O_6$. Processing of high-T_c superconductors to single-phase material requires elaborate processes, which we will not go in details about here. Let us just remark that as far as YBCO is concerned the processing route usually goes via making the non-superconducting $YBa_2Cu_3O_6$ first. This materials may be obtained under appropriate conditions in an oxygen atmosphere from a mixture of $Y_2O_3 + BaO + CuO_x$, heated to over $900\,^\circ C$. The resulting material is of tetragonal structure, with a lot of oxygen vacancies. On lowering the temperature to somewhere between $400\,^\circ C$ and $500\,^\circ C$ more oxygen is absorbed by the structure and a different oxygen ordering takes place, with Cu-O chains, as shown in Figure 2.10. The chain formation

Figure 2.12   Permanent magnet levitating above a chunk of high-T_c superconducting $Y_1Ba_2Cu_3O_7$.

breaks the tetragonal structure with different lattice parameters along the a- and b-directions. A subsequent quenching of the material traps the oxygen in the structure. When all available oxygen sites are filled in this structure we have the $YBa_2Cu_3O_7$ compound shown.

Doping of $SrTiO_3$ had been demonstrated to lead to superconductivity due to a reduction process, whereby oxygen was removed. The situation in YBCO is related, but more complicated, as can be seen in Figure 2.11. Starting with the $YBa_2Cu_3O_{6+x}$ structure at $x = 0$ and increasing the oxygen content via external oxygen supply at high temperatures the compound at first is an insulator, but becomes a superconductor at x-values grater than 0.4. From there on T_c increases until it goes through a maximum at $x \approx 0.93$, and diminishes slightly until $x = 1$ is reached. Optimal doping (corresponding to the highest T_c) of YBCO therefore is achieved in $Y_1Ba_2Cu_3O_{7-\delta}$ with $\delta \approx 0.07$. On inspecting the resulting structure, one plane at a time, one finds that the orthorhombic YBCO structure consists of layers as do all high-T_c compounds. Beginning at the bottom and proceeding upwards they are: CuO, BaO, CuO_2, Y, CuO_2, BaO, CuO. The CuO_2 layers are superconducting in YBCO like in all other high-T_c materials. But YBCO possesses the additional CuO chains, as mentioned. These can also be superconducting, depending on the degree of filling. In this respect YBCO is unique among high-T_c materials. Another aspect of this structure is that the oxygens are not in equivalent positions in the lattice, contrary to ABO_3 type structures. There are four inequivalent O sites and two inequivalent Cu sites.

2.6.3 Making high T_c materials

High-T_c materials, being doped oxides, are made by heating a mixture of appropriate amounts of metal oxides in powder form to temperatures in the range 850–950 °C. Materials that contain some percentage of superconductor are then easily obtained. But making materials of good quality, whether polycrystalline or single crystals is quite a different and highly non-trivial matter. Only quite elaborate procedures with attention paid to every detail of composition, temperature, pressure and atmosphere will bring out high quality materials. Regrinding, compressing and sintering repeatedly, is usually necessary. The resulting bulk material is then a fine-grained ceramic, brittle, and with high density of defects and grain boundaries. Making the big step towards single crystals is an even greater challenge. Growing large crystals of high quality has turned out to be extremely difficult. The main reason for this is to be found in the layered structure, which permits easy diffusion and growth along the (a, b)-planes, but allows only very slow diffusion and growth along the c-direction. Some of these materials end up quite 'flaky', like in the Bi-based materials where layer by layer may be peeled off quite easily. This is an intrinsic property of the material

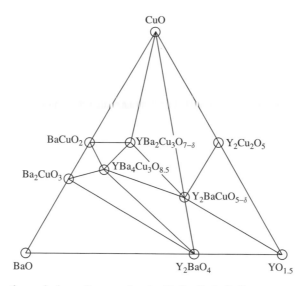

Figure 2.13 Isothermal phase diagram for the Y_2O_3-BaO-CuO_x system at 900 °C in 1 atm oxygen. Calculations by Rian [24]. Lines connect phases which exist in equilibrium at this temperature.

itself, reflecting the great anisotropy. Much of the difficulty in crystal growth and processing lies also in the fact that even the simplest of them are quaternary compounds. When 3 oxides are mixed, several different phases may result. Figure 2.13 shows a chemical equilibrium phase diagram relevant to YBCO.

Matters grew even more complex when it turned out that the road to higher T_c went through compounds like $Bi_2Sr_2Ca_2Cu_3O_{10+\delta}$ or $Bi_2Sr_2Ca_1Cu_2O_{8+\delta}$, with four metal ions, as can be seen in Figure 2.14, and several similar ones with Tl or Hg replacing Bi. In the process of making such substances a number of unwanted phases will normally be created. In this case perfection is practically impossible in bulk production, but single crystals may be grown. In order to study the intrinsic properties of new materials, access to single crystals is a necessity. Table 2.7 shows a list of high-T_c compounds, with corresponding T_c-values.

2.6.4 Phase diagrams and doping

Superconductivity in high T_c materials depends on appropriate doping; and the phase diagrams show a remarkable similarity: as a function of the appropriate doping parameter they all have an antiferromagnetic phase near zero doping, followed by an insulating phase, then the superconducting phase underneath a sort of metallic phase in a certain doping range, and finally T_c goes to 0 again. The typical situation is sketched in Figure 2.15, where the various phases

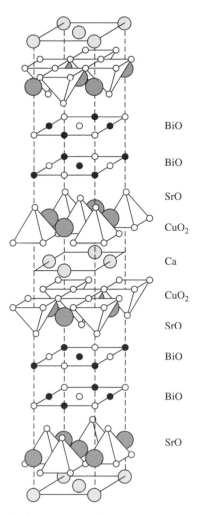

BiO

BiO

SrO

CuO$_2$

Ca

CuO$_2$

SrO

BiO

BiO

SrO

Figure 2.14 Structure of Bi$_2$Sr$_2$CaCu$_2$O$_8$ crystal.

are indicated. One finds the peculiar property that in every substance there exists a range of x-values, where superconductivity is observed, with optimal doping corresponding to the highest T_c, somewhere near the middle of that range. Above this dome-shaped curve is the 'normal' state, a kind of unconventional metallic region. To the left of the dotted line in Figure 2.15 is a region which is often referred to as that of a 'strange' metal, while to the right of the same line one finds more normal metallic properties. Much of the secret of high-T_c compounds may be found in these peculiar normal state properties. Often the question is asked: how can we understand high-T_c superconductivity if we don't even understand the normal state? Anderson is the scientist who most relentlessly has pursued this question [25].

Table 2.7   Some representative examples of high-T_c cuprate superconductors and their T_c's. The reported values of T_c will vary somewhat, depending on the processing conditions and resulting oxygen content and other deviations from stoichiometry

Compound	T_c [K]	Nicknames
$La_{1.85}Sr_{0.15}CuO_4$	39	LCCO or LaSCCO
$YBa_2Cu_3O_7$	92	Y123 or YBCO
$Bi_2Sr_2CaCu_2O_8$	84	Bi2212 or BiSCCO
$Bi_2Sr_2Ca_2Cu_3O_{10}$	110	Bi2223 or BiSCCO
$Tl_2Ba_2CuO_6$	90	
$Tl_2Ba_2CaCu_2O_8$	110	
$Tl_2Ba_2Ca_2Cu_3O_{10}$	125	Tl2223 or TBCCO
$TlBa_2CaCu_2O_7$	91	
$TlBa_2Ca_2Cu_3O_9$	116	
$TlBa_2Ca_3Cu_4O_{11}$	122	
$HgBa_2CuO_4$	95	
$HgBa_2CaCu_2O_6$	122	
$HgBa_2Ca_2Cu_3O_8$	133	Hg1223 or HBCCO
$Nd_{1.85}Ce0.15CuO_{4-y}$	25	NCCO

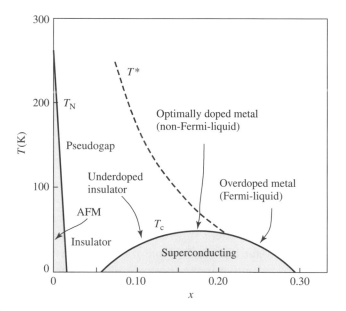

Figure 2.15   Typical overall phase diagram with doping in high-T_c cuprate superconductors (AFM = antiferromagnetic phase).

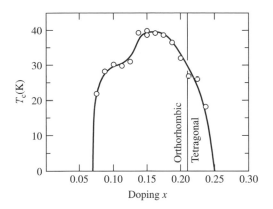

Figure 2.16 T_c versus doping in $La_{1-x}Sr_xCuO_4$. Data from Radelli *et al.* [26].

How doping is achieved, i.e. how the parameter x on the horizontal axis of Figure 2.15 is controlled, depends on the compound. In the case of the La-based, original high-T_c materials, the key is to replace trivalent La with divalent Ba or Sr (Figure 2.16). Because the basic compound La_2CuO_4 is charge neutral, replacing La^{3+} with Ba^{2+} and still demanding charge neutrality results in a freed positive charge, donated to the valence band. On the other hand, in the case of oxygen depletion, removing an oxygen atom leaves behind two incomplete bonds, which upon completion free two positive charges, or holes.

The charge structure of the cuprates can be described in a very simple block form by separating the superconducting atomic layers from the non-conducting ones, as shown in Figure 2.10 for the case of $YBa_2Cu_3O_7$. This case is special due to the existence of CuO chains. But the general charge structure scheme for all cuprate superconductors is as follows: A stack of alternating layers of charge reservoirs (like BaO or CuO) and conducting CuO_2 layers that receive the mobile charges (usually holes) from the charge reservoirs. This internal redistribution of charge occurs in a fine tuned balance between the valences (affinities) of participating atoms. The undoped $YBa_2Cu_3O_6$ has an insulating CuO_2 layer where the d^9 electrons of Cu^{2+} ions are antiferromagnetically bound. On admitting oxygen to $YBa_2Cu_3O_6$ the chains are formed at the expense of taking away the necessary two electrons per atom from the CuO_2-plane. This is equivalent to doping the CuO_2 plane with positive mobile charges, or holes. The result is a hole-like superconductor. The process is different in $La_{2-x}Ba_xCuO_4$, where Ba^{2+} replaces La^{3+}, but the result is similar. When Ba is brought into the insulating La_2CuO_4 unit cells, it delivers two negative charges to the bonds, but it needs one more to make up for three electrons donated by La, so an extra electron is taken from CuO_2 to complete all bonds, and thereby a hole is released in the CuO_2 plane, to give conduction. Although hole- doping is the rule in high-T_c superconductors there are some cases where the mobile charges are electrons. Figure 2.17 shows these possibilities for both types of carriers where the aspect

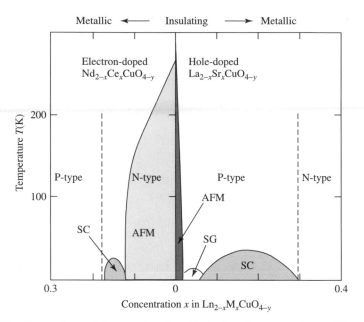

Figure 2.17 Comparison of electron-doped and hole-doped superconductor. The assignment SG refers to the possible existence of a spin glass phase.

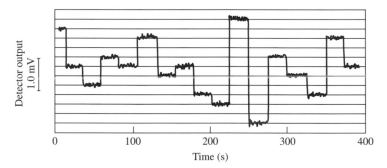

Figure 2.18 These measurements of flux jumps in a superconducting ring gave the first verification of the flux quantisation according to $\Phi_0 = \frac{2e}{h}$ in high-T_c superconductors. Adapted from [27].

of electron-hole symmetry is emphasized. Upon learning that these materials are dominated by hole carriers, one might ask: What about the pairing of carriers, is it still as we know it from the low-T_c metallic superconductors? The answer was found quite early by the Birmingham group [27] who demonstrated that flux quantization was intact in the new materials, and that the flux quantum was $\Phi_0 = \frac{h}{2e}$ as before. In this sense it did not matter what the sign of the charge is. The measurements which provided proof of pairing in high-T_c superconductors are shown in Figure 2.18.

2.6.5   Some remarks on the original idea which led to the discovery of cuprate superconductors

In spite of all efforts which have been spent on establishing a theoretical understanding of superconductivity in cuprate perovskites, a final conclusion on this matter has not yet been reached. It is nevertheless quite interesting to follow the reasoning which led Bednorz and Müller on the right track. We quote here a section taken directly from a paper given by Müller on the occasion of the celebration of the first 10 years of high-T_c superconductivity in 1996 [28]. (for simplicity we omit here the numerous references given):

> Strong electron-phonon interactions can occur in oxides, owing to polaron formation as well as mixed-valence states. This can go beyond the standard BCS theory. A phase diagram with a superconducting to bipolaronic insulator transition was proposed early by Chakraverty. A mechanism for polaron formation is the Jahn–Teller (JT) effect as studied by Höck *et al.* in a linear chain model. From it, one expects heavy polaron masses if the JT stabilization energy becomes comparable to or larger than the bandwidth of the degenerate orbitals, and thus localization. Intermediate polarons are expected if the JT energy is not too large compared to the bandwidth. We recall that the JT theorem states the following: A nonlinear molecule or a defect in a crystal lattice exhibiting an electron degeneracy will spontaneously distort in lowering its symmetry, thereby removing its degeneracy. Isolated Fe^{4+}, Mn^{3+}, Ni^{3+} and Cu^{2+} in an octahedral oxygen environment show strong JT effects because their incompletely occupied e_g orbitals, transforming as $3z^2 - r^2$ and $x^2 - y^2$, point towards the negatively charged oxygen ligands. Although $SrFe^{4+}O_3$ is a distorted perovskite insulator, $LaNiO_3$ is a JT undistorted metal in which the transfer energy b_π of the e_g electrons of the Ni^{3+} is large enough to quench the JT distortion. On the other hand, $LaCuO_3$ is a metal containing only the non-JT Cu^{3+}. Therefore, it was decided to investigate and 'engineer' nickel- and copper-containing oxides, with reduced bandwidth $\simeq b_\pi$, partially containing Ni^{3+} or Cu^{2+} states. The JT polaron proposed in 1983 was one envisaged to lead to a heavy mass of the particle; and intermediate, mobile polaron suited for superconductivity was not considered. However, in the ferromagnetic conductor $La_{1-x}Ca_xMnO_3$, a giant oxygen isotope effect has been discovered most recently and ascribed to the presence of intermediate JT polarons due to the Mn^{3+} ions in the oxide. This important and new evidence for the existence of intermediate JT polarons is quite in favor of the new original concept.
>
> In Rüsclikon, there was a tradition of more than two decades of research in insulating oxides that undergo structural and ferroelectric transitions, which was a strong motivation to pursue the program. Furthermore, in 1979 the present author had started to work in the field of granular superconductors in which small Al grains are surrounded by amorphous Al_2O_3. In these systems T_c's have been reported to be as high as 5 K, compared to pure Al with a T_c of 1.1 K. In our laboratory, the search for superconductivity was initiated together with J. G. Bednorz in mid-summer of 1983. our efforts first concentrated on Ni^{3+}-containing perovskites, such as mixed crystals of $LaNiO_3$ and $LaAlO_3$. In these unpublished efforts, the metallic behavior of the various synthesized double and triple oxides was measured, and at low temperatures

they exhibited localization upon cooling. This indicated the possible existence of JT polarons, however, without any signs of superconductivity. In Figure 2.9, results of these efforts are reproduced. In late summer of 1985, the efforts were shifted to copper-containing compounds, such as $LaCuO_3$. Because Cu^{3+} has two electrons in the e_g subshell, the latter is half-filled. Thus, its ground state is not degenerate. It was clear that an oxide with mixed Cu^{2+}/Cu^{3+} or Cu^{3+}/Cu^{4+} had to be tried.

At this stage, Bednorz became aware of a paper by Michael, Er-Rakho and Raveau on the mixed perovskite $BaLa_4Cu_5O_{13.4}$, exactly meeting the requirements of mixed valence. The French authors had shown that this mixed oxide, a metal at room temperature and above, contained Cu^{2+} and Cu^{3+}. Thus, we tried to reproduce it, at the same time continuously varying the Cu^{2+}/Cu^{3+} ratio by changing the Ba concentration on $Ba_xLa_{5-x}Cu_5O_{5(3-y)}$, and we looked for superconductivity. A representative and concise account of the discovery of superconductivity in $Ba_xLa_{5-x}Cu_5O_{5(3-y)}$ and the relevant superconducting phase present appeared in the September 4, 1987, issue of *Science* and, in more detail, in the first of the two Nobel lectures in 1987, and would exceed the scope of this contribution.

2.6.6 Thermal fluctuations of the superconducting condensate. A preliminary discussion

Already in 1988 two more significant series of high-T_c superconductors were synthesized: one Bi-based, and one Tl-based. Later, a Hg-based series was also found. These series gave T_c's higher than the 93 K of YBCO, as could already be seen from Table 2.7. At the same time, however, their anisotropy was found to be much stronger than in YBCO, and the coherence length is very short, both a disadvantage and a challenge for their practical use. Why the coherence length is so short can be seen from Eq. 2.1, which can be obtained from the BCS theory, and may be assumed to hold approximately even in high-T_c. Equation 2.1 predicts a very short coherence length in high-T_c due to the combination of small Fermi surfaces with low v_F, caused by the low concentration of holes, and the high T_c. The expression used here may also be derived from the uncertainty relation, and is not sensitive to the pairing mechanism.

This prediction for ξ_0 is approximately followed in real systems. While $\xi_0 = 1600$ nm in Al, it is only of the order of 1 nm in the high-T_c compounds. This gives the thermal energy of kT a chance to create fluctuations of large spatial extent in the order parameter since they will occur in a typical volume ξ^3. This volume is of the order of $(1/1600)^3 = 2.4 \times 10^{-10}$ times smaller in high T_c materials than in aluminium. The energy barrier ΔE against the creation of a fluctuation in the order parameter is proportional to the respective volumes ξ^3 in each case, and also proportional to the square of the thermodynamic field, i.e.

$$\Delta E \approx \frac{1}{2\mu_0}\xi^3 B_c^2 \tag{2.2}$$

The ratio of condensation energies in a volume of $\bar{\xi}^3$ in the two materials is therefore

$$\frac{\Delta E_{YBCO}}{\Delta E_{Al}} = \frac{(\bar{\xi}^3 B_c^2)_{YBCO}}{(\bar{\xi}^3 B_c^2)_{Al}} \tag{2.3}$$

where we have written $\bar{\xi}^3$ to indicate a geometric average of $\xi_1\xi_2\xi_3$ for the anisotropic cases. A numerical estimate of the barrier ratio expressed by Eq. 2.3 gives 1.3×10^{-6}, favouring fluctuations enormously in YBCO as compared to Al. The ratio of probabilities for thermal fluctuations at a temperature T can be written as

$$\frac{P_{YBCO}}{P_{Al}} = \frac{p_1 e^{-\left(\frac{\Delta E}{kT}\right)_{YBCO}}}{p_2 e^{-\left(\frac{\Delta E}{kT}\right)_{Al}}} = \frac{p_1 e^{-\left(\frac{\bar{\xi}^3 B_c^2}{2\mu_0 kT}\right)_{YBCO}}}{p_2 e^{-\left(\frac{\bar{\xi}^3 B_c^2}{2\mu_0 kT}\right)_{Al}}} \tag{2.4}$$

Let us try to gain some insight into the consequences of Eq. 2.4. Here the exponents are far more important than the unknown coefficients p_1 and p_2. From measurement, B_c in aluminium is 0.02 T, and B_c in YBCO may be taken to be about 1.5 T. The ξ_0-values for YBCO and Al, can be estimated from Eq. 2.1. For Al, $v_F = 2.03 \times 10^6$ and $T_c = 1.17$ K which gives the estimate $\xi_0 = 2.2 \times 10^3$ nm. This is not too far away from the measured value, $\xi_0 = 1.6 \times 10^3$ nm. For YBCO, v_F is about 100 times lower than in Al. This tells us that the ratio of coherence lengths $\xi_{Al}/\bar{\xi}_{YBCO}$ should be about 10^4, which is of the correct magnitude compared to measured values. Furthermore, the ratio between fluctuating volumes $(\xi_{Al})^3/(\bar{\xi}_{YBCO})^3$ is about 10^{12}. So, not only is the fluctuating volume in YBCO a factor 10^{-12} smaller than in Al, but since the fluctuations take place at temperatures near 100 K in YBCO, versus 1 K in Al, the available energy to drive the fluctuations in YBCO is $\approx 10^2$ greater. The effect of thermal fluctuations should therefore be expected to be hugely greater in YBCO than in Al. We can now estimate the exponents in Eq. 2.4 to gain an impression of the difference in probabilities for thermal fluctuations in the respective superconducting volumes of size $\bar{\xi}^3$. For Al we find near T_c:

$$\left(\frac{\Delta E}{kT}\right)_{Al} \approx \left(\frac{\bar{\xi}^3 B_c^2}{2\mu_0 kT_c}\right)_{Al} \approx 4.0 \times 10^7 \tag{2.5}$$

Clearly, this is an impossible barrier to overcome. Hence, thermally driven fluctuations in the density of Cooper pairs in a volume $\bar{\xi}^3$ in Al occur with practically vanishing probability.

Let us now compare this with the situation on YBCO. We find

$$\left(\frac{\Delta F}{kT}\right)_{\text{YBCO}} \approx \left(\frac{\bar{\xi}^3 B_*^2}{2\mu_0 kT_c}\right)_{\text{YBCO}} \approx 1 \qquad (2.6)$$

In this estimate, we have written $\bar{\xi}^3 = \xi_{ab}^2 \times \xi_c$, and taken a rough average of tabulated values to find $\bar{\xi}^3 = (1.5 \times 10^{-9}\,\text{m})^2 \times 0.5 \times 10^{-9}\,\text{m} \approx 2 \times 10^{-27}\,\text{m}^3$. With a resulting probability $P_{\text{YBCO}} \propto e^{-1}$ the conclusion is that thermally driven fluctuations that suppress superconductivity in a $\bar{\xi}^3$-volume must occur profusely. Conversely, the same argument says that in the normal state above T_c, fluctuations will occur with similar likelihood, in this case creating superconducting $\bar{\xi}^3$-volumes in the normal matrix. In conclusion, the superconducting transitions in YBCO and other high-T_c materials, occur with such violent fluctuation of the superconducting wavefunction near T_c that the behaviour is bound to be non-classical, non-mean-field like, quite different from low-T_c physics. At this point one may even wonder how a substance with such violent fluctuations can sustain superconductivity at all near T_c. This is a very valid question indeed that will be dealt with in Chapters 9 and 10. Clearly, the superconducting phase transition in YBCO and other high-T_c superconductors is fundamentally different from low-T_c materials. The important question is *how* it is different.

At this stage, let us also ask: What topological form will the fluctuations discussed above have? Will they always be in the shape of a $\bar{\xi}^3$-volume as defined above? Not necessarily. There is sufficient thermal energy that the volume could be extended in one coordinate, for instance in a tubular shape, and still have a very good chance for the fluctuation to occur. As is now well established by computer-simulation there is firm prediction for a high rate of generation of *thermally induced vortex loops*. According to the prediction, this behaviour is accompanied by a vortex 'blow-out' phase transition within a vortex liquid phase [29, 30]. Figure 2.19 attempts to sketch such possibilities. These issues are discussed more precisely in Advanced Topics.

It should be kept in mind for comparison, that on lowering the temperature of a high-T_c compound to liquid helium temperatures, fluctuations even in these compounds would occur with a low probability. It is the combination of high temperature, short coherence length and anisotropy which make the high-T_c compounds so different in terms of fluctuation properties at elevated temperatures. Figure 2.20 shows a highly unusual phenomenon in a high-T_c superconductor. The characteristic rule of superconductivity, that diamagnetic response is found only below T_c, is broken; diamagnetism is observed above T_c. The effect is explained by the presence of Cooper-pairs already formed above T_c.

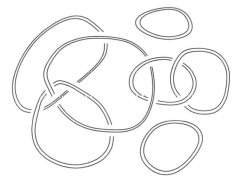

Figure 2.19 Sketch of possible configurations of thermally generated flux loops in high-T_c superconductors in the ordered phase as discussed by Tesanovic [31, 32], and Nguyen and Sudbø [29, 30].

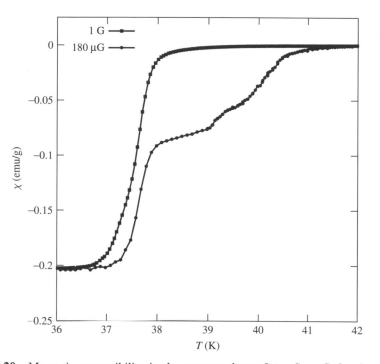

Figure 2.20 Magnetic susceptibility in the superconductor La$_{1.85}$Sr$_{0.15}$CuO$_4$ showing the presence of diamagnetic response (upper curves) even above T_c due to the presence of significant concentration of Cooper pairs above the superconducting phase transition. T_c is known to be 38 K from other independent measurements. The numbers given in the figure, 1 G and 180 μG, refer to the ac-field used in the measurements. Only of the lowest field did the weak superconducting resparse show up above T$_c$. After Thisted *et al.* [33].

2.7 Heavy fermion superconductors

The term 'heavy fermion superconductors' refers to a class of superconductors characterized by very high effective electron mass, m^*, 2-3 orders of magnitude larger than the free electron mass. The first discovery of such superconductors was made by Stoglich and coworkers in 1979 [34]. The compound was $CeCu_2Si_2$. This discovery was later followed by UBe_{13} and UPt_3. The main characteristics of some of these superconductors are given in Table 2.8. Figure 2.21 shows the structure of the material where superconductivity was first discovered.

Typical for these superconductors is that they contain elements like the rare earth Ce with two 4f electrons, and the actinide U with 5f electrons. When these f electrons mix or hybridize with conduction electrons a sharp band with high density of states $D(E)$ arises at the Fermi energy. Since $D(E) \propto (m^*)^{3/2}$

Table 2.8 Heavy fermion superconductor properties.

Superconductor	$T_c(K)$	$H_{c1}(mT)^{\dagger}$	$H_{c2}(T)$	$\lambda(\mu m)$	$\xi(nm)$	κ	m^*/m_0
UPt_3	0.48	0.7	$1.8//c$ $2.3//a,b$	1	$12//a,b$ $14//c$	90	180
UBe_{13}	0.87	6.7	10.2	~1	9.5	55	260
URu_2Si_2	1.2	0.2	$2.5//c$ $10.5//a$	~1	$7.5//a$ $13//c$	260	140
$CeCu_2Si_2$	~1	~2.5	$2.1//a$ $2.5//c$	0.5	9		380

Reproduced from the website of Zbigniew Koziol http://www.iyp.org/vp/super/heavy/reference.html.
†the value is estimated.

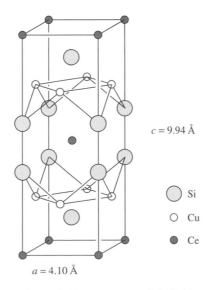

$c = 9.94\ \text{Å}$

◯ Si
○ Cu
● Ce

$a = 4.10\ \text{Å}$

Figure 2.21 Structure of CeCuSi.

one ascribes the high density of states to the high effective mass. Hence the name referred to above. Clearly, this property carries over to other physical characteristics as well. The electronic specific heat is directly proportional to the effective mass. Hence the electronic specific heat is hugely increased compared to that of usual metals. In the superconducting state the penetration depth λ is proportional to $(m^*)^{1/2}$. Hence this quantity is also very large, as seen in Table 2.8. Other heavy fermion superconducting compounds not included in the table are $Ce_3Bi_4Pt_3$, $CeNi_2Ge_2$, UPd_2Al_3 and YBiPt.

Recent research in heavy fermion superconductors indicate that they are a quite diverse group, showing distinctly non-universal behaviour. Some, like UPd_2Al_3 exhibit coexistence of long-range antiferromagnetic order and a Landau Fermi-liquid phase in a range $T_c < T < T_N$. Another group, like $CeCu_2Si_2$ (see Figure 2.21), are intermetallics that show pronounced deviations from Landau Fermi-liquid behaviour, due to fluctuations in the local magnetization. For these and other reasons heavy fermion superconductors are still an active area of research. It is still not clear whether a generalized Fermi-liquid theory can be found to describe all superconductors in this class.

2.8 MgB₂ superconductor

Magnesium diboride, MgB_2, is a metallic compound known since the 1950s. It is a hard and brittle material with a hexagonal structure, as seen in Figure 2.22, with magnesium located in the external corners and in the face centre positions of the upper and lower hexagon, and with boron located in the inner hexagon, rotated by 60° with respect to the Mg hexagons.

Superconductivity in MgB_2 was discovered as late as 2001 [35, 36], with T_c at 39 K, a record by far in an ordinary metallic compound. This value of T_c is close to what has been considered the maximum possible by pairing caused by electron–phonon interaction. MgB_2 has several interesting physical properties.

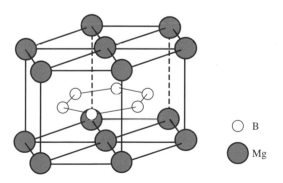

○ B

● Mg

Figure 2.22 Structure of MgB₂ superconductor.

Table 2.9 Special properties of some superconductor categories

Superconductor	Category	T_c(K)	B_{c2}(T)	ξ(nm)	λ(nm)	Special properties
MgB$_2$	Layered non-cuprate	39	19–40	2–5	85–180	Record T_c for BCS-type metal
Nb$_3$Ge	A15-phase	23.2	39	3	90	High T_c, lattice instability
Nb$_3$Sn		17.9	24	3	65	
V$_3$Si		17.0	23	3	60	
PbMo$_6$S$_8$	Chevrel phase: Mo-based chalcogenides	15.2	60	2.2	215	High T_c, high B_{c2}
LaMo$_6$Se$_8$		11	5			
UPd$_2$Al$_3$	Heavy fermion compounds	2.0	~40			High carrier mass: $m \sim 10^2 m_e$
CeCu$_2$Si$_2$			~2			
UPt$_3$		0.46	1.9	12–14	600	
K$_3$C$_{60}$	Fullerene based	19.3	17–32	~3	240	Organic, high T_c, molecular based
Rb$_3$C$_{60}$		29.6	38	~2		
Pb$_{2.7}$Tl$_{2.2}$C$_{60}$		45				
$\kappa-$(BEDT$-$TTF)$_2$Cu(NCS)$_2$	Charge transfer salt Bechgaard salts, spin density wave	8.7–10.5		~1–10	980	Organic, quasi-2D
(TMTSF)$_2$X X = ClO$_4$, PF$_6$, ReO$_4$		0.9–1.4				Organic, quasi-1D
Ba$_{1-x}$KBiO$_3$; x = 0.4	Non-cuprate perovskite	30				Record T_c for non-cuprate perovskite

	T_c(K)	B_{c2}(T)	ξ_{ab}	ξ_c	λ_{ab}	λ_c	Special properties
La$_{2-x}$Ba$_x$CuO$_4$; x = 0.2	30		3.3		290		Perovskite high-T_c cuprate superconductors
YBa$_2$Cu$_3$O$_{7-\delta}$	93	115	2.5	0.5	150	500	High T_c, high B_{c2}, extreme Type II, anisotropic layered
Bi$_2$Sr$_2$Ca$_2$Cu$_3$O$_{10}$	110	198	2.9	0.09			
TlBa$_2$Ca$_2$Cu$_3$O$_9$	123				173	480	
HgBa$_2$Ca$_2$Cu$_3$O$_{8+\delta}$	133	190	1.3		130	3500	

In addition to the high T_c-value, it has been found to have two quite distinct energy gaps caused by pairing involving electrons from different parts of the Fermi surface. The high T_c value is a compromise between the two gap values. The high T_c is favoured both by the light element boron and by strong covalent binding. The isotope effect reflected in the dependence of T_c on isotope mass is a strong indication of BCS type electron–phonon pairing mechanism. The most important phonon modes involved in the pairing are in the B-planes.

The upper critical magnetic field B_{c2} has not been determined with any accuracy yet. Values as high as 32 T have been reported, but much lower values have also been found.

MgB_2 shows promise for several applications including transport of high currents and Josephson junctions. Several other compounds of related composition have been found, but all with lower T_c than MgB_2.

2.9   Summarizing remarks

In this chapter we have attempted to review some basic facts about the most important materials in superconductivity research and applications today. Necessarily such a review will be sketchy, considering the vast literature available on these subjects. Essentially, this chapter can only be an appetizer for a deeper look into this immensely rich area. Still, it is hoped that it may be sufficient for those students who want to get a broad overview, but who do not necessarily intend to pursue the subjects at a much deeper level.

Finally, in Table 2.9 we summarize some special properties associated with different categories of superconductors, so that the reader can in one glance get an idea about how they differ in some important respects.

3

Fermi-Liquids and Attractive Interactions

3.1 Introduction

In solid state physics, *The Theory of Everything* is given by a Hamiltonian of electrons moving more or less freely through a lattice of ions. Electrons and ions have a kinetic energy, and they interact with each other via the Coulomb interaction. This Hamiltonian is in some sense trivial, it is simply the sum of potential and kinetic energies. However, as we know, the amazing number of different phenomena taking place in condensed matter physics goes to show that trivial Hamiltonians need not exhibit trivial ground states. Superconductivity is an example of this. Another notable example is the fractional quantum Hall effect. The main difficulty is to be able to identify correctly the manybody renormalized low-energy excitations in various circumstances. This is what is non-trivial, and the excitations vary enormously, ranging from bosonic spin-waves in Mott–Hubbard insulators and quantum Heisenberg antiferromagnets to fermionic bogoliubons in superconductors via anyonic excitations with fractional charge in the fractional quantum Hall effect. These systems all have the same underlying microscopic Hamiltonian, and the diversity of the possible low-energy excitations basically reflect that they are *emergent* long-wavelength properties whose basic characteristics to some extent are a result of the initial conditions in which we have prepared the underlying electron-ion system.

The grand unified theory of solid state physics is given by the Hamiltonian

$$H = H_{e-e} + H_{ion-ion} + H_{e-ion} \tag{3.1}$$

where H_{e-e} is the Hamiltonian for electrons moving through the solid, $H_{ion-ion}$ is the Hamiltonian for the ions in the solid, and H_{e-ion} is the Hamiltonian describing the coupling between the electrons and the ions. For electrons with

Superconductivity: Physics and Applications   Kristian Fossheim and Asle Sudbø
© 2004 John Wiley & Sons, Ltd   ISBN 0-470-84452-3

mass m and ions with mass M, we have

$$H_{e-e} = \sum_i \frac{p_i^2}{2m} + \sum_{i,j} V_{\text{Coulomb}}^{e-e}(\mathbf{r_i} - \mathbf{r_j})$$

$$H_{\text{ion}-\text{ion}} = \sum_i \frac{P_i^2}{2M} + \sum_{i,j} V_{\text{Coulomb}}^{\text{ion}-\text{ion}}(\mathbf{R_i} - \mathbf{R_j})$$

$$H_{e-\text{ion}} = V_{\text{Coulomb}}^{e-\text{ion}}(\mathbf{r_i} - \mathbf{R_j}) \tag{3.2}$$

Here $\mathbf{r_i}$ denotes an electron-position, while $\mathbf{R_i}$ denotes an ion-position. The dominant effect of the Coulomb interactions between the ions is to freeze them into a lattice. Once we have assumed the existence of this ion-lattice ground state, $V_{\text{Coulomb}}^{\text{ion}-\text{ion}}(\mathbf{R_i} - \mathbf{R_j})$ has performed its principle task and we no longer need to consider it. In a perfect solid with no lattice defects and no lattice vibrations, the electrons may be considered as moving in an external periodic potential set up by the ions. Ignoring electron–electron interactions, this is basically a one-body problem of an electron moving in a periodic external potential. The exact eigenstates are Bloch-states. These states are adjusted to the perfect lattice and are not scattered at all, a perfect crystal is transparent to Bloch-electrons. Including lattice vibrations around equilibrium positions of the ions, but still no lattice defects, the Bloch-states will couple to the lattice vibrations, whose quantized excitations are bosonic phonons. We then need to include coupling of Bloch-electrons to phonons. Finally, we should include Coulomb-interactions between electrons. Hence, a sensible way of viewing the above Hamiltonian in the absence of lattice defects (we shall ignore such site-disorder) is that of Bloch-electrons with Coulomb-interactions coupled to phonons. This is the situation we shall consider. We start by describing electrons ignoring interaction effects altogether, then modify the description in the necessary fashion when $V_{\text{Coulomb}}^{e-e}(\mathbf{r_i} - \mathbf{r_j})$, and finally include the electron-phonon interaction.

In this chapter, we will give a brief introduction to the concept of Landau Fermi-liquid theory, which forms the cornerstone of our current understanding of ordinary metals. We will also present the effective Hamiltonian for phonon-mediated superconductivity, which forms the basis for the celebrated BCS theory of conventional low-temperature superconductivity arising out of good metals. The Fermi-liquid theory is a spectacularly successful framework which transforms the daunting problem of interacting electron systems into a theory which conceptually is almost as simple as the non-interacting electron gas. We will also give a brief review of *why* such a drastic simplification of the problem actually works, even in extreme cases like the heavy fermion compounds. The accepted theory of conventional phonon-mediated superconductivity we have available today for superconductivity (namely the BCS theory, named after the originators of the theory, Bardeen, Cooper, and Schrieffer [7]), basically assumes that the interacting electron gas forming the metal that ultimately becomes a superconductor, exhibits well defined *electron-like* excitations that can form Cooper-pairs. Theories for superconductivity based on metallic states which do *not* exhibit such electron-like excitations have been proposed, notably in the context of attempting to explain high-temperature superconductivity in copper oxides. None of them have so far achieved what the Fermi-liquid-based BCS theory did in such a brilliant manner–to make predictions for a complicated many-body problem, with right numbers. The Fermi-liquid theory, and the BCS theory, which we will introduce the reader to in this chapter, are certainly two of the most successful theories in condensed matter physics ever.

3.2   The non-interacting electron gas

The language we will use to formally describe manybody systems in this book, is second quantization, mainly due to its convenience in the context of manybody systems with non-definite number of particles. This is precisely the situation which is forced upon us for superconductors. The basic building blocks of this formalism, are the *annihilation-* and *creation-*operators for electrons with linear momentum $\mathbf{k}$ and spin σ. The operator $c_{\mathbf{k},\sigma}^{\dagger}$ creates a state of an electron in a plane-wave (or Bloch-state) and spin state with quantum numbers $(\mathbf{k}, \sigma)$, while $c_{\mathbf{k},\sigma}$ destroys such a state. The fact that electrons are fermions obeying the Pauli principle is is taken into account by postulating *anticommutation* relations for these operators, namely

$$c_{\mathbf{k},\sigma}^{\dagger} c_{\mathbf{k}',\sigma'} + c_{\mathbf{k}',\sigma'} c_{\mathbf{k},\sigma}^{\dagger} = \delta_{\mathbf{k},\mathbf{k}'}\delta_{\sigma,\sigma'} \tag{3.3}$$

where the δ-symbols are Kronecker-deltas, $\delta_{\alpha,\beta} = 1$ if $\beta = \alpha$, and $\delta_{\alpha,\beta} = 0$ otherwise. In this language, the operator that counts the number of electrons in a given state defined by the quantum numbers $(\mathbf{k}, \sigma)$ is given by the operator $n_{\mathbf{k},\sigma} = c_{\mathbf{k},\sigma}^{\dagger} c_{\mathbf{k},\sigma}$. The Hamiltonian of a non-interacting electron system of states $(\mathbf{k}, \sigma)$ where the energy of such states is given by $\varepsilon_{\mathbf{k},\sigma}$, is simply the energy per state multiplied by the number operator of such states, and finally summed over all states. Normally, we assume that the energies are independent of σ, i.e. we have $\varepsilon_{\mathbf{k}}$. Then, the Hamiltonian is given by

$$H = \sum_{\mathbf{k},\sigma} \varepsilon_{\mathbf{k}} c_{\mathbf{k},\sigma}^{\dagger} c_{\mathbf{k},\sigma} \tag{3.4}$$

The ground state of this system is given by filled states $(\mathbf{k}, \sigma)$ up to the Fermi-momentum $\mathbf{k}_{\mathrm{F}}$, where we denote the energy of the uppermost filled state by ε_{F}.

This Fermi-sea is completely inert in the absence of interactions, and we will denote this ground state by $|\phi_0\rangle$. It has the important property that

$$n_{\mathbf{k},\sigma}|\phi_0\rangle = \Theta(\varepsilon_F - \varepsilon_{\mathbf{k}}) \tag{3.5}$$

where the Heaviside step function Θ has the property $\Theta(x) = 1, x > 0, \Theta(x) = 0, x < 0$. Physically, the above is simply a statement that in the ground state of the non-interacting electron gas, all levels below the Fermi-level are occupied with (maximal) occupation number equal to 1 per spin state in accordance with the Pauli principle, all states above the Fermi-level are unoccupied.

An extremely important quantity in a many-particle system is the probability amplitude of finding a particle in a state $(\mathbf{k}', \sigma')$ at time t', given that it was in state $(\mathbf{k}, \sigma)$ at time t. This quantity is the so-called *single-particle* propagator, or Green's function, of the system. For the non-interacting case we will denote it by

$$G_0(\mathbf{k}, \mathbf{k}', t - t') = G_0(\mathbf{k}, t - t')\delta_{\mathbf{k}, \mathbf{k}'} \tag{3.6}$$

It plays a central role in describing the electronic structure of any manybody electron system. The right hand side of the relation above follows from the fact that in a non-interacting system, a particle cannot be scattered out of a plane-wave state defined by a wavenumber $\mathbf{k}$. Note also that G_0 only depends on t and t' only via the combination $t - t'$ when H is time-independent. Mostly, it is convenient to work with the Fourier-transform of the above quantity, namely

$$G_0(\mathbf{k}, \omega) = \int_{-\infty}^{\infty} dt e^{i\omega t} G_0(\mathbf{k}, \omega)$$

$$= \frac{1}{\omega - \varepsilon_{\mathbf{k}} + i\delta_{\mathbf{k}}} \tag{3.7}$$

where $\delta_{\mathbf{k}} = \delta sign(\varepsilon_{\mathbf{k}} - \varepsilon_F)$, and δ is an infinitesimal positive quantity. The pole in the propagator is given by $\omega = \varepsilon_{\mathbf{k}} - i\delta_{\mathbf{k}}$, which defines the spectrum of the single-particle excitations of the system. The imaginary part plays the role of a damping term for the excitations, i.e. an inverse lifetime. In the free-electron case, this damping term is infinitesimal in the free-electron gas, i.e. the excitations of energy $\varepsilon_{\mathbf{k}}$ are infinitely long lived. The imaginary part of $G_0(\mathbf{k}, \omega)$ gives the single-particle spectral weight

$$A(\mathbf{k}, \omega) = -\frac{1}{\pi}\Im[G_0(\mathbf{k}, \omega)] = \frac{1}{\pi}\frac{\delta_{\mathbf{k}}}{(\omega - \varepsilon_{\mathbf{k}})^2 + \delta_{\mathbf{k}}^2} = \delta(\omega - \varepsilon_{\mathbf{k}}) \tag{3.8}$$

which provides direct information on the occupation of a plane-wave state $|\mathbf{k}\rangle$ of energy ω. Hence, the momentum distribution $n(\mathbf{k})$ of the free electron gas is given by $A(\mathbf{k}, \omega)$ integrated over all frequencies ω, while the $\mathbf{k}$-space sum of

$A(\mathbf{k}, \omega)$ gives the density of states $D(\omega)$. At zero temperature $n(\mathbf{k})$ is given by the Fermi-Dirac distribution function

$$n(\mathbf{k}) = \Theta(\varepsilon_F - \varepsilon_\mathbf{k})$$

Note that the magnitude of the step-discontinuity in this momentum distribution function is 1 at the Fermi-level. Going back to Eq. 3.7, we see that this value of the discontinuity happens to be the same as the value of the residue at the pole of the free-electron propagator. As we shall see below, this is not a coincidence.

3.3 Interacting electrons, quasiparticles and Fermi-liquids

When spin-independent pairwise-interactions between electrons are taken into account (we ignore multibody interactions, as usual), the Hamiltonian includes a two-particle scattering term where two incoming electrons in states $\mathbf{k}, \sigma$ and $\mathbf{k}', \sigma'$ are scattered to states $\mathbf{k} + \mathbf{q}, \sigma$ and $\mathbf{k}' - \mathbf{q}, \sigma'$ with a matrix element V which in principle depends on the initial scattering states and the momentum transfer $\mathbf{q}$ in the process. We assume that the two-particle scattering is elastic. Thus, we may under quite general conditions write the total Hamiltonian as follows

$$H = \sum_{\mathbf{k},\sigma} \varepsilon_\mathbf{k} c^\dagger_{\mathbf{k},\sigma} c_{\mathbf{k},\sigma} + \sum_{\mathbf{k},\mathbf{k}',\mathbf{q},\sigma,\sigma'} V_{\mathbf{k},\mathbf{k}',\mathbf{q}} c^\dagger_{\mathbf{k}+\mathbf{q},\sigma} c^\dagger_{\mathbf{k}'-\mathbf{q},\sigma'} c_{\mathbf{k},\sigma} c_{\mathbf{k}',\sigma'} \qquad (3.9)$$

A typical term in the interaction term of the Hamiltonian is illustrated in Figure 3.1.

For such an interacting system, a rather remarkable fact is that we can again write down an exact expression for the single-particle propagator, denoted by $G(\mathbf{k}, t)$, i.e. the probability amplitude that if a particle is found in state $|\mathbf{k}, \sigma\rangle$ at time 0, it will still be found in this state at time t. This expression is known

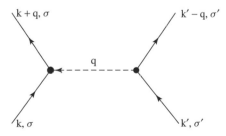

Figure 3.1 A typical two-body scattering event contributing to the interaction part of the Hamiltonian, Eq. 3.9. The dashed line connecting the electrons is an illustration of $V_{\mathbf{k},\mathbf{k}',\mathbf{q}}$ denoting some *effective* interaction, such as Coulomb-interaction, phonon-mediated electron-electron interaction, and soforth.

as Dysons equation, and is given by, when expressed in terms of the Fourier-transform of $G(\mathbf{k}, t)$

$$G^{-1}(\mathbf{k}, \omega) = G_0^{-1}(\mathbf{k}, \omega) - \Sigma(\mathbf{k}, \omega) \tag{3.10}$$

where $\Sigma(\mathbf{k}, \omega)$ is the so called one particle irreducible self-energy. In general, this is a complex quantity. It has the following physical interpretation: The real part of $\Sigma(\mathbf{k}, \omega)$, $\Sigma_R = \Re(\Sigma(\mathbf{k}, \omega))$, gives a reactive shift of the excitation spectrum of the interacting electrons, such that this spectrum is given by $\tilde{\varepsilon}_{\mathbf{k}} = \varepsilon_{\mathbf{k}} + \Sigma_R$. The imaginary part, $\Sigma_I = \Im(\Sigma(\mathbf{k}, \omega))$ essentially gives the inverse lifetime of the interacting electron in the plane-wave state $\mathbf{k}, \sigma$, i.e. $\Sigma_I \sim \tau_{\mathbf{k}}^{-1}$, where $\tau_{\mathbf{k}}^{-1}$ is the lifetime of the electron in the state $|\mathbf{k}, \sigma\rangle$. Let us work this out in more detail.

Inverting Eq. 3.10, we have

$$
\begin{aligned}
G(\mathbf{k}, \omega) &= \frac{G_0(\mathbf{k}, \omega)}{1 - G_0(\mathbf{k}, \omega)\Sigma(\mathbf{k}, \omega)} \\
&= \frac{1}{\omega - \varepsilon_{\mathbf{k}} - \Sigma(\mathbf{k}, \omega)}
\end{aligned} \tag{3.11}
$$

where we have ignored $\delta_{\mathbf{k}}$ in the free electron propagator compared to the imaginary part of Σ, which we anticipate to be finite. We now want to bring this onto a form such that it resembles, as much as possible, Eq. 3.4 for the free-electron propagator. We imagine that, in some sense, the interactions are not 'too strong', and write the complex self-energy on the form $\Sigma = \Sigma_R + i\Sigma_I$, and assume that the imaginary part (the damping-piece) is small compared to the real part (the reactive-shift piece), $|\Sigma_I|/|\Sigma_R| \ll 1$. The pole in $G(\mathbf{k}, \omega)$ determines the exact single-particle excitation spectrum of the interacting system. This spectrum is therefore determined from the equation

$$\omega - \varepsilon_{\mathbf{k}} - \Sigma_R(\mathbf{k}, \omega) - i\Sigma_I(\mathbf{k}, \omega) = 0 \tag{3.12}$$

To zeroth order, we ignore the imaginary part of Σ. This gives the self-consistency equation for the reactively shifted single-particle spectrum

$$\omega = \tilde{\varepsilon}_{\mathbf{k}} = \varepsilon_{\mathbf{k}} + \Sigma_R(\mathbf{k}, \tilde{\varepsilon}_{\mathbf{k}}) \tag{3.13}$$

Taking into account a small imaginary part of Σ means that the frequency giving the pole in the propagator is shifted, and this in turn implies that we have to Taylor expand the real part of Σ around the frequency $\omega = \tilde{\varepsilon}_{\mathbf{k}}$, as is evident from Eq. 3.13. Thus, we consider the real part of Σ Taylor-expanded to first order around $\omega = \tilde{\varepsilon}_{\mathbf{k}}$, as follows

$$\Sigma_R(\mathbf{k}, \omega) = \Sigma_R(\mathbf{k}, \tilde{\varepsilon}_{\mathbf{k}}) + (\omega - \tilde{\varepsilon}_{\mathbf{k}})\frac{\partial\Sigma_R}{\partial\omega}\Big|_{\omega=\tilde{\varepsilon}_{\mathbf{k}}} \tag{3.14}$$

On the other hand, since we are computing to first order in Σ_I, we consider the imaginary part of Σ evaluated at $\omega = \tilde{\varepsilon}_k$. Let us now insert these expressions into Eq. 3.12. We obtain

$$\omega - \underbrace{[\varepsilon_k + \Sigma_R(k, \tilde{\varepsilon}_k)]}_{=\tilde{\varepsilon}_k} - (\omega - \tilde{\varepsilon}_k) \frac{\partial \Sigma_R}{\partial \omega}\Big|_{\omega = \tilde{\varepsilon}_k} - i \Sigma_I(k, \tilde{\varepsilon}_k) = 0$$

$$(\omega - \tilde{\varepsilon}_k)\left[1 - \frac{\partial \Sigma_R}{\partial \omega}\Big|_{\omega = \tilde{\varepsilon}_k}\right] - i \Sigma_I(k, \tilde{\varepsilon}_k) = 0 \qquad (3.15)$$

We now define the lifetime τ_k of the interacting electrons in state $|k, \sigma\rangle$, as well as the quantity z_k by the following equations

$$\frac{1}{\tau_k} = -\frac{\Sigma(k, \tilde{\varepsilon}_k)}{1 - \frac{\partial \Sigma_R}{\partial \omega}\big|_{\omega = \tilde{\varepsilon}_k}}$$

$$z_k = \frac{1}{1 - \frac{\partial \Sigma_R}{\partial \omega}\big|_{\omega = \tilde{\varepsilon}_k}} \qquad (3.16)$$

Inserting all of this in the exact expression for G, we find

$$G(k, \omega) = \frac{z_k}{\omega - \tilde{\varepsilon}_k + \frac{i}{\tau_k}} \qquad (3.17)$$

If we compare Eq. 3.17 with Eq. 3.7, we see that the main effects of interactions between electrons are to produce a shift of the single-particle spectrum as well as a finite lifetime for the interacting electrons in a given plane-wave state. In addition, the *residue* of the propagator, the quantity z_k, is reduced from the value 1 which it had in the non-interacting case. This is seen directly from Eq. 3.16, as Σ_R on physical grounds is expected to be a decreasing function of frequency: At high energies, the effect of interactions is expected to be smaller than at low energies, because rapidly moving electrons have less time to interact with their surroundings than slow electrons. Hence, $\partial \Sigma_R / \partial \omega < 0$, and thus $z_k < 1$. Roughly speaking, it is as if the electrons in the non-interacting case have been degraded in the interacting case, only a remainder (a residue) less than one remains. What we are particularly interested in, is if this residue is non-zero on the Fermi-surface, i.e. if there is anything left at all of the single-particle excitations we had in the free case. More precisely, are there *any remnants at all* left of the non-interacting electrons, at low energies close to the Fermi-surface, once interactions have been turned on? If there is, i.e. if $z_{k_F} > 0$, then we have some amount of well-defined low-energy single-particle excitations similar to the ones that are exact eigenstates in the non-interacting case. Such free-electron-like excitations are called *quasiparticles*. An interacting fermionic system with such quasiparticles is called a *Fermi-liquid*. Such single-particle

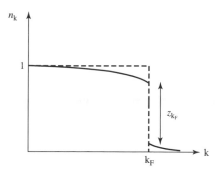

Figure 3.2   The zero-temperature momentum distribution n_k of an interacting electron system. The magnitude of the discontinuity on the Fermi surface is precisely the quasiparticle residue. The dashed line is the momentum distribution of the non-interacting system, which has a discontinuity of magnitude 1 on the Fermi-surface.

like excitations are no longer exact eigenstates of the system, since interaction terms permit scattering in and out of Bloch-states. However, in a Fermi-liquid the quasiparticles are nevertheless defined well enough at low energies, at least, to enable us to consider them as bona fide electron-like excitations of the system.

The momentum distribution which we in principle may obtain from the *quasiparticle propagator* given in Eq. 3.17 is illustrated in Figure 3.2. The important feature to note is the discontinuity of magnitude z_{k_F} on the Fermi-surface, the hallmark of a Fermi-liquid.

The concept of quasiparticles in interacting fermion systems was first introduced by Landau in 1950. It has proven to be an enormously successful and important paradigm, basically forming the cornerstone of our understanding of metallic systems. The basic idea which underlies the Landau Fermi-liquid picture is that there is a one-to-one correspondence between the quantum numbers of non-interacting electrons and those of interacting electrons. This means that we may imagine that we start out with some simple Hamiltonian like Eq. 3.4, of which we know the eigenstates and eigenexcitations, and then perturbatively introduce interactions in the problem. When this is done, the resulting quasiparticles of the theory are obtained adiabatically from the excitations of the non-interacting system, and in particular the quantum numbers of the excitations in the interacting and non-interacting case are in one-to-one correspondence with each other. Although this may naively seem like a hopeless approach, it turns out to be a surprisingly robust framework when considering metals. Even in heavy fermion compounds, where interaction effects are so strong that one gets a renormalization of the bare electron mass by factors of order 1000, one still has not obliterated completely the residue of the electrons that exist in the noninteracting case.

Examples of systems that do not conform to the Fermi-liquid paradigm exist, however. Notably, quasi-one dimensional organic conductors are not Fermi-liquids. This is because interactions in one-dimensional systems constitute a *singular* perturbation to the free-electron case as a consequence of severely restricted Fermi-surface kinematics. As a result, forward scattering, which is innocuous in higher dimensions, is singular in one dimension and all vestiges of free-electron systems are destroyed as soon as interactions are switched on. The long-lived low-energy excitations in this case turn out to be bosonic in character, and are collective excitations in terms of the constituent electrons of the underlying system. Under such circumstances, there can be no one-to-one correspondence between the quantum numbers of the interacting case and of the non-interacting case.

How can it be that an approximation that has all the appearances of a non-starter, works so spectacularly well in most cases? Understanding the micro-scopics of the Fermi-liquid theory is not the main purpose of this book, so we mention it only briefly. It is however an important problem to understand, and for more details the reader is referred to the vast literature on the subject. What is absolutely crucial when one does any sort of perturbation theory, is to start with the right reference system. Why, then, is the non-interacting electron system a good and sensible starting point to perturb about, when the goal is to describe an interacting electron system? The reason is embedded in the Pauli principle. The number of states that an added particle to the system can scatter into, is severely limited by the Pauli principle, and this effectively suppresses the effect of switching on interactions. As a result, the free case is a good stable system to perturb around. It has the feature of having a finite compressibility in the ground state, and therefore collective modes, such as zero-sound, as well as thermodynamic quantities will change in a smooth manner when interactions are introduced. In contrast, this is not true for a bosonic system. For instance, non-interacting bosons do not support zero-sound, whereas the interacting system does. Hence, quantities do not evolve in a smooth manner as some interaction parameter is varied. Quite the contrary, the limit of zero interactions is a singular limit. So, the non-interacting boson system is not a good starting point to perturb around. *The principal reason for why Fermi-liquid theory works is the kinematic restrictions imposed on the interacting system through the Pauli principle.* This is all we shall have to say about this, for a more detailed exposition the reader may for instance consult the very nice review article by Varma *et al.* [37].

It should be emphasized here that in all of the above, we have assumed inter-actions to mean a *repulsive* interaction between electrons. This is very natural, in that the dominant interaction between electrons in condensed matter systems is the Coulomb interaction, whose Fourier-transform is given by a matrix element V of the type we considered in connection with Eq. 3.9, namely

$$V_{\mathbf{k},\mathbf{k}',\mathbf{q}} = \frac{1}{4\pi\varepsilon_0} \frac{2\pi e^2}{q^2} \qquad (3.18)$$

In this case, V only depends on the momentum transfer $\mathbf{q}$ in the scattering process, not on the initial scattering states. However, as we shall see in the next section, this is not the only source of interaction between electrons in a solid. For instance, electrons obviously interact with the ions of a solid as they move through it causing scattering of electrons and ions. These scattered ions may in turn interact with other electrons, thus mediating an effective electron–electron interaction. This phonon-mediated interaction turns out to be of a quite different nature than the photon-mediated Coulomb-interaction above. In particular, it has the property that in a thin energy-shell around the Fermi-surface, it is *attractive*. This feature alone suffices to produce an instability of the Fermi-liquid. *Attractive electron–electron interactions are singular perturbations to the non-interacting case.*

3.4   Instability due to attractive interactions

In the previous section, we established that the Fermi-liquid picture of interacting electrons is stable to repulsive interactions. In this section, we will start by investigating what happens to the Fermi liquid when an effective attractive interaction between two electrons is introduced as a perturbation to the non-interacting system. Later, we will show how a weak phonon-mediated interaction between electrons may produce a net attraction, even when the *a priori* dominant repulsive Coulomb interaction is accounted for.

3.4.1   Two electrons with attractive interaction

In this subsection, we will consider a rather artificial, but nonetheless extremely instructive problem, first considered by Leon Cooper. It exhibits in a lucid manner the dramatic effect an arbitrarily weak attractive interaction between electrons has, in stark contrast to a repulsive interaction. The Cooper problem is defined as follows. Consider the situation where we have an inert Fermi-sea, in which the electrons are treated as non-interacting. To this Fermi-sea we add two electrons above the Fermi-surface (by necessity). These two extra electrons do not interact with the inert Fermi-sea. They do however interact with each other, but in a very peculiar manner. The interaction is such that the two electrons attract if they both are within a small energy ω_0 from the Fermi-surface, and *on opposite sides of the Fermi-surface*, otherwise they do not interact at all. In the absence of interactions, one electron occupies a plane-wave state $|\mathbf{k}\rangle$, the other occupies the plane-wave state $|-\mathbf{k}\rangle$. Let us denote the two-particle state of the two extra electrons in the absence of such a peculiar interaction by $|\mathbf{k}, -\mathbf{k}\rangle$, and in the presence of the interaction by $|1, 2\rangle$. Denoting the Hamiltonian of the system by

$$H = H_0 + V_{\text{eff}} \qquad (3.19)$$

then we have

$$H_0|\mathbf{k}, -\mathbf{k}\rangle = 2\varepsilon_{\mathbf{k}}|\mathbf{k}, -\mathbf{k}\rangle \qquad (3.20)$$

where $\varepsilon_{\mathbf{k}}$ is the single-particle excitation energy of the non-interacting fermion system (see Eq. 3.4). Note that due to the Pauli principle, we must have $2\varepsilon_{\mathbf{k}} > 2\varepsilon_F$ in the presence of the inert Fermi-sea. This sea plays the role of providing blocking-factors for added states. Adding interactions, the exact Schrödinger equation for the (admittedly peculiar) two-particle problem defined above is given by

$$H|1, 2\rangle = E|1, 2\rangle \qquad (3.21)$$

where E is the exact two-particle energy of the two electrons above the Fermi-surface, in the presence of an attractive interaction. We assume that the states $|\mathbf{k}, -\mathbf{k}\rangle$ form a complete set such that the exact two-particle eigenstate can be expanded in this basis, thus

$$|1, 2\rangle = \sum_{\mathbf{k}} a_{\mathbf{k}}|\mathbf{k}, -\mathbf{k}\rangle \qquad (3.22)$$

Our task is to obtain the expansion coefficients $a_{\mathbf{k}}$ and finding the value of E. This will solve the Schrödinger-equation for the problem. Let us insert Eq. 3.22 into Eq. 3.21, obtaining

$$(H_0 + V_{\text{eff}}) \sum_{\mathbf{k}} a_{\mathbf{k}}|\mathbf{k}, -\mathbf{k}\rangle = E \sum_{\mathbf{k}} a_{\mathbf{k}}|\mathbf{k}, -\mathbf{k}\rangle \qquad (3.23)$$

We next project the resulting equation down on the adjoint states $< \mathbf{k}', -\mathbf{k}'|$, where we have $\langle \mathbf{k}', -\mathbf{k}'|\mathbf{k}, -\mathbf{k}\rangle = \delta_{\mathbf{k},\mathbf{k}'}$, in order to obtain a Schrödinger equation for the coefficients $a_{\mathbf{k}}$. Using the above orthogonality relation, we thus find

$$a_{\mathbf{k}}[2\varepsilon_{\mathbf{k}} - E] = - \sum_{\mathbf{k}'} a_{\mathbf{k}'} \langle \mathbf{k}', -\mathbf{k}'|V_{\text{eff}}|\mathbf{k}, -\mathbf{k}\rangle \qquad (3.24)$$

The quantity $\langle \mathbf{k}', -\mathbf{k}'|V_{\text{eff}}|\mathbf{k}, -\mathbf{k}\rangle$ is basically a two-particle scattering matrix element for particles on opposite sides of the Fermi surface, from a two-particle state $|\mathbf{k}, -\mathbf{k}\rangle$ to a two-particle state $|\mathbf{k}', -\mathbf{k}'\rangle$. This scattering process is illustrated in Figure 3.3.

So far, this is quite general, and we now specialize to the case where this scattering matrix element is of the somewhat peculiar nature described above, namely attractive in a thin shell around the Fermi-surface, and zero elsewhere. That is, we have

$$\langle \mathbf{k}', -\mathbf{k}'|V_{\text{eff}}|\mathbf{k}, -\mathbf{k}\rangle = -V; \ |\varepsilon_{\mathbf{k}} - \varepsilon_F| < \omega_0$$

$$= 0; \text{ otherwise} \qquad (3.25)$$

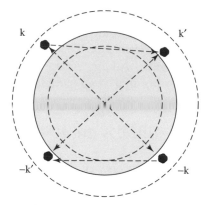

Figure 3.3 Scattering event of two particles maintained on opposite sides of the Fermi-surface, scattered by the matrix element $\langle \mathbf{k}', -\mathbf{k}' | V_{\text{eff}} | \mathbf{k}, -\mathbf{k} \rangle$ which is operative in the thin shell around the Fermi-surface indicated by the dashed circles.

Then Eq. 3.24 takes the form

$$a_{\mathbf{k}}[2\varepsilon_{\mathbf{k}} - E] = V \sum_{\mathbf{k}'} a_{\mathbf{k}'} \, \Theta(\varepsilon_{\mathbf{k}} - \varepsilon_{\mathrm{F}}) \Theta(\omega_0 - |\varepsilon_{\mathbf{k}} - \varepsilon_{\mathrm{F}}|) \qquad (3.26)$$

The first Θ-function ensures that the single-particle energies ε are above the Fermi-surface, the second one ensures that the effective attraction only is operative in a thin shell around the Fermi-surface. At this stage it is more convenient to go over to energy integrals instead of $\mathbf{k}$-space integrals, by introducing the density of states $D(\varepsilon)$, and viewing the expansion coefficients as functions of energy ε. Introducing the density of states $D(\varepsilon)$ (the $\mathbf{k}$-space sum of the spectral weight $A(\mathbf{k}, \varepsilon)$ obtained from $G_0(\mathbf{k}, \varepsilon)$ (see Eq. 3.8), as follows

$$D(\varepsilon) = -\frac{1}{\pi} \sum_{\mathbf{k}} \Im[G_0(\mathbf{k}, \varepsilon)] = \frac{1}{\pi} \sum_{\mathbf{k}} \frac{\delta_{\mathbf{k}}}{(\varepsilon - \varepsilon_{\mathbf{k}})^2 + \delta_{\mathbf{k}}^2}$$

$$= \sum_{\mathbf{k}} \delta(\varepsilon - \varepsilon_{\mathbf{k}}) \qquad (3.27)$$

where $\delta(\varepsilon - \varepsilon_{\mathbf{k}})$ is a Dirac delta function, and we have used that $\lim_{\eta \to 0} \eta/(x^2 + \eta^2) = \pi \delta(x)$. Noting that $a_{\mathbf{k}}$ only depends on $\mathbf{k}$ via $\varepsilon_{\mathbf{k}}$, we may write Eq. 3.26 as follows

$$a(\varepsilon_{\mathbf{k}})[2\varepsilon_{\mathbf{k}} - E] = V \int_{-\infty}^{\infty} d\varepsilon \sum_{\mathbf{k}'} a(\varepsilon)\delta(\varepsilon - \varepsilon_{\mathbf{k}})\Theta(\varepsilon - \varepsilon_{\mathrm{F}})\Theta(\omega_0 - |\varepsilon - \varepsilon_{\mathrm{F}}|)$$

$$= V \int_{-\infty}^{\infty} d\varepsilon D(\varepsilon)\Theta(\varepsilon - \varepsilon_{\mathrm{F}})\Theta(\omega_0 - |\varepsilon - \varepsilon_{\mathrm{F}}|) \qquad (3.28)$$

We then obtain, upon renaming variables

$$a(\varepsilon)[2\varepsilon - E] = V \int_{\varepsilon_F}^{\varepsilon_F + \omega_0} d\varepsilon' D(\varepsilon') a(\varepsilon') \tag{3.29}$$

The right hand side of Eq. 3.29 is seen to be independent of ε, and hence we immediately conclude that $a(\varepsilon)$ must have the form

$$a(\varepsilon) = \frac{C}{2\varepsilon - E} \tag{3.30}$$

where C is some normalization constant. Its precise value will not be important, and we do not specify it further. Inserting the Ansatz Eq. 3.30 for $a(\varepsilon)$ into Eq. 3.29, we find the self-consistent equation for the eigenvalue E

$$1 = V \int_{\varepsilon_F}^{\varepsilon_F + \omega_0} \frac{D(\varepsilon') d\varepsilon'}{2\varepsilon - E} \tag{3.31}$$

Utilizing the fact that we are considering a thin shell around the Fermi-surface, and assuming that the density of state $D(\varepsilon)$ varies slowly, we may simply replace it by its value on the Fermi surface, $D(\varepsilon_F)$. Introducing the dimensionless coupling constant $\lambda \equiv V D(\varepsilon_F)$, we obtain

$$1 = \lambda \int_{\varepsilon_F}^{\varepsilon_F + \omega_0} \frac{d\varepsilon'}{2\varepsilon - E}$$
$$= \lambda \ln \left[\frac{2(\varepsilon_F + \omega_0) - E}{2\varepsilon_F - E} \right] \tag{3.32}$$

We then introduce the energy difference between the states of two non-interacting particles on the Fermi-surface, and the exact energy eigenvalue E, i.e. $\Delta = 2\varepsilon_F - E$. In terms of this variable, Eq. 3.32 may be written

$$\frac{1}{\lambda} = \ln \left[1 + \frac{2\omega_0}{\Delta} \right] \tag{3.33}$$

Note that the way V is defined in Eq. 3.25, it is a positive quantity, hence λ is a positive quantity. Thus we see that in order for Eq. 3.33 to have a solution, $\Delta > 0$ is required. This means that $E < 2\varepsilon_F$! This is a a quite dramatic result, since the basic starting supposition was that the inert Fermi-sea was completely filled up to the Fermi-level, and the two extra electrons lived above the Fermi-surface. Now, we have just reached the rather startling conclusion that an arbitrary weak attractive interaction leads to an exact two-particle state with a total energy lower than twice the Fermi-energy. What this indicates, is that the inert Fermi-sea is suffering some sort of collapse such that energy states below the top of the Fermi-sea become accessible. At this stage, such an interpretation of the above result is merely a plausible conjecture. Confirmation of its physical soundness will have to await the analysis in the next chapter, where the full self-consistent treatment of *all* electrons in a thin shell around the Fermi-surface will be completed.

Let us solve Eq. 3.33 for Δ. We find

$$\Delta = \frac{2\omega_0}{e^{1/\lambda} - 1} \approx 2\omega_0 e^{-1/\lambda} \tag{3.34}$$

where the last approximation follows if $\lambda \ll 1$, i.e. if the effective electron–electron attraction is in some sense weak.

Several remarks are now in order. First, we see that Δ has an essential singularity as $\lambda \to 0$. This means that the above result for Δ could never have been obtained to any finite order in perturbation theory in λ. Hence, an arbitrarily weak attraction is a singular perturbation to a free electron gas. It destabilizes a Fermi-liquid. Secondly, we see that $\lambda > 0$ only if $D(\varepsilon_F) > 0$. For a parabolic band $\varepsilon_k = \hbar^2 k^2 / 2m$ in three dimensions, we have $D(\varepsilon) \sim \sqrt{\varepsilon}$. Hence, if the Fermi sea vanishes by draining electrons out of the system, $\lambda \to 0$ and $\Delta \to 0$. In other words, the tendency to pairing of electrons that we see above vanishes when the Fermi-sea vanishes. We should not expect a two-electron bound state for arbitrarily weak attraction in vacuum. Thirdly, if we were to reinstate $\hbar$ in the problem, we would find

$$\Delta = 2\hbar\omega_0 e^{-1/\lambda} \tag{3.35}$$

This shows that the pairing is also a *quantum effect*, since the bound state energy gap Δ vanishes when $\hbar \to 0$, i.e. it vanishes in the classical limit. Last, but by no means least, we note that to the extent we choose to take the above results seriously (and we should), the fact that the two-electron bound state resides at an energy which apparently is prohibited by the Pauli principle, in a vague way hints at the fact that perhaps bound pairs of electrons are not as much slaves to the Pauli principle as are individual electrons. In fact, this is precisely true, as we will come back to in Chapter 4. By forming Cooper-pairs, electrons can shed their fermion statistics. The resulting quantum fluid of Cooper-pairs is not as protected from interactions as are the Pauli principle loyal fermionic fluids. This is basically the root cause of why attractive interactions are able to destabilize the Fermi-liquid while repulsive interactions are virtually incapable of doing so in metallic states. Recall that when we briefly analyzed the microscopic foundations of Landau Fermi-liquid theory, we stated that the Pauli principle severely suppressed the effects of (repulsive) interactions, thus saving the electron-like quasi-particles from obliteration. For attractive interactions, the Pauli principle is essentially sidelined and no longer acts as a suppressor of interaction effects. This opens up the possibility of ground states qualitatively different from the free electron gas.

The resulting electron pair that we have described the genesis of in the above, is referred to as a Cooper-pair, named after Leon Cooper who had the audacity to consider the then outrageous problem of two electrons interacting attractively in the peculiar manner described above. Cooper's calculation is truly remarkable in that when he performed it, there was no known mechanism by which electrons could interact attractively in solid states physics. As it turns out, the phenomenon of Cooper-pairing of electrons is at the heart of the phenomenon of superconductivity.

3.4.2   Phonon-mediated attractive interactions

In this section we will start with the Hamiltonian for electrons interacting with each other through the Coulomb interaction and add the coupling of electrons to phonons. A canonical transformation reduces this to an effective Hamiltonian with a modified electron-electron interaction that under certain circumstances could be attractive, despite the apparent complete dominance of the Coulomb-interaction compared to the weak electron–phonon couling. It is very much like David's triumph over Goliath–intelligence and a sit of luck wins over brute force.

Let us, before we start with formalities, give an intuitive picture of what is going on. The crucial point which facilitates the above rather remarkable result, is that heavy lattice ions are slow movers around their equilibrium positions compared to fast and light electrons. Hence, dynamic local lattice distortions arising from electron-ion collisions, remain long after the electron that caused them have passed. This phenomenon will be called *retardation*. A local lattice distortion gives rise to a small dipole moment that can interact with electrons. As a result, another electron in the vicinity of the local lattice distortion can be drawn towards it long after the initial electron has left the scene. As the electrons are far removed from each other, the repulsive Coulomb interaction is reduced. It is this 'wait a little and see' tactics that allows the electron–phonon coupling to circumvent the apparent insurmountable obstacles of the Coulomb-repulsion. On the other hand, the second electron is not allowed to wait for a too long time before taking advantage of the local lattice distortion, since the latter eventually will decay as the ion settles around its equilibrium again. Hence the second electron has a limited amount of time at its disposal before moving in on the lattice distortion caused by the first electron. It therefore stands to reason that in order to reduce as much as possible the (for pairing purposes) adverse effects of the Coulomb-repulsion, the two electrons should put as much distance as possible between themselves per unit time. The best two-particle state for such purposes is a state where the two electrons occupy plane-wave states of opposite wavenumbers, i.e. they move in opposite directions. (Anybody who has ever been on an airliner between two major central European cities has experienced this by looking out the window of the aircraft and seeing another airliner whizzing past in the opposite direction. They are out of sight in seconds, in contrast to those that fly in more or less the same direction, which can be followed for minutes.) This crude real-space picture already gives us a hint that Cooper's calculation discussed in the previous subsection, while appearing quite artificial, in fact is precisely the right problem to consider. We now proceed to derive an effective electron–electron interaction, mediated by photons and phonons, which gives rise to the physics discussed above.

The starting point is the second-quantized version of the interacting electron gas coupled to quantized lattice vibrations, or phonons. It is given by

$$H = \sum_{\mathbf{k},\sigma} \varepsilon_{\mathbf{k}} c^{\dagger}_{\mathbf{k},\sigma} c_{\mathbf{k},\sigma} + \sum_{\mathbf{k},\mathbf{k}',\mathbf{q},\sigma,\sigma'} \frac{1}{4\pi\varepsilon_0} \frac{2\pi e^2}{q^2} c^{\dagger}_{\mathbf{k}+\mathbf{q},\sigma} c^{\dagger}_{\mathbf{k}'-\mathbf{q},\sigma'} c_{\mathbf{k},\sigma} c_{\mathbf{k}',\sigma'} + V_{\text{e–ph}}$$

$$V_{\text{e–ph}} = \sum_{\mathbf{k},\mathbf{q},\sigma} M_{\mathbf{q}}(a^{\dagger}_{-\mathbf{q}} + a_{\mathbf{q}}) c^{\dagger}_{\mathbf{k}+\mathbf{q}} c_{\mathbf{k}} \tag{3.36}$$

where $M_{\mathbf{q}}$ is the matrix element for electron–phonon coupling.

The operators $(a^{\dagger}_{\mathbf{q}}, a_{\mathbf{q}})$ are creation and destruction operator for phonons, and satisfy the following commutation relations reflecting that phonons are bosons

$$a^{\dagger}_{\mathbf{q}'} a_{\mathbf{q}} - a_{\mathbf{q}} a^{\dagger}_{\mathbf{q}'} = \delta_{\mathbf{q},\mathbf{q}'} \tag{3.37}$$

Phonons are not material particles, and their number is not conserved in a solid. A typical electron-phonon scattering event is shown in Figure 3.4.

In Eq. 3.36, we have assumed that one particular phonon-mode dominates the coupling to the electrons. Had we had several phonon-modes of equal coupling strength, we would have had to include a sum over phonon modes λ as well, in $V_{\text{e–ph}}$. The matrix element $M_{\mathbf{q}}$ may be expressed in terms of the eigenfrequencies $\omega_{\mathbf{q},\lambda}$ of harmonic lattice vibrations and the Fourier-transform $\tilde{V}_{\text{e–ion}}(\mathbf{q})$ of

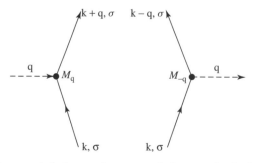

Figure 3.4   Scattering events between electrons and phonons. In the left process, a phonon is absorbed by an electron causing recoil. In the right process, an electron emits a phonon causing recoil. In the latter case, the electron sets up a local dynamic lattice distortion. The vertex is denoted by a black dot and its strength is given by $M_{\mathbf{q}}$.

the electrostatic electron-ion interaction as follows (when also a phonon-mode index λ is included)

$$M_{\mathbf{q},\lambda} = i(\mathbf{q} \cdot \boldsymbol{\xi}_\lambda)\sqrt{\frac{\hbar}{2M_{\text{ion}}\omega_{\mathbf{q},\lambda}}}\, \tilde{V}_{\text{e-ion}}(\mathbf{q}) \tag{3.38}$$

where M_{ion} is the ion-mass and $\boldsymbol{\xi}_\lambda$ is a unit vector giving the direction of the lattice-vibration mode λ. It is difficult to compute $M_{\mathbf{q},\lambda}$ with any precision from first principles, and we will simply take this matrix element as some parameter. *This coupling is much weaker than electron–photon coupling.* Note also that $\lim_{\mathbf{q}\to 0} M_{\mathbf{q},\lambda} = 0$. The fact that $M_{\mathbf{q},\lambda} \sim \mathbf{q}$ reflects the fact that the electron-phonon coupling is a coupling between a point charge and a dipole moment.

Let us see how we can transform the electron-phonon coupling into an effective electron-electron interaction in a fairly simple manner. Note that the interaction in Figure 3.1, when viewed as a Coulomb-interaction, may be considered as obtained from two scattering events as those depicted in Figure 3.4, but where the dashed lines are photons and not phonons, and where the scattering vertex has strength e, the electron charge, and not strength $M_{\mathbf{q}}$. In Figure 3.1, we then interpret the dashed line between the electrons as a *free-photon propagator*. Since electrons are virtually at a standstill compared to photons, this interaction is non-retarded and we may ignore the frequency dependence of the photon-propagator. Under such circumstances, the photon propagator is nothing but the static Coulomb-interaction.

We may proceed precisely along the same lines when phonons are involved. We obtain an effective electron–electron interaction by taking the two scattering events in Figure 3.4 and forming a two-particle scattering term as in Figure 3.1. We then interpret the dashed line connecting the electrons as a *free phonon-propagator*. Free phonons have the Hamiltonian

$$H = \sum_{\mathbf{q}} \omega_{\mathbf{q}} a_{\mathbf{q}}^\dagger a_{\mathbf{q}} \tag{3.39}$$

where $\omega_{\mathbf{q}}$ is the eigenvalue of a phonon in a plane-wave state $|\mathbf{q}\rangle$. The explicit expression for the free phonon-propagator $D_0(\mathbf{q}, \omega)$, which has the same interpretation as the propagator for electrons, is given by

$$D_0(\mathbf{q}, \omega) = \frac{2\omega_{\mathbf{q}}}{\omega^2 - \omega_{\mathbf{q}}^2 + i\eta} \tag{3.40}$$

where η is a positive infinitesimal quantity. Using this, we obtain the phonon-mediated part of the electron–electron interaction

$$V_{\text{eff}}(\mathbf{q}, \omega) = \frac{2|M_{\mathbf{q}}|^2 \omega_{\mathbf{q}}}{\omega^2 - \omega_{\mathbf{q}}^2} \tag{3.41}$$

In the above, $\mathbf{q}$ is the momentum transfer in the electron-electron scattering process, while ω is the energy transfer. Electrons are by no means at a standstill in solids compared to the velocity of sound waves. Thus, we cannot ignore the frequency dependence of this part of the electron–electron interaction. The result of all of the above is that Eq. 3.36 may be written on the form of a purely electronic system where the trace of the phonons is manifest in the additive contribution they give rise to in the electron-electron interaction, as follows

$$H = \sum_{\mathbf{k},\sigma} \varepsilon_{\mathbf{k}} c_{\mathbf{k},\sigma}^{\dagger} c_{\mathbf{k},\sigma} + \sum_{\mathbf{k},\mathbf{k}',\mathbf{q},\sigma,\sigma'} \tilde{V}_{\text{eff}}(\mathbf{q},\omega) c_{\mathbf{k}+\mathbf{q},\sigma}^{\dagger} c_{\mathbf{k}'-\mathbf{q},\sigma'}^{\dagger} c_{\mathbf{k},\sigma} c_{\mathbf{k}',\sigma'}$$

$$\tilde{V}_{\text{eff}}(\mathbf{q},\omega) = \frac{2|M_{\mathbf{q}}|^2 \omega_{\mathbf{q}}}{\omega^2 - \omega_{\mathbf{q}}^2} + \frac{1}{4\pi\varepsilon_0} \frac{2\pi e^2}{q^2} \qquad (3.42)$$

Note that when $\omega^2 < \omega_{\mathbf{q}}^2$, the phonon-mediated piece of the electron–electron interaction is negative, i.e. it is *attractive*. This in itself does not mean all that much, the issue is if this attractive piece can overcome the manifestly repulsive piece originating in the Coulomb-interaction. In fact, it is easy to see that it *can*, at least for some momentum transfer and energy transfer. We note that as ω^2 approaches $\omega_{\mathbf{q}}^2$ from below, the phonon-mediated piece of the interaction formally diverges! Hence, it does not matter how small $|M_{\mathbf{q}}|^2$ is, the phonon-piece will always be able to overcome the Coulomb-piece. The physics of this is precisely what was alluded to in the introduction to the section. Namely, by waiting for a time which is of order the time-scale of the lattice vibrations, an electron may come in and be dragged towards a local lattice distortion caused by an electron passing the site at some earlier time. This minimum waiting time corresponds to the upper frequency cutoff.

In Figure 3.5, a sketch of $\tilde{V}_{\text{eff}}(\mathbf{q},\omega)$ is given as a function of frequency ω at fixed finite momentum $\mathbf{q}$, the figure on the left. Note that when ω is close enough to typical phonon-frequencies $\omega_{\mathbf{q}}$ the total effective interaction is positive. For all $\omega^2 > \omega_{\mathbf{q}}^2$ it is repulsive, and this is also the case for too small ω^2. This latter fact is what we alluded to in the introduction to this subsection, namely that the electron can not wait for *too* long a time before taking advantage of the local lattice distortion. Thus, the effective attraction we see for not too small and not too large frequencies, is basically the Fourier-spacetime version of the "wait and see" tactics described at the start of the section.

The figure on the right in Figure 3.5 is an idealized and simplified version of $\tilde{V}_{\text{eff}}(\mathbf{q},\omega)$. We ignore entirely the repulsive part of the interaction, since we expect the effect of repulsive terms to be suppressed by the Pauli principle as discussed in the previous section. The attractive part is extended to all frequencies $\omega^2 < \omega_{\mathbf{q}}^2$, and is moreover taken to be a constant independent of ω in this range. This simplification is the one that was originally considered by Bardeen,

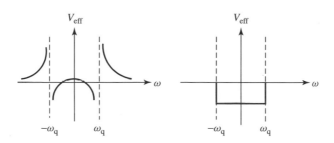

Figure 3.5   The left figure shows $\tilde{V}_{\mathrm{eff}}(\mathbf{q}, \omega)$ at fixed $\mathbf{q}$ as a function of ω. The right figure shows a much simplified version of it.

Cooper, and Schrieffer in their seminal 1957 work on phonon-mediated super-conductivity [7]. It is an extreme simplification, but nonetheless yields results that are in spectacular agreement with experiments. While the agreement with experiments (to within a few percent for studiously chosen quantities) may have been a source of some surprise at the time, given the crudeness of the approximation of $\tilde{V}_{\mathrm{eff}}(\mathbf{q}, \omega)$, we now understand why this is so. The point is that the simplified interaction contains the essential feature of the more real-istic $\tilde{V}_{\mathrm{eff}}(\mathbf{q}, \omega)$, namely attraction in a certain frequency range close to typical phonon-frequencies. What other details are left out are not all that important, the validity of the results obtained is virtually guaranteed by the principle of *universality*: Certain physical properties do not depend on tiny details of the microscopic Hamiltonian. All we need to do is to derive some *effective* descrip-tion and compute properties from this, and we get the right results (if we focus on the right properties!). In the next chapter, we shall compute some results that come out right due to universality, and we shall also compute one notable quantity which definitely does not.

3.4.3   Reduction of the effective Hamiltonian

Let us next proceed to further simplifying the Hamiltonian Eq. 3.42. The simpli-fications we will make are expected to be sensible provided that the frequency range where $\tilde{V}_{\mathrm{eff}}(\mathbf{q}, \omega)$ is attractive in a tiny energy range compared to the Fermi-energy, i.e. the shell around the Fermi-surface in Figure 3.3 is exceed-ingly thin compared to the radius of the Fermi-sphere. The implication of this is that when the frequency range of attraction is narrow, the phase-space for scat-tering such that the attractive interaction is felt, is maximized by having both the initial and final two-particle states within the thin shell. The energy transfer in the scattering process caused by $\tilde{V}_{\mathrm{eff}}(\mathbf{q}, \omega)$ is $\omega = \varepsilon_{\mathbf{k}} - \varepsilon_{\mathbf{k}+\mathbf{q}} = \varepsilon_{\mathbf{k}'} - \varepsilon_{\mathbf{k}'-\mathbf{q}}$. All energies $(\varepsilon_{\mathbf{k}}, \varepsilon_{\mathbf{k}'}, \varepsilon_{\mathbf{k}+\mathbf{q}}, \varepsilon_{\mathbf{k}'-\mathbf{q}})$ must lie within a thin energy shell around the Fermi-surface if the initial and final two-particle states are to contain elec-trons that attract each other. What values of $(\mathbf{k}, \mathbf{k}', \mathbf{q})$ in Eq. 3.42 will ensure

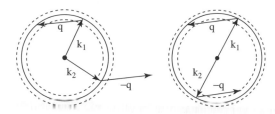

Figure 3.6   Phase-space of two-particle scattering within a thin shell of the Fermi-surface. There is only one situation where both the initial and scattered states are within the thin shell, provided one of the initial and one of the scattered states are, and that is if the initial momenta are on opposite sides of the Fermi-surface.

this? The interaction term describes scattering of initial state $|\mathbf{k}, \sigma\rangle$ to state $|\mathbf{k} + \mathbf{q}, \sigma >$ and scattering of the state $|\mathbf{k}', \sigma'\rangle$ to state $|\mathbf{k}' - \mathbf{q}, \sigma'\rangle$. What relation between $(\mathbf{k}, \mathbf{k}')$ ensures that if $|\mathbf{k} + \mathbf{q}, \sigma\rangle$ is a state within the energy shell, then so is $|\mathbf{k}' - \mathbf{q}, \sigma'\rangle$ for all $\mathbf{q}$? The answer is that this occurs if $\mathbf{k}' = -\mathbf{k}$ (see Figure 3.3), and this maximizes the phase-space for scattering of attractively interacting states. Hence, we will keep only such terms in Eq 3.42.

The situation is also illustrated in Figure 3.6. Consider two electrons in momentum states $\mathbf{k}_1$ and $\mathbf{k}_2$, both within a thin energy shell around the Fermi-surface. Suppose one electron is scattered into momentum state $\mathbf{k}_1 + \mathbf{q}$ and the other into $\mathbf{k}_2 - \mathbf{q}$. If the total momentum of the initial and scattered states is $\mathbf{k}_1 + \mathbf{k}_2 = \mathbf{Q}$, then we see from the left figure that in general, one of the scattered states will not be within the thin shell even if the other is. There is only one situation that guarantees that both the scattered states are within the thin shell if one of them are, and that is if $\mathbf{k}_1 + \mathbf{k}_2 = 0$. This situation will tend to dominate the contribution to the scattering phase space, given that the two-particle scattering matrix element is operative only if both of the initial and final states are within this thin shell. This domination becomes progressively more pronounced the thinner the shell is. If we dispensed with retardation altogether, and considered the scattering matrix element to be operative over the whole electronic band, then the above reduction into opposite momentum states is more questionable. Nonetheless, with the recent renaissance of purely electronic pairing mechanisms invoked in attempts of explaining such phenomena as superconductivity in high-T_c cuprates, the reduction is often used. Mostly out of old habit and with little or no justification. In the case of phonon-mediated attraction, the approximation is however on quite firm ground.

Moreover, we expect the attractive interaction mediated by phonons to be of short range. Thus, electrons should be allowed to approach each other closely in space (although they avoid each other in time). Due to the Pauli principle, it is therefore plausible to suggest that the electrons that can approach each other in space most easily are such electrons that are in opposite spin states. Therefore, we also set $\sigma' = -\sigma$. Setting $\mathbf{k}' = -\mathbf{k}$, $\sigma' = -\sigma$, and redefining $\mathbf{k} + \mathbf{q} \rightarrow \mathbf{k}$, $\mathbf{k} \rightarrow \mathbf{k}'$ in the interaction term of the effective Hamiltonian, Eq. 3.42 may be written on the form

$$H = \sum_{\mathbf{k},\sigma} \varepsilon_{\mathbf{k}} c^{\dagger}_{\mathbf{k},\sigma} c_{\mathbf{k},\sigma} + \sum_{\mathbf{k},\mathbf{k}',\sigma} V_{\mathbf{k},\mathbf{k}'} c^{\dagger}_{\mathbf{k},\sigma} c^{\dagger}_{-\mathbf{k},-\sigma} c_{-\mathbf{k}',-\sigma} c_{\mathbf{k}',\sigma} \qquad (3.43)$$

where $V_{\mathbf{k},\mathbf{k}'}$ is an attractive two-particle scattering matrix element which is operative if both $\mathbf{k}$ and $\mathbf{k}'$ are within a thin shell around the Fermi-surface. Eq. 3.43 is the celebrated BCS-reduced Hamiltonian forming the starting point for discussing phonon-mediated superconductivity.

In Figure 3.7, we also provide a real-space illustration of a Cooper-pair of two electrons moving in a lattice, where we have in mind a phonon-assisted pairing mechanism (the only really known pairing mechanism to date). The electrons move approximately linearly in opposite directions and in opposite spin-states. Each electron leaves behind a trail of lattice distortions as it moves through the lattice, which the other electron can feel and be attracted towards via the matrix element Eq. 3.38. In order to maximize the attraction, it is advantageous for the electrons to move co-linearly, since this maximizes the time the electrons spend close to lattice distortions while at the same time being mobile. Moreover, moving in precisely opposite directions puts maximum distance between the two electrons per unit time, thus minimizing the adverse effects, for pairing-purposes, of the repulsive Coulomb-interaction. The above is a rough real-space version of the $\mathbf{k}$-space arguments given in the previous paragraph. It clearly illustrates that Cooper-pairing is a $\mathbf{k}$-space phenomenon, not a pairing phenomenon in real-space.

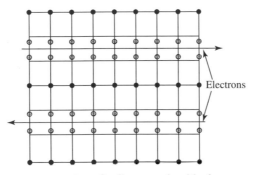

Figure 3.7   Real-space representation of a Cooper-pair with phonon-mediated superconductivity. The black dots are lattice ions unperturbed by the motion of the electron through the lattice. The hatched dots are lattice ions that have been displaced slightly by the electrons passing close by. Since the ions are heavy and slow, the local lattice distortions remain long after the electrons have passed. An electron moving in precisely the opposite direction can now take full advantage of the extended line of distortions, more so than an electron crossing this line. Thus, phonon-mediated electron–electron attraction is maximized for electrons in opposite linear momentum states. Moreover, such states also minimize the adverse effects of Coulomb-repulsion, since the two electrons then put a maximum amount of distance between themselves in a minimum amount of time.

In metals, electrical resistivity has many sources, among them electron–phonon coupling, but also scattering of electrons off impurities. In very clean metals with low density of lattice defects or other sources of point-like or extended impurities, one might expect the electron–phonon coupling to dominate the contribution to the scattering. Provided the pairing mechanism we have described in previous sections is the source of superconductivity, the source of resistivity and superconductivity is basically one and the same scattering mechanism. This immediately gives some insight into a fact which seems to be borne out in most cases, namely that good metals become superconductors only at low temperatures. While mercury has a room temperature resistivity of 98×10^{-8} Ω m and critical temperature of about 4.2 K, copper, the archetypical GOOD metal, has a room temperature resistivity of 1.7×10^{-8} Ω m and critical temperature which is in the millikelvin range. There really is no theory available to us that allows us to compute T_c from known experimentally accessible normal state properties with any great precisions. What we have, as we shall see in the next chapter, are rough estimates in terms of an electron–phonon coupling constant for which one has some crude estimate. It will turn out, however, that the critical temperature is exponentially *sensitive* to this *couling* constant, and hence great variation in T_c are expected. Paradoxically, therefore, we would expect the largest superconducting critical temperature in a poor metal! This goes to show that superconductivity is much more than just perfect conductivity. We already know this, of course, because the Meissner effect is a defining property of a superconductor which obviously has no counterpart in a perfect free electron gas with zero resistivity. The physical consequences of the Meissner effect have been mentioned, and will be elaborated on in much more detail in the chapters to follow.

4

The Superconducting State – an Electronic Condensate

4.1 BCS theory: a magnetic analogue

We have seen in Chapter 3 that the effective Hamiltonian for electrons interacting attractively (by some mechanism) when two electrons are in opposite spin-states and states of opposite linear momentum, is given by the BCS reduced Hamiltonian

$$H = \sum_{\mathbf{k},\sigma} \varepsilon_{\mathbf{k}} c^{\dagger}_{\mathbf{k},\sigma} c_{\mathbf{k},\sigma} + \sum_{\mathbf{k},\mathbf{k}',\sigma} V_{\mathbf{k},\mathbf{k}'} c^{\dagger}_{\mathbf{k},\sigma} c^{\dagger}_{-\mathbf{k},-\sigma} c_{-\mathbf{k}',-\sigma} c_{\mathbf{k}',\sigma} \tag{4.1}$$

It is possible to recast the effective BCS Hamiltonian into a language reminiscent of what we are acquainted with for the Heisenberg model for magnetic insulators, using the so-called Anderson pseudospin formalism. Such pseudospin formalisms have subsequently been used extensively in the context of a number of physical phenomena, such as the bi-layer quantum Hall effect, to mention one example.

We have seen in the previous chapter, that the *effective* Hamiltonian capturing the essentials of conventional phonon-assisted low-T_c superconductivity, is given by

$$H = \sum_{\mathbf{k},\sigma} \varepsilon_{\mathbf{k}} c^{\dagger}_{\mathbf{k},\sigma} c_{\mathbf{k},\sigma} + \sum_{\mathbf{k},\mathbf{k}'} V_{\mathbf{k},\mathbf{k}'} c^{\dagger}_{\mathbf{k},\uparrow} c^{\dagger}_{-\mathbf{k},\downarrow} c_{-\mathbf{k}',\downarrow} c_{\mathbf{k}',\uparrow} \tag{4.2}$$

where the interaction term represents scattering of two-particle spin-singlet states on opposite sides of the Fermi-surface. We now recast this into the form of a (pseudo)spin-Hamiltonian, where it will become evident that superconducting off-diagonal long-range order is equivalent to pseudo-magnetic ordering, which is more familiar to us in that it at least has a classical counterpart. (Throughout

Superconductivity: Physics and Applications Kristian Fossheim and Asle Sudbø
© 2004 John Wiley & Sons, Ltd ISBN 0-470-84452-3

this chapter, we work with units such that $\hbar = 1$). We introduce the standard Pauli matrices τ_1, τ_2, τ_3, as well as the spin-lowering and raising operators $\tau^{\pm} = \tau_1 \pm i\tau_2$. Next, we introduce the operators $\tau_{\mathbf{k}}^1$, $\tau_{\mathbf{k}}^2$, $\tau_{\mathbf{k}}^3$, as follows

$$\tau_{\mathbf{k}}^3 = c_{\mathbf{k},\uparrow}^{\dagger} c_{\mathbf{k},\uparrow} - c_{-\mathbf{k},\downarrow}^{\dagger} c_{-\mathbf{k},\downarrow}$$

$$\tau_{\mathbf{k}}^+ = 2c_{\mathbf{k},\uparrow}^{\dagger} c_{-\mathbf{k},\downarrow}^{\dagger} \tag{4.3}$$

Inserting this into the effective Hamiltonian (Eq. 4.2), we obtain

$$H = \sum_{\mathbf{k}} \varepsilon_{\mathbf{k}}(\tau_{\mathbf{k}}^3 + 1) + \frac{1}{4} \sum_{\mathbf{k},\mathbf{k}'} V_{\mathbf{k},\mathbf{k}'}(\tau_{\mathbf{k}}^1 \tau_{\mathbf{k}'}^1 + \tau_{\mathbf{k}}^2 \tau_{\mathbf{k}'}^2) \tag{4.4}$$

Note that this has the form of an *XY*-Heisenberg Hamiltonian of spins in an external magnetic field in the z-direction. However, it should be noted that the usual labels for spin-sites on a real-space lattice is replaced by indices in Fourier-space. This just underscores that point that Cooper-pairing is a k-space phenomenon. The point that enormously simplifies the treatment of the above pseudo-spin Hamiltonian is that the original effective interaction matrix element $V_{\mathbf{k},\mathbf{k}'}$ is essentially a contact interaction in real-space, and hence a long-range potential in k-space. Thus, van der Waals type mean-field arguments can be brought to bear on the problem, and we expect that a mean-field decomposition of the pseudo-spin Hamiltonian will yield excellent results. With longer range interaction in real-space, and hence shorter range interactions in k-space, fluctuation effects are expected to become more prominent. Moreover, we have tacitly assumed that the majority of the two-particle scattering events take place between electron pairs of zero total linear momentum, an approximation that is expected to be excellent as long as the phase-space for attraction is limited to a thin shell around the Fermi-surface. Hence, Eq. 4.4 is expected to be an effective Hamiltonian which is reasonable for the case where the interaction responsible for Cooper-pairing is retarded. Spontaneous symmetry breaking in this pseudomagnet occurs for the case where the pseudomagnetic field vanishes, i.e. for the case where we ignore the kinetic energy term. This magnetic analog leads us to choose the magnetization to be oriented in the $1-2$-plane, and thus define an order parameter as follows

$$\langle \tau_{\mathbf{k}}^1 \rangle = \langle c_{\mathbf{k},\uparrow}^{\dagger} c_{-\mathbf{k},\downarrow}^{\dagger} \rangle + \langle c_{-\mathbf{k},\downarrow} c_{\mathbf{k},\uparrow} \rangle \tag{4.5}$$

The task now will be to derive a self-consistent mean-field equation for this pseudomagnetic order parameter.

Before doing this, let us make a few general remarks. Note that if the order parameter exists, it implies that the particle number is not conserved in the ground state of the system. Global conservation of particle number is reflected by a global $U(1)$-symmetry of the original effective Hamiltonian. Thus, we see that superconductivity really amounts to breaking a global $U(1)$-symmetry in the ground state of the interacting electron gas. In physical terms, this means that the superconducting ground state is a *coherent superposition* of states with different number of particles, however with a majority of the states containing a number of particles which is close to the number of particles one would have in the $U(1)$-symmetric normal metallic state. Going back to the magnetic analogue, we see that superconductivity is equivalent to obtaining spontaneous pseudomagnetization in the $1-2$-plane. It is as if an XY-magnet has ordered. For this reason, and since the symmetry group $U(1)$ essentially is isomorphic to the symmetry group $O(2)$, it is often said that the normal metal-superconductor transition is in the universality class of the XY-model. As we shall see later, while this statement is correct for the superfluid to normal fluid transition in He^4, it is strictly speaking not correct in the case of a superconductor, which has a charged condensate. If the condensate has a charge, it means that it couples to an electromagnetic vector potential, a gauge-field. This elevates the symmetry of the system to a *local* $U(1)$ symmetry, as opposed to the *global* $U(1)$ symmetry of the XY model describing a superfluid. The presence of a gauge-field in the problem in principle alters the interactions, and hence also some critical properties. In many cases, the differences are, however, unobservably small.

Not only is the normal metal–superconductor transition strictly speaking not in the XY-universality class, but depending on the value of the Ginzurg–Landau order parameter, the transition may not even be a critical phenomenon, but a first order phase transition! What will transpire is that the demarcation between first order phase transition and second order phase transition is given by the tricritical value of the Ginzburg–Landau parameter $\kappa_{\text{tri}} = 0.79/\sqrt{2}$. It also turns out that this value is the fluctuation-corrected demarcation between type-I and type-II superconductivity!

4.2 Derivation of the BCS gap equation

In this section, we compute the free energy of the superconductor in the mean-field approximation. The superconducting gap is found by minimizing the expression for the free energy as a function of the gap-parameter. We first proceed by reducing the effective Hamiltonian to a self-consistent one-body problem by first introducing expectation values of the creation- and annihilation operators for Cooper-pairs. We have justified this mean-field procedure above, in the context of the pseudo-spin formalism. We first define the c-numbers

$$b_{\mathbf{k}} = \langle c_{-\mathbf{k},\downarrow} c_{\mathbf{k},\uparrow} \rangle$$
$$b_{\mathbf{k}}^{\dagger} = \langle c_{\mathbf{k},\uparrow}^{\dagger} c_{-\mathbf{k},\downarrow}^{\dagger} \rangle \qquad (4.6)$$

such that we have

$$c_{-\mathbf{k},\downarrow}c_{\mathbf{k},\uparrow} = b_{\mathbf{k}} + \underbrace{c_{-\mathbf{k},\downarrow}c_{\mathbf{k},\uparrow} - b_{\mathbf{k}}}_{\delta b_{\mathbf{k}}} . \tag{4.7}$$

We next insert this into the effective Hamiltonian and retain terms up to linear in $\delta b_{\mathbf{k}}$, ignoring terms of order $\mathcal{O}(\delta b_{\mathbf{k}}^2)$. This gives

$$H = \sum_{\mathbf{k},\sigma} \varepsilon_{\mathbf{k}} c_{\mathbf{k},\sigma}^\dagger c_{\mathbf{k},\sigma} + \sum_{\mathbf{k},\mathbf{k}'} V_{\mathbf{k},\mathbf{k}'} [b_{\mathbf{k}}^\dagger c_{-\mathbf{k}',\downarrow} c_{\mathbf{k}',\uparrow} + b_{\mathbf{k}'} c_{\mathbf{k},\uparrow}^\dagger c_{-\mathbf{k},\downarrow}^\dagger - b_{\mathbf{k}}^\dagger b_{\mathbf{k}'}] \tag{4.8}$$

This Hamiltonian is immediately simplified by introducing the superconducting *gap-parameter*, defined as follows

$$\Delta_{\mathbf{k}'}^\dagger \equiv -\sum_{\mathbf{k}} V_{\mathbf{k},\mathbf{k}'} b_{\mathbf{k}}^\dagger$$

$$\Delta_{\mathbf{k}} \equiv -\sum_{\mathbf{k}'} V_{\mathbf{k},\mathbf{k}'} b_{\mathbf{k}'} . \tag{4.9}$$

We the obtain an effective one-body Hamiltonian on the following form

$$H = \sum_{\mathbf{k},\sigma} \varepsilon_{\mathbf{k}} c_{\mathbf{k},\sigma}^\dagger c_{\mathbf{k},\sigma} - \sum_{\mathbf{k}} [\Delta_{\mathbf{k}}^\dagger c_{-\mathbf{k},\downarrow} c_{\mathbf{k},\uparrow} + \Delta_{\mathbf{k}} c_{\mathbf{k},\uparrow}^\dagger c_{-\mathbf{k},\downarrow}^\dagger - b_{\mathbf{k}}^\dagger \Delta_{\mathbf{k}}] \tag{4.10}$$

Note that because of presence of pair-production and pair-annihilation terms, Eq. 4.10 is not yet on the standard form of an effective non-interacting electron gas, and hence we cannot at this stage read off what the quasiparticle spectrum of Eq. 4.10 is. In order to achieve this, we perform a rotation of the basis to a set of new fermion operators that diagonalize this Hamiltonian, as follows. We introduce two new species of fermion annihilation operators ($\eta_{\mathbf{k}}$, $\gamma_{\mathbf{k}}$) along with their corresponding creation operators, which are related to the original electron operators via the unitary transformation

$$c_{\mathbf{k},\uparrow} = \cos(\theta)\eta_{\mathbf{k}} - \sin(\theta)\gamma_{\mathbf{k}}$$

$$c_{-\mathbf{k},\downarrow}^\dagger = \sin(\theta)\eta_{\mathbf{k}} + \cos(\theta)\gamma_{\mathbf{k}} \tag{4.11}$$

and their corresponding adjoint operators. Inserting this into the Hamiltonian we obtain

$$H = \sum_{\mathbf{k},\sigma} [\varepsilon_{\mathbf{k}} + \Delta_{\mathbf{k}} b_{\mathbf{k}}^\dagger]$$

$$+ \sum_{\mathbf{k}} [\varepsilon_{\mathbf{k}} \cos(2\theta) - \sin(\theta)\cos(\theta)(\Delta_{\mathbf{k}} + \Delta_{\mathbf{k}}^\dagger)] \eta_{\mathbf{k}}^\dagger \eta_{\mathbf{k}}$$

$$- \sum_{\mathbf{k}} [\varepsilon_{\mathbf{k}} \cos(2\theta) - \sin(\theta)\cos(\theta)(\Delta_{\mathbf{k}} + \Delta_{\mathbf{k}}^\dagger)] \gamma_{\mathbf{k}}^\dagger \gamma_{\mathbf{k}}$$

$$-\sum_{\mathbf{k}}[\Delta_{\mathbf{k}}\cos^2(\theta) - \Delta_{\mathbf{k}}^\dagger\sin^2(\theta) - 2\varepsilon_{\mathbf{k}}\sin(\theta)\cos(\theta)]\eta_{\mathbf{k}}^\dagger\gamma_{\mathbf{k}}$$

$$-\sum_{\mathbf{k}}[\Delta_{\mathbf{k}}^\dagger\cos^2(\theta) - \Delta_{\mathbf{k}}\sin^2(\theta) - 2\varepsilon_{\mathbf{k}}\sin(\theta)\cos(\theta)]\gamma_{\mathbf{k}}^\dagger\eta_{\mathbf{k}} \qquad (4.12)$$

We next proceed by choosing the value θ in such a way as to precisely diagonalize the Hamiltonian, i.e. such that only terms like $\eta_{\mathbf{k}}^\dagger\eta_{\mathbf{k}}$ and $\gamma_{\mathbf{k}}^\dagger\gamma_{\mathbf{k}}$ appear. This is achieved by choosing the following value of θ, upon ignoring phase-fluctuations in the complex order parameter entirely and setting $\Delta_{\mathbf{k}} = \Delta_{\mathbf{k}}^\dagger$

$$\tan(2\theta) = -\frac{\Delta_{\mathbf{k}}}{\varepsilon_{\mathbf{k}}} \qquad (4.13)$$

For such values of θ, we have

$$\sin^2(\theta) \equiv v_{\mathbf{k}}^2 = \frac{1}{2}\left[1 - \frac{\varepsilon_{\mathbf{k}}}{E_{\mathbf{k}}}\right]$$

$$\cos^2(\theta) \equiv u_{\mathbf{k}}^2 = \frac{1}{2}\left[1 + \frac{\varepsilon_{\mathbf{k}}}{E_{\mathbf{k}}}\right] \qquad (4.14)$$

where

$$E_{\mathbf{k}} \equiv \sqrt{\varepsilon_{\mathbf{k}}^2 + |\Delta_{\mathbf{k}}|^2} \qquad (4.15)$$

Before we go further, it is appropriate to mention the factors $v_{\mathbf{k}}^2$ and $u_{\mathbf{k}}^2$ which were introduced in Eq. 4.14. These are normally referred to as *coherence factors*. They play an extremely important role in the following sections where we will study ultrasound attenuation and nuclear magnetic resonance (NMR). Their existence was a prediction that came out of the BCS theory. It turns out that observed results for NMR and ultrasound attenuation in superconductors depend sensitively on these coherence factors, since in these two cases they conspire subtly and in entirely different ways. An even more subtle point is that, as we shall see, they conspire in different ways in transverse and longitudinal ultrasound attenuation! Bardeen, Cooper and Schrieffer [7] came up with predictions with correct numbers for these effects in a complicated manybody problem. This is undoubtedly one of the major triumphs of condensed matter physics of the 20th century.

Using Eqs 4.14 in 4.12, we finally obtain the simple fermionic Hamiltonian which formally has the form of a free fermion quasiparticle gas

$$H = \sum_{\mathbf{k},\sigma}[\varepsilon_{\mathbf{k}} + \Delta_{\mathbf{k}}b_{\mathbf{k}}^\dagger] + \sum_{\mathbf{k}} E_{\mathbf{k}}[\eta_{\mathbf{k}}^\dagger\eta_{\mathbf{k}} - \gamma_{\mathbf{k}}^\dagger\gamma_{\mathbf{k}}] \qquad (4.16)$$

The above Hamiltonian now has the form of a constant (mean-field) term $H_0 \equiv \sum_{\mathbf{k}}[\varepsilon_{\mathbf{k}} + \Delta_{\mathbf{k}}b_{\mathbf{k}}^\dagger]$ plus the Hamiltonian for a spinless fermion system with two species of fermions, described by the operators $(\eta_{\mathbf{k}}^\dagger, \eta_{\mathbf{k}})$ and $(\gamma_{\mathbf{k}}^\dagger, \gamma_{\mathbf{k}})$, where we can identify the quasiparticle excitation energies as $(E_k, -E_k)$ for the two species of fermions. Note that the two species of quasiparticle fermions are spinless since they are linear combinations of electrons and holes with opposite spins. Rather than having $\uparrow$-spin electrons and $\downarrow$-spin electrons as our degrees of freedom as in the metallic phase, we have two distinct linear combinations of particle-hole spin-singlets as our relevant degrees of freedom. Thus, the total number of degrees of freedom is conserved.

The expression in Eq. 4.15 immediately provides a physical interpretation of Δ_k defined in Eq. 4.9. It appears as a *gap* in the quasiparticle spectrum on the Fermi-surface (where $\varepsilon_k = 0$), a gap which is present because the expectation value $< c_{-k,\downarrow} c_{k,\uparrow} >$ is non-zero. This expectation value is essentially the order parameter of the superconducting state, as was made clear from the pseudo-magnet analog above, and hence we will refer to the gap Δ_k as the *superconducting gap*. Note that the gap strictly speaking is *not* precisely the same as the superconducting order parameter, by the definition (Eq. 4.9)! For one thing, the k-space structure of the two objects are different. However, they are both non-zero or zero simultaneously, and for this reason the gap in the quasiparticle spectrum may effectively serve as a *bona fide* order parameter, and in most cases this is in fact what is done.

Since this now has the form of a free fermion gas, plus a constant term which does not contain fermionic degrees of freedom (the first term), it is elementary to write down the grand canonical partition function for the problem, which is given by

$$Z_g = e^{-\beta H_0} \prod_k (1 + e^{-\beta E_k})(1 + e^{\beta E_k}) = e^{-\beta F}$$

$$H_0 \equiv \sum_k [\varepsilon_k + \Delta_k b_k^\dagger]. \tag{4.17}$$

This implies that the free energy F of the system is given by

$$F = H_0 - \frac{1}{\beta} \sum_k [\ln(1 + e^{-\beta E_k}) + \ln(1 + e^{\beta E_k})] \tag{4.18}$$

which has the form of a pure mean-field term plus the contribution to the free energy from a two-component effective non-interacting fermion gas. At zero temperature $(\beta \to \infty)$, the entropic contribution to the free energy vanishes, and Eq. 4.18 takes the form

$$F = H_0 + \sum_k [E_k \Theta(-E_k) + E_k \Theta(E_k)]$$

$$= H_0 + \sum_k [E_k (1 - \Theta(E_k)) - E_k \Theta(E_k)]$$

$$= \sum_k [\varepsilon_k + \Delta_k b_k^\dagger - E_k] \tag{4.19}$$

where we have used the Heaviside step-function $\Theta(x) = 1 - \Theta(-x)$, $\Theta(x) = 0$, $x < 0$, $\Theta(x) = 1$, $x > 0$.

The gap $\Delta_{\mathbf{k}}$, which at this stage has the status of a variational parameter, is self-consistently determined by minimizing the free energy with respect to variations in $\Delta_{\mathbf{k}}$. We therefore demand that the following (equivalent) sets of equations be satisfied for the superconducting gap

$$\frac{\partial F}{\partial \Delta_{\mathbf{k}}} = 0$$

$$\frac{\partial F}{\partial \Delta_{\mathbf{k}}^{\dagger}} = 0 \tag{4.20}$$

This mean-field requirement on the gap applies regardless of what the details of the $\mathbf{k}$-space structure of the attractive interaction $V_{\mathbf{k},\mathbf{k}'}$ is. One must remember that the above is a functional derivation such that only one term in the $\mathbf{k}$-space sums are picked out when the derivation is performed. Applying the first of Eqs 4.20 to 4.18, we obtain

$$b_{\mathbf{k}}^{\dagger} + \frac{\partial E_{\mathbf{k}}}{\partial \Delta_{\mathbf{k}}} \left(\frac{e^{-\beta E_{\mathbf{k}}}}{1 + e^{-\beta E_{\mathbf{k}}}} - \frac{e^{\beta E_{\mathbf{k}}}}{1 + e^{\beta E_{\mathbf{k}}}} \right) = 0 \tag{4.21}$$

or equivalently

$$b_{\mathbf{k}}^{\dagger} = \Delta_{\mathbf{k}} \underbrace{\frac{\tanh(\beta E_{\mathbf{k}}/2)}{2 E_{\mathbf{k}}}}_{\equiv \chi(\mathbf{k})} \tag{4.22}$$

where $\chi(\mathbf{k})$ may be interpreted as a *pair-susceptibility*, i.e. the ability the system has to form Cooper-pairs. Using this explicit self-consistent relation between the order parameter $b_{\mathbf{k}}$ and the gap parameter $\Delta_{\mathbf{k}}$ in the definition (Eq. 4.9), we finally arrive at a self-consistent equation for $\Delta_{\mathbf{k}}$, namely the celebrated BCS *gap-equation*

$$\Delta_{\mathbf{k}} = -\sum_{\mathbf{k}'} V_{\mathbf{k},\mathbf{k}'} \Delta_{\mathbf{k}'} \frac{\tanh(\beta E_{\mathbf{k}'}/2)}{2 E_{\mathbf{k}'}} \tag{4.23}$$

As it is written here, it applies at the mean-field level to any superconductor, and as long as we do not specify the origin of the attraction, it is not limited to phonon-mediated superconductivity. Physically, we expect that in reality the detailed frequency spectrum of the 'glue'-bosons does play some role in the phenomenon, but such details are left out of the above treatment. If we were to take such details into account, then we would arrive at a description which would be quite specific to the precise pairing mechanism we have in mind. The lack of such details in the above mean-field theory is usually referred to as a weak-coupling approximation. The more sophisticated (but still mean-field!) description where the details of the specific spectra of the 'glue'-bosons are taken into account, is usually referred to as Eliashberg theory, or strong coupling theory. We will not go into the details of this theory here, but simply state that the above picture of the superconductor as an electronic condensate is the basic picture in weak-coupling as well as strong-coupling theory.

One remark is appropriate here. The phenomenon of Bose–Einstein condensation is well known, both in the context of cold dilute inert gases, as well as in the superfluid transition of He^4. The basis for Bose–Einstein condensation even in ideal quantum gases is very much dependent on the fact that the basic constituents which condense are bosons that experience a statistical bosonic symmetry attraction. Electrons, on the other hand, obey the Pauli principle, which roughly may be viewed as a very brutal hard-core repulsion, rendering the pressure of even the ideal fermionic quantum gas finite at zero temperature, $P = (4\pi^2/3m)(3/8\pi)^{2/3}\rho^{5/3}$, where ρ is the particle density. How can electrons then condense? The answer lies in the fact that by first forming Cooper-pairs, which roughly may be considered to be bosons, the electrons effectively shed the shackles of the Pauli principle, and subsequently undergo a condensation into a superconducting condensate.

Therefore, we reach the important conclusion that *in principle*, superconductivity arises as a result of some sort of condensation in a quantum fluid of Cooper-pairs. We will return to a more detailed description of precisely what this means, but it is important to understand that the formation of Cooper pairs occurs, in principle, at a temperature which is above the true superconducting transition temperature. It is only as a result of a mean-field approximation that we find that the temperature at which we get Cooper-pairs is the same as the temperature where superconductivity occurs. In three dimensional systems of good metals, fluctuation effects that lead to a separation of the Cooper-pair formation temperature and the superconductivity, are tiny and almost immeasurably low. However, in systems like high-T_c superconductors, which first of all are quasi-two-dimensional and secondly arise out of metals with very low carrier density, but nonetheless have very respectable transition temperatures, fluctuation effects are much more prominent, and ignoring them therefore becomes a much more dubious endeavour.

The important fluctuations that destroy superconductivity, but do not dissociate Cooper-pairs, are transverse phase-fluctuations of the complex two-component superconducting gap function Δ_k. The stiffness of the phase of this gap-function, which is a measure of how difficult it is to excite transverse phase-fluctuations, is given by the superfluid density of the system, which is limited by the carrier density. Poor metals thus are expected to give rise to superconductivity where phase-fluctuations play a prominent role. Good metals are expected to give superconductivity where mean-field theory should work very well. In superconductors, the Ginzburg–Landau parameter κ is determined by the superfluid density and the critical temperature, $\kappa = \lambda/\xi \sim T_c/\sqrt{\rho_s}$. Hence, in type-I and moderate type-II superconductors, we can ignore fluctuation effects. In extreme type-II superconductor, like the high-T_c compounds, they cannot be ignored. We will give a detailed description of phase-fluctuations of the superconducting order parameter in Chapters 9 and 10.

4.3   Transition temperature T_c and the energy gap Δ

The minimum of the free energy is determined self-consistently by having the gap-parameter satisfy the equation

$$\Delta_{\mathbf{k}} = -\sum_{\mathbf{k}'} V_{\mathbf{k},\mathbf{k}'} \Delta_{\mathbf{k}'} \frac{\tanh(\beta E_{\mathbf{k}'}/2)}{2E_{\mathbf{k}'}} \tag{4.24}$$

At this stage, this is a mean-field equation with a quite general interaction matrix element $V_{\mathbf{k},\mathbf{k}'}$. Assuming that the potential is separable, i.e.

$$V_{\mathbf{k},\mathbf{k}'} = \sum_{\eta} \lambda_\eta g_\eta(\mathbf{k}) g_\eta(\mathbf{k}') \tag{4.25}$$

where $g_\eta(\mathbf{k})$ are basis functions for the irreducible representations of the crystal symmetry group of the lattice which the electrons move through, then the gap can be written on the general form

$$\Delta_{\mathbf{k}}(T) = \sum_{\eta} \Delta_\eta(T) g_\eta(\mathbf{k}) \tag{4.26}$$

The indices η denote symmetry channels that are represented in the total gap-function. We will take a particularly simple form of this matrix element, namely

$$V_{\mathbf{k},\mathbf{k}'} = -V \tag{4.27}$$

for such pairs of wavenumbers $\mathbf{k}, \mathbf{k}'$ that lie in a thin shell around the Fermi-surface of thickness ω_0, and zero otherwise. The physical interpretation of ω_0 is that it is an upper frequency cutoff on the spectrum of the bosons that provide the 'glue' between the electrons forming the Cooper-pairs. For phonon-assisted superconductivity, we have $\omega_0 = \omega_D$, where ω_D is the Debye-frequency. Inserting this into the gap equation (Eq. 4.24), we conclude that the gap $\Delta_{\mathbf{k}}$ is k-independent, i.e. it is the simplest case of a gap which transforms as the identity under the allowed symmetry operations of the crystal which the electrons move through. Hence, the gap-equation can be written in the form

$$1 = V \sum_{\mathbf{k}'} \frac{\tanh(\beta E_{\mathbf{k}'}/2)}{2 E_{\mathbf{k}'}} \tag{4.28}$$

There are two cases where this equation is easy to solve, namely $T = T_c$ and $T = 0$, and we consider the former first.

Letting $T \to T_c^-$, i.e. in the limit of a vanishing gap, the above equation simplifies in that we ignore the gap in the quasiparticle spectrum $E_{\mathbf{k}}$. Moreover, we replace the k-sum by an energy integral by introducing the normal-state density of states $D_n(\varepsilon)$. The k-space sum is by assumption limited to a narrow region around the Fermi-surface. We therefore need to consider the density of states near the Fermi-surface. Assuming that the Fermi-level is not located close to energies where $D_n(\varepsilon)$ has a rapid variation, e.g. not located close to van Hove singularities, we may then simply replace $D_n(\varepsilon)$ with $D_n(0)$, i.e. the density of states at the Fermi-level when all energies are measured relative to the Fermi-surface. Finally, upon introducing the *dimensionless* coupling constant $\lambda \equiv V D_n(0)$, we get the following equation determining T_c

$$1 = \lambda \int_0^{\omega_D} d\varepsilon \frac{\tanh(\beta\varepsilon/2)}{\varepsilon}$$

$$= \lambda \left[\tanh(\beta\omega_D/2) \ln(\beta\omega_D/2) - \int_0^{\beta\omega_D/2} dx \frac{\ln(x)}{\cosh^2(x)} \right], \tag{4.29}$$

where we have performed a partial integration to isolate the logarithmic singularity in the integral. For phonon-assisted superconductivity (more generally, for weak-coupling superconductivity), $\lambda \ll 1$. This means that in order to satisfy the above equation, we require that $\beta\omega_D/2 \gg 1$. Hence, since the second integral above is rapidly convergent, the above simplifies to

$$1 = \lambda \left[\ln\left(\frac{\beta\omega_D}{2}\right) - \int_0^\infty dx \frac{\ln(x)}{\cosh^2(x)} \right]$$

$$= \lambda \left[\ln\left(\frac{\beta\omega_D}{2}\right) - \ln\left(\frac{\pi}{4e^\gamma}\right) \right]$$

$$= \lambda \ln\left(\frac{2e^\gamma \beta\omega_D}{\pi}\right) \tag{4.30}$$

where $\gamma \equiv \lim_{m \to \infty} \left(\sum_{l=1}^{m} 1/l - \ln(m) \right) = 0.5772156649 \ldots$ is the Euler–Mascheroni constant. This equation may be solved to give

$$k_B T_c = \frac{2e^{\gamma}}{\pi} \omega_D e^{-\frac{1}{\lambda}} \approx 1.13 \omega_D e^{-\frac{1}{\lambda}} \tag{4.31}$$

On the other hand, at $T = 0$, the gap-equation (Eq. 4.28) takes the form

$$1 = V \sum_{\mathbf{k}'} \frac{1}{2E_{\mathbf{k}'}} = \lambda \int_0^{\omega_D} \frac{d\varepsilon}{\sqrt{\varepsilon^2 + \Delta^2}}$$

$$= \lambda \int_0^{\omega_D/\Delta} \frac{dx}{\sqrt{x^2 + 1}} = \lambda \sinh^{-1} \left(\frac{\omega_D}{\Delta} \right) \tag{4.32}$$

Again, when $\lambda \ll 1$, this requires $\omega_D/\Delta \gg 1$ for solution of the equation to exist, and hence $\sinh^{-1}(\omega_D/\Delta) \approx \ln(2\omega_D/\Delta)$. Thus we obtain

$$\Delta(T = 0) = 2\omega_D e^{-\frac{1}{\lambda}} \tag{4.33}$$

Note that both T_c and $\Delta(T = 0)$ depend sensitively on the coupling constant λ, and that both are functions of λ with an essential singularity as $\lambda \to 0$. This reflects the fact that an attractive interaction is a singular perturbation to the free electron gas, in contrast to the repulsive Coulomb-interaction. There is no way we can obtain the above results for T_c and $\Delta(T = 0)$ to any finite order in perturbation theory in the dimensionless coupling constant λ. Moreover, since the dimensionless coupling constant λ is not easy to compute reliably from first principles even in simple metals like aluminum, it is extremely hard to compute values of T_c and $\Delta(T = 0)$ reliably. This reflects the fact that these quantities depend on all details of the system, and really should not be the focus of any computations, when the goal is to gain some qualitative understanding of the problem. This is actually a lesson that applies to all types of phase transitions. However, what was noted by Bardeen, Cooper and Schrieffer was the following remarkable fact. If we form the ratio of $\Delta(T = 0)$ and $k_B T_c$, then we obtain

$$\frac{2\Delta(T = 0)}{k_B T_c} = \frac{2\pi}{e^{\gamma}} \approx 3.52. \tag{4.34}$$

This is a universal amplitude ratio, all materials-dependent quantities precisely cancel out!

4.4 Generalized gap equation, s-wave and d-wave gaps

As the BCS Hamiltonian is formulated in Eq. 4.1, it does not make specific references to any particular pairing mechanism, although originally the founders of the theory definitely had in mind a phonon-assisted pairing mechanism. What is central, is the idea that the basic physical phenomenon underlying superconductivity is *pairing* of electrons into Cooper-pairs, followed by a condensation of the electron system into a macroscopically phase-coherent matter wave. In a sense, superconductivity is the matter-wave counterpart of laser-light, which is basically a phase-coherent lightwave. The electron-pairing picture will be maintained. However, as a slight generalization of the above approach, we may ignore the issue of the origin of some attractive effective interaction between electron, leaving open even the possibility that it may not be phononic in origin, but could also be purely electronic. Thus, we may generalize the interaction to the case where it is not limited to be operative in a thin shell around the Fermi-surface, i.e. we include the possibility of a non-retarded attractive interaction. We also propose a slightly more complicated form of the matrix element $V_{\mathbf{k},\mathbf{k}'}$ than what we suggested in the previous section, based on the physics of phonon-mediated attractive interactions between electrons.

Assume that an electron moves through a crystal with some symmetry. Any physical quantity will ultimately reflect the symmetry of the crystal in some way. It can therefore be expanded in a complete set of functions that all reflect the particular crystal symmetry that the electrons experience. For instance, if the system is approximately isotropic we can expand any physical quantity in the complete set of the well-known spherical harmonics, if the crystal has simple cubic symmetry we can expand in the complete set of cubic harmonics, and so on. Let us denote the set of such complete basis functions by $\{g_\eta(\mathbf{k})\}$, where η is an index that denotes which particular one of the basis functions is meant. Now, let us assume further that the matrix element $V_{\mathbf{k},\mathbf{k}'}$ is factorizable. By this we mean that it can be factorized into two factors, where one factor only depends on $\mathbf{k}'$ and the other one only depends on $\mathbf{k}$. Now, since the set $\{g_\eta(\mathbf{k})\}$ is complete, by assumption, we may write

$$V_{\mathbf{k},\mathbf{k}'} = \sum_\eta \lambda_\eta g_\eta(\mathbf{k}) g_\eta(\mathbf{k}') \tag{4.35}$$

where λ_η is the strength of the interaction in the η-channel.

To be specific, consider an example. Assume that the lattice is square (two-dimensional, with lattice constant set to unity) and that the electron-electron interactions are such that when two electrons are on the same site, on nearest-neighbor sites (horizontal and vertical bonds), and on next-nearest-neighbour sites (diagonal bonds) they experience interactions $U/2$, $2V$, and $4W$, respectively. Fourier-transforming this to obtain $V_{\mathbf{k},\mathbf{k}'}$ we find

$$V_{\mathbf{k},\mathbf{k}'} = \frac{U}{2} + 2V[\cos(k_x)\cos(k_x') + \sin(k_x)\sin(k_x')$$
$$+ \cos(k_y)\cos(k_y') + \sin(k_y)\sin(k_y')]$$

$$+ 4W[\cos(k_x)\cos(k_x') + \sin(k_x)\sin(k_x')]$$
$$\times [\cos(k_y)\cos(k_y') + \sin(k_y)\sin(k_y')] \qquad (4.36)$$

Now, we would like to re-express this as a sum of factorized terms using the complete set of *square lattice harmonics*. As the interaction given above is of finite range, $V_{k,k'}$ has a finite number of terms and thus a finite number of square lattice harmonics will suffice to completely express the matrix element. It turns out that the five lowest order square lattice harmonics will do the job. They are given by

$$g_1(\mathbf{k}) = \frac{1}{2\pi}$$

$$g_2(\mathbf{k}) = \frac{1}{2\pi}[\cos(k_x) + \cos(k_y)]$$

$$g_3(\mathbf{k}) = \frac{1}{2\pi}\cos(k_x)\cos(k_y)$$

$$g_4(\mathbf{k}) = \frac{1}{2\pi}[\cos(k_x) - \cos(k_y)]$$

$$g_5(\mathbf{k}) = \frac{1}{2\pi}\sin(k_x)\sin(k_y) \qquad (4.37)$$

Re-expressing Eq. 4.36 in terms of the above basis functions, we find an expression of the form of Eq. 4.35, with $\lambda_1 = 2U\pi^2$, $\lambda_2 = \lambda_4 = 4V\pi^2$, $\lambda_3 = \lambda_5 = 4W\pi^2$, and $\lambda_\eta = 0$, $\eta \geq 6$.

We note in passing the following fact, which will figure prominently in what follows. The symmetry group of the square lattice is C_{4v}. What is important to note about the five basis functions listed above, is that the first two *transform as the identity* under all symmetry operations included in C_{4v}. The latter three change sign under $\pi/2$ rotations. In this respect, they have common properties with $l = 0$ and $l = 2$ spherical harmonics functions that are the basis functions for the isotropic case. $l = 0$ spherical harmonics are commonly referred to as s-wave functions, while $l = 2$ spherical harmonics are referred to as d-wave functions. Hence, the first two basis functions above are s-wave basis function, $g_2(\mathbf{k})$ is often called an extended s-wave function. The latter three are d-wave functions. Of these, $g_4(\mathbf{k})$ is the most well-known one in the context of high-T_c superconductivity, since the gap in the cuprate superconductors to a very good approximation appears to be of the form $\Delta_\mathbf{k} = \Delta_0(T)g_4(\mathbf{k})$. Note that $l = 1$ square lattice harmonics do not enter in the above. The reason is that in all of the above, we have implicitly considered only such cases where we have *spin-singlet pairing*. This means that the spin-part of the Cooper-pair wave functions is antisymmetric so the space-part of it must be symmetric. This requires even-l basis functions. Spin-triplet pairing would be required in order to make use of odd-l basis functions.

Returning now to the general case, as far as the superconducting gap is concerned, it is clear that when $V_{\mathbf{k},\mathbf{k}'}$ takes the form of Eq. 4.35, we obtain the gap-equation

$$\Delta_{\mathbf{k}} = -\sum_{\mathbf{k}} V_{\mathbf{k},\mathbf{k}'} \Delta_{\mathbf{k}'} \chi_{\mathbf{k}'} = -\sum_{\eta=1}^{5} \lambda_\eta g_\eta(\mathbf{k}) \underbrace{\sum_{\mathbf{k}'} g_\eta(\mathbf{k}') \Delta_{\mathbf{k}'} \chi_{\mathbf{k}'}}_{\equiv -\Delta_\eta/\lambda_\eta}$$

$$= \sum_{\eta=1}^{5} \Delta_\eta g_\eta(\mathbf{k}) \tag{4.38}$$

which shows that the gap itself is also expressible as a linear combination of basis functions, as it should be, being a physical quantity. The coefficients Δ_η are not $\mathbf{k}$-dependent, but do of course depend on temperature. Using the expansion of the gap in terms of basis function on both sides of the gap-equation, we find the set of coupled self-consistent equations for that gap in the different channels Δ_η

$$\Delta_\eta = \sum_{\eta'=1}^{5} \Delta_{\eta'} M_{\eta,\eta'}$$

$$M_{\eta,\eta'} = -\lambda_\eta \sum_{\mathbf{k}} g_\eta(\mathbf{k}) g_{\eta'}(\mathbf{k}) \chi_{\mathbf{k}} \tag{4.39}$$

In general, these equations require numerical efforts in order to be solved.

Considering, once again, the above square lattice example, and setting $U > 0$, $V < 0$, and $W = 0$, we find that there is no attraction in the λ_1-channel, and attraction in the λ_2-channel (extended s-wave) and the λ_4-channel (d-wave). Hence, we see that the gap function, which is closely related to the Fourier-transform of the Cooper-pair wave function, adjusts itself so as to be zero in channels where the wavefunction is non-zero for zero separation between the electrons. By choosing extended s-wave or d-wave, the electrons in the Cooper-pair avoid the onsite Coulomb interaction. Recall that in phonon-assisted Cooper-pairing, the electrons did not avoid each other in space, but used retardation to eliminate the adverse effects of Coulomb-interactions. I.e. they avoided each other in time. In the present non-retarded example, the electrons avoid each other in space, but not in time. Avoiding each other in space (higher-angular momentum pairing) or in time (retardation) are the basic two ways electrons can avoid the repulsive Coulomb-interaction.

In the BCS gap-equation, the presence or lack of retardation effects is manifest in the presence or lack of an energy cutoff around the Fermi-surface. In the absence of such a cutoff, i.e. in the absence of retardation, the entire Brillouin-zone contributes to the gap. Since $\Delta(\mathbf{r}) = \sum_{\mathbf{k}} \Delta_{\mathbf{k}} e^{i\mathbf{k}\cdot\mathbf{r}}$, this means that $\Delta(\mathbf{0}) = \sum_{\mathbf{k}} \Delta_{\mathbf{k}}$. Hence, $\Delta_{\mathbf{k}}$ must be chosen so as to obtain $\Delta(\mathbf{0}) = 0$ in order to avoid

Coulomb-repulsion. Gaps of the form $\Delta_{\mathbf{k}} = \Delta_0(T)g_2(\mathbf{k})$, and in particular $\Delta_{\mathbf{k}} = \Delta_0(T)g_4(\mathbf{k})$, have this property. The latter gap-function has a form which is believed to be very close to the form of the gap-function in high-T_c cuprates.

As an important application and in preparation for later sections, let us obtain the superconducting density of states $D_s(E)$, which may now be found as follows. Imagine that we have some function $g(\mathbf{k})$ that depends on $\mathbf{k}$ only through $E_{\mathbf{k}} = \sqrt{\varepsilon_{\mathbf{k}}^2 + \Delta_{\mathbf{k}}^2}$, and assume $\Delta_{\mathbf{k}}$ to be independent of $\mathbf{k}$. Then we have

$$
\sum_{\mathbf{k}} g(\mathbf{k}) = \int d\varepsilon\, D_n(\varepsilon) g(\sqrt{\varepsilon^2 + \Delta^2})
$$

$$
= \int dE\, D_n(\varepsilon) \frac{d\varepsilon}{dE} g(E)
$$

$$
= \int dE\, D_s(E) g(E) \tag{4.40}
$$

where we have introduced $E = \sqrt{\varepsilon^2 + \Delta^2}$ and $D_s(E) = D_n(\varepsilon)\frac{d\varepsilon}{dE}$, where $D_n(\varepsilon)$ is the normal state density of states. The above is exact provided we have

$$
D_n(\varepsilon) = \frac{1}{N} \sum_{\mathbf{k}} \delta(\varepsilon - \varepsilon_{\mathbf{k}}) \tag{4.41}
$$

Here, N is the number of Fourier modes of the wave-number, equivalently the number of lattice sites in the problem. The normal state density of states is now assumed to be a slowly varying function of ε around the Fermi surface such that we approximate it by $D_n(\varepsilon) \approx D_n(0)$ when we measure energies relative to the Fermi surface energy. Implicit in this assumption is that sharp features in $D_n(\varepsilon)$ are absent near the Fermi surface, i.e. there are no van Hove singularities near the Fermi surface. Then we have

$$
\frac{D_s(E)}{D_n(0)} = \frac{E}{\sqrt{E^2 - \Delta^2}} \Theta(E - |\Delta|) \tag{4.42}
$$

where $\Theta(x)$ is the Heaviside step-function $\Theta(x) = 1, x > 0, \Theta(x) = 0, x < 0$.

When the gap function $\Delta_{\mathbf{k}}$ is not a constant in $\mathbf{k}$-space, we derive a density of states more formally from the spectral weight $A(\mathbf{k}, \omega)$ introduced in Section 3.2, as follows. Introduce, in analogy to Eq. 3.6, the Green's function $G(k, t)$ for the superconducting condensate. Formally, this Green's function is given by

$$
G(\mathbf{k}, t; \sigma) = -i \langle 0 | c_{\mathbf{k},\sigma}^+(t) c_{\mathbf{k},\sigma}(0) | 0 \rangle \tag{4.43}
$$

where $|0\rangle$ is the BCS superconducting groundstate. It is an eigenstate of the Hamiltonian Eq. 4.16. Moreover, we have

$$
G(\mathbf{k}, \omega; \sigma) = \frac{1}{2\pi} \int dt\, e^{i\omega t} G(\mathbf{k}, t; \sigma) \tag{4.44}
$$

Without loss of generality we choose one spin direction, $\sigma = \uparrow$ say, and find

$$G(\mathbf{k}, t; \uparrow) = -i \langle 0|(\cos(\theta)\eta_{\mathbf{k}}^{+}(t) - \sin(\theta)\gamma_{\mathbf{k}}^{+}(t))(\cos(\theta)\eta_{\mathbf{k}}(0)$$

$$- \sin(\theta)\gamma_{\mathbf{k}}(0))|0\rangle \tag{4.45}$$

Now, using the fact that $(\eta_{\mathbf{k}}, \gamma_{\mathbf{k}})$ diagonalize the BCS Hamiltonian, we have that

$$\langle 0|\eta_{\mathbf{k}}^{+}\gamma_{\mathbf{k}}^{+}|0\rangle = 0$$

$$\langle 0|\eta_{\mathbf{k}}\gamma_{\mathbf{k}}|0\rangle = 0 \tag{4.46}$$

Hence, we obtain

$$G(\mathbf{k}, t; \uparrow) = -i \cos^2(\theta)\langle 0|\eta_{\mathbf{k}}^{+}(t)\eta_{\mathbf{k}}(0)|0\rangle$$

$$- i \sin^2(\theta)\langle 0|\gamma_{\mathbf{k}}^{+}(t)\gamma_{\mathbf{k}}(0)|0\rangle \tag{4.47}$$

Each bracket is now a free fermion propagator in the superconducting state, hence upon Fourier-transforming we find, introducing the coherence factors in Eq. 4.14,

$$G(\mathbf{k}, \omega; \uparrow) = \frac{u_{\mathbf{k}}^2}{\omega - E_{\mathbf{k}} + i\delta_{\mathbf{k}}} + \frac{v_{\mathbf{k}}^2}{\omega + E_{\mathbf{k}} + i\delta_{\mathbf{k}}} \tag{4.48}$$

where $\delta_{\mathbf{k}}$ is infinitesimal. The spectral weight is therefore given by, using Eq. 3.8

$$A(\mathbf{k}, \omega; \uparrow) = u_{\mathbf{k}}^2 \delta(\omega - E_{\mathbf{k}}) + v_{\mathbf{k}}^2 \delta(\omega + E_{\mathbf{k}}) \tag{4.49}$$

which should be compared to Eq. 3.8, and which is seen to be spin-independent. Due to the fact that we are considering spin-singlet pairing, we obtain the same result for $\sigma = \downarrow$. Thus, we finally obtain the superconducting density of states per spin channel as

$$D_s(\omega) = \frac{1}{N}\sum_{\mathbf{k}} A(\mathbf{k}, \omega)$$

$$= \frac{1}{N}\sum_{\mathbf{k}}[u_{\mathbf{k}}^2 \delta(\omega - E_{\mathbf{k}}) + v_{\mathbf{k}}^2 \delta(\omega + E_{\mathbf{k}})] \tag{4.50}$$

which again is seen to be independent of the spin index σ due to the fact that we are here considering spin-singlet pairing.

4.5   Quasi-particle tunnelling and the gap

4.5.1   Introductory remarks

The field of superconductivity made two important steps forward at the experimental discovery of tunneling of single charge carriers ('quasiparticles') by Giaever in 1960 [38], and the subsequent predictions of tunnelling of Cooper pairs by Josephson in 1962 [8]. These discoveries set the stage for a new era in superconductivity on the small scale. Structures in which the thickness dimension was as small as a few nanometers could be manufactured in laboratories, and their properties controlled reproducibly. Avenues were opened for entirely new research and insights, and for electronic devices of a kind that could not even in principle be achieved by conventional methods, all due to the unique properties of the superconducting state, and its wave-function.

At the time of the discovery of these new phenomena quantum mechanical tunnelling was a well-established concept in atomic and nuclear matter, and had recently been demonstrated by Esaki to occur between semiconductors separated by a thin barrier. It took the curiosity of two young and independent student minds to divert from conventional thinking, and venture into the new landscapes:

The scientific world was startled at the simplicity and profoundness of their discoveries. Giaever pointedly describes his discovery as a method of 'measuring the superconducting energy gap with a voltmeter'. This apparent simplicity, and the importance of the results led to immediate acceptance. Josephson's theory, on the other hand, met some initial skepticism. But within a year, experimental evidence began to accumulate in favour of Josephson's ideas, and further theoretical work by many groups expanded our understanding of the physics involved, and contributed to the foundation of present-day sophisticated devices, leading on to what we call today 'Josephson technology'. Here again initial discoveries in pure physics found their way to applications in electrical engineering sciences and technical products.

4.5.2   The tunnelling principle

The tunnelling of a small particle through a barrier in vacuum is a standard problem in atomic physics books, and solvable by relatively simple mathematics using the Schrödinger equation, on requiring continuity of the wave-function through the barrier. Usually, the height of the barrier is an unknown parameter, which may eventually be determined by comparison of experiment and theory.

When tunnelling is to take place between two conductors, separated by an insulating barrier, several possibilities arise. If we let N symbolize a normal metal, and S a superconductor, and have the substances separated by a thin insulator I, then we may combine these substances in various configurations: NIN, SIS, NIS, N_1IN_2, S_1IS_2 etc.

Some of these possible structures are illustrated schematically in Figure 4.1. The symbols ϕ, ϕ_l, ϕ_r in superconductors indicate the phases of the wavefunctions. In each case the tunnelling particle leaves the substance to the left, say,

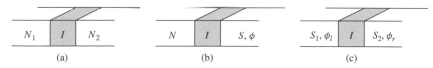

N_1 | I | N_2 N | I | S, ϕ S_1, ϕ_l | I | S_2, ϕ_r

(a) (b) (c)

Figure 4.1   Schematic illustration of various possible configurations for tunneling between two normal metals (a), between normal metal and superconductor (b), and between two superconductors (c).

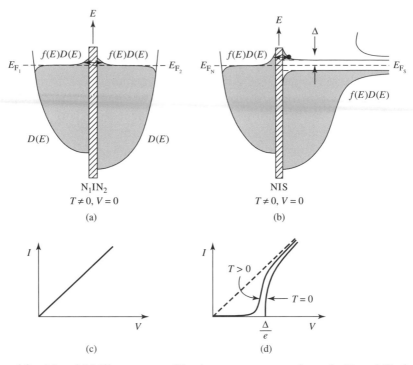

Figure 4.2   (a) and (c) Illustrate tunnelling between two normal metals N_1 and N_2 through a thin insulating barrier I as seen in a density of states versus energy diagram, with the Fermi–Dirac function factored in. In (a) $T \neq 0$ and applied voltage $V = 0$. In (c) the $I - V$ characteristic is shown as linear, as a consequence of displacing the electron distributions in (a) along the vertical axis by application of an increasing voltage. The right hand figures illustrate similarly the situation for tunneling between a normal metal and a superconductor with an energy gap Δ, at $T \neq 0$ and $V = 0$ in (b), and that of varying voltage applied, in (d). The consequence of varying T is indicated in (d). Note the onset of tunnelling at $V = \Delta/e$.

where it occupies an allowed state in the conduction band of free carriers, and, after tunnelling enters into an allowed state in a band on the right side of the insulating barrier. These processes are restricted by the energy spectra and the statistics governing the bands from which they tunnel, and into which they arrive. Normal metals are well understood so that the appropriate mathematical description of the process between normal metals is straightforward. Furthermore, the BCS theory for superconductivity gave an accurate description of the energy bands in typical low-T_c metallic superconductors. Because this theory already existed by the time of the experimental work by Giaever and coworkers, the appropriate integrals could be written down. Figure 4.2 shows sketches of bands for two of the interesting cases. Figure 4.2a shows the $N_1 I N_2$ case, while Figure 4.2b illustrates the NIS case with no external voltage applied. Figure 4.2c and d illustrate the corresponding current–voltage characteristics for the NIN case (c), and the NIS case (d), respectively. None of these represent the Josephson case, however. In the Josephson effect the phase properties of the wavefunction come into full play. This requires a different approach, as will be discussed later. The Giaever case is a simpler one. Let us look at the terms that contribute to the tunneling of single carriers, beginning with the NIN case where only normal electrons are involved.

4.5.3   Single-particle NIN tunnelling

For this discussion we refer to Figure 4.2a. The total number of electrons travelling across the barrier from left to right is proportional to the number of filled electron states on the left-hand side (l) and to the number of states that are unoccupied on the right-hand side (r). The tunnelling current $i_{l \to r}$ at a particular energy level E and in an interval dE is then

$$i_{l \to r} = C_n D_l(E) f_l(E) \times D_r(E) f_{r,h}(E) dE \qquad (4.51)$$

where f_l and $f_{r,h}$ are the probabilities that the energy level E to the left is occupied, and that states at the same level to the right are unoccupied, respectively. According to quantum theory, when the transitions take place from left to right, i.e. when the transitions are completed as $i_{l \to r}$, the unknown constant C_n may be written as

$$C_n = \frac{2\pi}{\hbar} |T_n|^2 \qquad (4.52)$$

where $|T_n|^2$ is the tunnelling probability, given by the squared matrix element for the transition of an electron from the left electrode to the right electrode.

We have, in addition, to factor in the number of available states (holes, h) to the right in the same energy interval by use of the relation

$$f_{r,h}(E) = 1 - f_r(E) \qquad (4.53)$$

The total tunnelling current, when adding processes taking place at all energy levels is then

$$I_{l \to r} = \int_{-\infty}^{\infty} i_{l \to r}(E) dE = \frac{2\pi}{\hbar} \int_{-\infty}^{\infty} |T_n|^2 D_l(E) f_l(E) D_r(E) (1 - f_r(E)) \, dE \qquad (4.54)$$

In equilibrium, of course, a similar integral applies to transitions from right to left. These transitions have to be subtracted to get the net current across the barrier. Finally, therefore: The net current I_{NIN} between two normal metals found upon subtraction of $I_{r \to l}$ from $I_{l \to r}$ becomes:

$$I_{NIN} = I_{l \to r} - I_{r \to l} = \frac{2\pi}{\hbar} \int_{-\infty}^{\infty} |T_n|^2 D_l(E) D_r(E) (f_l(E) - f_r(E)) \, dE \qquad (4.55)$$

In both equations above we have allowed integration from $-\infty$ to $+\infty$. This is permissible due to the difference term $(f_l(E) - f_r(E))$. The properties of these functions makes the integrand extremely small outside a narrow energy window of order kT.

Now, without a voltage applied, the difference $(f_l(E) - f_r(E))$ is simply zero, as may be seen from Figure 4.2a and b: No net charge transport can occur without an applied voltage. The claim that $f_l(E) - f_r(E) = 0$ may seem like an oversimplification if we consider two metals with different Fermi energies to the left and to the right. A very important condition has to be fulfilled for this equality to hold: The two electron gases must constitute one thermodynamic system. This requires particle exchange across the barrier, in fact vigorous tunneling must take place. It is a basic result of thermodynamics that two systems which come into equilibrium due to exchange of particles, will do so by spilling over particles until they have adjusted their chemical potentials to the same level. In the present context we will approximate the chemical potential μ by the Fermi-energy E_F. So $E_{F,l}$ and $E_{F,r}$ become exactly equal at equilibrium. Now clearly $f_l(E) - f_r(E)$ becomes zero since what really goes into the Fermi-functions is always the difference $E - E_F$. This argument also points to the fact that *initially* there must be a current to establish equilibrium. But this process is fast, and happens during deposition of the films. We need not worry about such transient properties.

We can break this symmetry by applying a voltage V, raising the electron band on one side with respect to the other side. Now current will flow since we have to put $E + eV$ or $E - eV$ as the argument in one or the other of the f-functions, and $f_l(E) - f_r(E + eV)$ becomes non-zero. The current is now

$$I_{NIN} = \frac{2\pi}{\hbar} |T_n|^2 \int_{-\infty}^{\infty} D_l(E) D_r(E + eV)(f_l(E) - f_r(E + eV)) \, dE \qquad (4.56)$$

We may if $eV \ll E_f$, to a very good approximation, make the following simplification for calculation of the integral for I_{NIN}:

$$(f(E) - f(E + eV)) \, dE \approx -\frac{\partial f}{\partial E} eV \, dE \qquad (4.57)$$

Under the integral $-\frac{\partial f}{\partial E}$ acts like a δ-function; hence we obtain

$$I_{NIN} = \frac{2\pi}{\hbar} |T_n|^2 D_l(E_F) D_r(E_F) eV \equiv G_{NIN} V \equiv \frac{1}{R_{NIN}} V \qquad (4.58)$$

where G_{NIN} is the normal conductance, or the inverse of the resistance R_{NIN} of the entire NIN contact. We have taken D_l and D_r outside the integration, replacing them by the constants $D_l(E_F)$ and $D_r(E_F)$ since under normal circumstances E will be near E_F where $D(E)$ varies slowly with E. This result is shown in Figure 4.2c where the simple linear characteristic of one NIN contact is illustrated.

4.5.4 NIS quasiparticle tunnelling

The analysis given above proves that the NIN tunneling contact acts as an ohmic element in a circuit, a fact worth noting, but as such not particularly exciting or useful since for most purposes we have more practical ways of making resistors. However, the situation is substantially altered once we allow one, or both, of the metals to become superconducting. This can in many cases be achieved simply by cooling the contact. If the two metals were, for instance, Nb with a superconducting transition temperature $T_c = 9\,$K and Pb with $T_c = 7\,$K, we could implement the following contacts by lowering the temperature: N_1IN_2 at all $T > 9\,$K; NIS between $7\,$K and $9\,$K; and S_1IS_2 below $7\,$K.

What makes the NIS or SIS configurations so interesting, and useful, is (i) that they provide us with radically different, nonlinear current-voltage I–V characteristic both for single electron tunneling and for Cooper pair tunnelling, and (ii) that entirely new phenomena arise in SIS contacts in particular, reflecting deeper aspects of the superconducting wavefunction. These aspects are exceedingly interesting from a physics point of view, and harbor the promise for a multitude of important applications.

On changing our system under investigation to an NIS-contact, as long as we are discussing single-particle tunnelling, the integrals introduced above do not change in principle, only the density of states D_s in the superconductor replaces one of D_n functions in the normal state. The Fermi–Dirac functions remain unaltered. Again we have to apply a voltage to set up a net current in one of the two directions (see Figure 4.3.)

In Section 4.4 we have shown that the density of states of quasiparticles in the superconductor $D_s(E)$ is related to the one in the normal metal $D_n(0)$ near

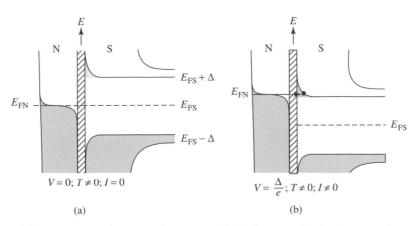

$V = 0; T \neq 0; I = 0$

(a)

$V = \dfrac{\Delta}{e}; T \neq 0; I \neq 0$

(b)

Figure 4.3 (a) $T \neq 0$. Almost no electrons available for tunneling leads to very low tunnel current. (b) Bands displaced by eV. Optimal condition for quasiparticle tunnelling from N to S. In both cases E_{FN} and E_{FS} refer to Fermi energies of normal and superconducting metals, respectively.

the Fermi energy by

$$\frac{D_s(E)}{D_n(0)} = \begin{cases} \dfrac{E}{(E^2 - \Delta^2)^{1/2}}; & |E| \geq \Delta \\[2mm] 0; & |E| < \Delta \end{cases} \tag{4.59}$$

where energy E is now measured from the Fermi-level E_{FS}. On insertion of $D_s(E)$ the current can be expressed as

$$I^q_{NIS} = \frac{2\pi}{\hbar} |T_n|^2 \int_{-\infty, |E| > \Delta}^{\infty} D_l(0) D_r(E_F) \frac{|E|}{\left|E^2 - \Delta^2\right|^{\frac{1}{2}}} (f(E) - f(E + eV)) \, dE \tag{4.60}$$

where the superscript q reminds us that only quasiparticles (single electrons) are to be counted.

The notation $|E| > \Delta$ under the integral sign indicates that integration over the interval $|E| > \Delta$ is excluded. On comparison with the NIN-case we discover that I^q_{NIS} can be simplified to

$$I^q_{NIS} = \frac{G_{NIN}}{e} \int_{-\infty}^{\infty} \frac{|E|}{\left|E^2 - \Delta^2\right|^{\frac{1}{2}}} (f(E) - f(E + eV)) \, dE \tag{4.61}$$

This integral can only be solved analytically in the $T = 0$ limit, giving

$$I^q_{NIS} = \begin{cases} 0; & |eV| < \Delta \\[2mm] \dfrac{G_{NIN}}{e} \left[(eV)^2 - \Delta^2)\right]^{\frac{1}{2}}; & |eV| \geq \Delta \end{cases} \tag{4.62}$$

The quasiparticle current increases sharply from zero at $V < \frac{\Delta}{e}$ like the square root of the voltage in excess of $\frac{\Delta}{e}$ when $V > \frac{\Delta}{e}$.

A crucial point, which was eloquently exploited by Giaever and coworkers, is the ensuing relation for the differential conductance, defined as $G^d_{NIS} = dI_{NIS}/dV$ (rather than $G_{NIS} = I_{NIS}/V$ used before). One quickly verifies that

$$G^d_{NIS} \equiv \frac{dI_{NIS}}{dV} = G_{NIN} \int_{-\infty}^{\infty} \frac{D_s(E)}{D_n(E_F)} \frac{\partial f(E + eV)}{\partial(eV)} \, dE \tag{4.63}$$

Here the derivative of f approaches a δ-function when $kT \to 0$. At finite values it assumes a form similar to a Gaussian distribution, and with a width of the order of kT. In the $T = 0$ limit the result is then exactly

$$G^d_{NIS} = G_{NIN} \frac{eV}{\left[(eV)^2 - \Delta^2\right]^{\frac{1}{2}}} = G_{NIN} D_s(E = eV) \tag{4.64}$$

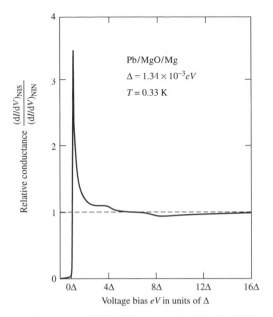

Figure 4.4 Relative conductance of Pb/MgO/Mg contact in the NIS configuration relative to that of the corresponding NIN contact. The structure above the gap is caused by phonon processes during tunnelling. Adapted from Giaever *et al.* [39].

This means that a measurement of dI_{NIS}/dV, indeed a measurable quantity, probes the superconducting quasiparticle density of states function *directly*. Most clearly we see this in the experimentally accessible quantity

$$G_{NIS}^d / G_{NIN} = D_s(E = eV) \tag{4.65}$$

An example of an early manifestation of this relationship is seen in Figure 4.4, except for some additional phonon structure.

The expected property that the relative conductance diverges like D_s at $eV = \Delta$, is therefore confirmed. A physical measurement can of course never truly measure a divergence in the strict mathematical sense. Neither should one take such a prediction literally in the real world. After all, we are talking about a model, the BCS theory, which has built into it certain idealizations that are not to be regarded as exact. Still, the model predicts a remarkably sharp onset of quasiparticle current *vs.* V that has been confirmed by experiments, even at temperatures well above $T = 0$, and the overall behaviour is very close to the BCS prediction.

In the so-called *excitation representation* electron-like and hole-like excitations are depicted in a positive energy continuum above the superconductor gap energy Δ. This is illustrated in Figure 4.5.

The superconducting energy gap is introduced by joining the hole-like branch and the electron-like branch at a finite gap energy Δ, and with consequences for the density of states, proportional to $|\nabla E(\mathbf{k})|^{-1}$.

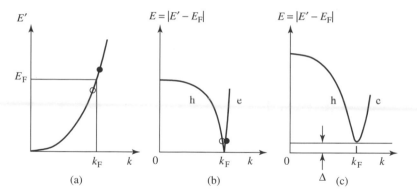

Figure 4.5 (a) Usual free-electron gas representation of $T = 0$ in a normal metal; holes ○ below E_F, electrons • above. (b) Excitation representation of a normal electron gas at $T = 0$. Hole branch: ○h, electron branch •e. (c) Excitation representation in a superconductor with a gap Δ. The two branches join at Δ, and the density of states proportional to $\left(\frac{\partial E}{\partial k}\right)^{-1}$ near the minimum point is strongly modified from that of the normal state in (b).

4.5.5   SIS quasiparticle tunnelling

With the construction of an a SIS contact the possibility arises to establish both a quasiparticle current and a Cooper-pair current between the two superconductors. Both of these can be treated fully by quantum theory, on a similar footing to that used in NIN and NIS contacts above. However, only the SIS-case is reasonably straightforward in the established language of single particle transitions. When Cooper-pairs tunnel the addition of matrix elements for the two constituent particles requires heavier formalism which belongs in a more advanced treatment. The familiar problem in quantum mechanics of adding before squaring comes in. We will attack this case with a different and simpler method in the next section. And, having established a method for calculation of NIS -quasiparticle current, one needs only extend that procedure to the S_1IS_2 case directly. The resulting integral can be guessed on the basis of the earlier discussion:

$$I_{SIS}^q = \frac{G_{NIN}}{e} \int_{-\infty}^{\infty} \frac{|E|}{\left|E^2 - \Delta_1^2\right|^2} \frac{|E + eV|}{((E + eV)^2 - \Delta_2^2)^{\frac{1}{2}}} ((f(E) - f(E + eV)) \, dE$$

$$(4.66)$$

Again, it is understood that integration is not to be carried out in the gaps. This current can only be calculated numerically. Instead of working out this problem we resort to simple physical argument. In four successive sketches, Figure 4.6, a qualitative picture is developed for quasiparticle tunneling between two superconductors with different gaps, Δ_1 and Δ_2.

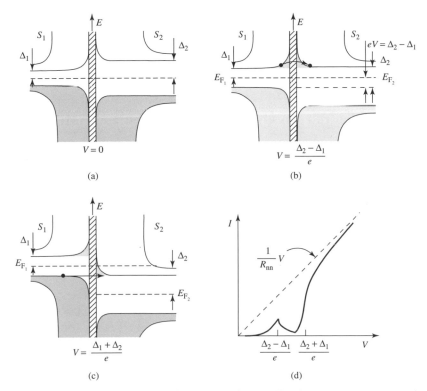

Figure 4.6 Qualitative illustration of quasiparticle tunneling between two superconductors with different gaps, (a), (b) and (c) at different values of bias field, (d) showing the resulting overall I–V characteristic to be expected.

4.6 BCS coherence factors versus quasiparticle-effects: ultrasound and NMR

4.6.1 Introductory remarks

As we have seen in Section 4.2 the BCS theory predicts an energy gap $\Delta(T) = \Delta_o(T_c - T)^{1/2}$ associated with the superconducting state due to an effectively attractive electron-electron interaction in the presence of an ionic lattice. This picture needs experimental confirmation. In the preceding section we already discussed a possible method, quasi-particle tunnelling. There are several other methods as well. In fact, specific heat measurements and infrared absorption had already shown, even before the BCS theory was worked out, that a gap had to be present. This was particularly evident well below T_c. In this sense the BCS theory explained already existing results. However, a more complete and satisfactory confirmation of the BCS theory requires measurement of both the gap size, the temperature dependence of the gap, the density of states, and confirmation of more subtle effects like the so-called coherence factors. In this section we will illustrate both some theoretical approaches, and some experimental procedures that confirm central aspects of the BCS theory for metallic, low-temperature superconductors.

4.6.2 Transition rates in ultrasound propagation and NMR

Ultrasonic attenuation and NMR have turned out to be excellent tools for testing of the predictions from the BCS theory. Some experimental results for ultrasound appeared already in the BCS paper, and lent immediate support to this important work. Longitudinal phonons interact with the quasi-particles in the same way as with electrons in the normal state. The BCS theory for the attenuation of longitudinal phonons gives a very simple expression to test, as will be outlined below, with $\Delta(T)$ as the only unknown quantity. Early measurements at Brown University [40] showed overall consistency with the predictions of BCS, although the experimental data had low precision near T_c.

We will now investigate how ultrasonic attenuation may be used to gain insight, possibly verifying more fully the BCS result. Let us follow the main line of reasoning leading to a prediction for ultrasonic attenuation and NMR relaxation rate. Up to a certain point these physical effects can be described by the same formalism. In quantum mechanical terms we need to calculate transition rates α_s. These are the net transition rates between energy levels E and $E + \hbar\omega$, in fact quite similar to tunnelling which we discussed previously, in Section 4.5. In the present context the energy $\hbar\omega$ refers to the exchange of a phonon in the case of ultrasound, and to the exchange of a quantum of electromagnetic radiation in the case of NMR. With the tunnelling effect as a reference, the expression for the desired quasiparticle transition rate in the superconducting state may be logically constructed as

$$\alpha_s = \int |M|^2 \, C(\Delta, E, E + \hbar\omega) D_s(E) D_s(E + \hbar\omega) \left[f(E) - f(E + \hbar\omega) \right] dE$$

(4.67)

Here, M is a matrix element for the appropriate electronic transition, the same one as in the normal state due to the close correspondence between quasiparticles and normal state electrons. The D_s-factors are the same density of states we used under the discussion of tunnelling in Section 4.5. However, there is a fundamental difference between the normal state and the superconducting state: In the latter case there is coherence between occupied one-electron states, making scattering events interdependent. This is an interference effect which is expressed by the coherence factors contained in the function $C(\Delta, E, E + \hbar\omega)$[1].

---

[1] See for instance de Gennes: *Superconductivity of metals and alloys*, Chapter 4. W. A. Benjamin, Inc. New York, 1966, for a thorough discussion of coherence factors in superconductivity.

The choice of sign in the coherence factor is essential, and distinguishes two important cases. We quote here the form valid at low frequencies, $\hbar\omega \ll \Delta$.

$$C(\Delta, E, E + \hbar\omega) = \frac{1}{2}\left|1 \mp \frac{\Delta^2}{E(E + \hbar\omega)}\right| = \frac{1}{2}\frac{|E(E + \hbar\omega) \mp \Delta^2|}{E(E + \hbar\omega)} \qquad (4.68)$$

Using previously developed expressions for D_s we now obtain the following integral for the transition rate in the superconducting state

$$\alpha_s = |M|^2 D_n^2(0) \int_{-\infty}^{\infty} \frac{|E(E + \hbar\omega) \mp \Delta^2| \left[f(E) - f(E + \hbar\omega)\right]}{(E^2 - \Delta^2)^{1/2}\left[(E + \hbar\omega)^2 - \Delta^2\right]^{1/2}} \, \mathrm{d}E \qquad (4.69)$$

The upper sign here refers to so-called case I processes, and the lower one to case II. The first of these involves a scalar potential as in longitudinal ultrasound, the second one corresponds to the presence of a vector potential A as in transverse ultrasound and in NMR.

The notation for the integration limits has been simplified. It is understood that no integration is to be carried out in the range $|E| < \Delta$, since no quasi-particles are present there. Next, to normalize the superconducting scattering rate (Eq. 4.69), against the rate in the normal state we set $\Delta = 0$ to obtain α_n corresponding to the normal state transitions rate. Integration of (4.69) gives

$$\alpha_n = |M|^2 D_n^2(0)\hbar\omega. \qquad (4.70)$$

Performing the normalization α_s/α_n we find an integral which subsequently may be worked out for different cases:

$$\alpha_s/\alpha_n = \frac{1}{\hbar\omega} \int_{-\infty}^{\infty} \frac{|E(E + \hbar\omega) \mp \Delta^2| \left[f(E) - f(E + \hbar\omega)\right]}{(E^2 - \Delta^2)^{1/2}\left[(E + \hbar\omega)^2 - \Delta^2\right]^{1/2}} \, \mathrm{d}E \qquad (4.71)$$

By calculating and measuring the ratio α_s/α_n the matrix element $|M|^2$ drops out of the problem, as it did here, and we need not be concerned with its actual value; similarly for $D_n^2(0)$. We are now in a position to discuss two distinct cases: ultrasound and NMR.

4.6.3   Longitudinal ultrasonic attenuation

The attenuation coefficient for ultrasound, α^u, is the exponent by which the intensity of an ultrasonic wave at frequency ω decays with distance. Its value is measured in dB/m or in Nepers/m. The quantity α^u is also closely related to the mechanical quality factor, the Q-factor, of the sample under study when treated as an acoustic resonator. It represents the ratio of dissipated energy to supplied energy. For this reason the normalized transition rate also corresponds to the normalized ultrasound attenuation, i.e. $\alpha_s^u/\alpha_n^u = \alpha_s/\alpha_n$. These relative rates can for instance be evaluated in the limit when $\hbar\omega$ approaches zero, an approach which applies in the usual ultrasonic phonon energy range – typically at 100 MHz – where $\hbar\omega$ is very small, on the scale of the energy gap Δ as well as on the scale of kT in most cases. Upon cancellation of the coherence factors against a density of states factor in Eq. 4.71 one finds an integral which can be simplified as follows in the low ω limit ($\omega \to 0$).

$$\int \frac{1}{\hbar\omega} \left[f(E) - f(E + \hbar\omega) \right] dE = - \int \frac{\partial f}{\partial E} dE \qquad (4.72)$$

Integration in the allowed range, i.e. excluding the range $-\Delta$ to $+\Delta$ by integration from $-\infty$ to $-\Delta$, and from Δ to ∞, one finds directly for longitudinal attenuation

$$\alpha_s^l / \alpha_n^l = 2f(\Delta) = \frac{2}{1 + e^{\Delta(T)/kT}}, \qquad (4.73)$$

a remarkably simple result, in form very similar to the Fermi–Dirac function. This result should not come as a complete surprise: Rather, the transition rate and therefore also the attenuation, should be expected to depend on the gap in this manner, since the Fermi–Dirac function determines the rate at which quasi-particles are created above the gap, and hence also the relative attenuation. In spite of this, the result (Eq. 4.73) is not entirely obvious. Without the accidental cancellation of a factor in the function C against a density of states factor in D_s, this simplicity would have been lost. Figure 4.7 shows data obtained on indium and analysed to give the limiting value of Δ at low temperatures, $\Delta(0)$ [41]. The early measurements quoted above, gave a good overall confirmation of the theory, but did not test this result fully, especially not near T_c where the temperature dependence is very strong, with a vertical tangent at T_c. Because this is also the region where the difference between the two-fluid model and the BCS model could be expected to be most distinct, ultrasound provides an important testing ground for the predictions of the combined effects of the energy gap and the coherence factors. In the case just discussed, only the gap remains as the controlling factor. This is quite different for transverse ultrasound, as will be discussed below.

The BCS prediction for attenuation (Eq. 4.73), was tested by careful measurements using longitudinal ultrasound at 210 MHz very close to T_c in superconducting aluminium [42], (Figure 4.8). The agreement with BCS theory is seen to be remarkably good. We note the almost vertical slope in ultrasonic attenuation at T_c, and the fact that the attenuation drops by 25% over only 40 mK, all of this in full agreement with the BCS prediction. We conclude, that in the case of aluminium the BCS predictions for the temperature exponent of the gap, i.e. the exponent 1/2 in $(T_c - T)^{1/2}$, as well as its zero degree value, $\Delta_0 = 3.52\,kT$, are in agreement with the experiment. Recalling that BCS is a mean field theory, this constitutes a demonstration that the BCS approach is good in typical low T_c superconductors even in the close vicinity of T_c. No traces of fluctuation effects are seen, in accord with the arguments given in Section 2.6.6.

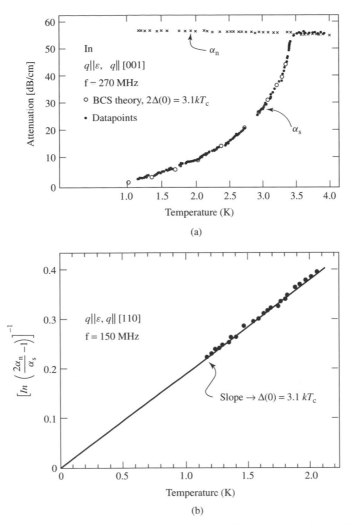

Figure 4.7   (a) Measurements of attenuation of longitudinal ultrasound in indium in the normal and superconducting states. The normal state data were obtained by application of a magnetic field $H > H_c$. (b) Analysis of data at 150 MHz using the function (Eq. 4.73), leads to determination of $\Delta(0)$[41].

However, theoretical work by Halperin and colleagues [43], and later work by Bartholomew [44] and by Mo and co-workers [45], has shown that when $\kappa = \lambda/\xi \ll 1$ the superconducting phase transition is expected to be *first order*. Aluminium is extreme Type I with a κ-value of order 10^{-2}, so the theory should apply in this case. Evidently the experiment quoted above would have to be carried out with even much higher than mK temperature resolution to reveal the predicted effect. Judging by the data presented here, the predicted first order jump has to be very small in Al, but its existence even here cannot be excluded on the basis of the presented data.

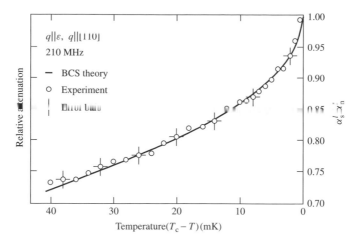

Figure 4.8 Measurements of the attenuation of longitudinal polarized ultrasound in super-conducting Al, compared to BCS prediction in Eq. 4.73. Notice the temperature scale in mK. Adapted from Fossheim *et al.* [42].

4.6.4 Transverse ultrasound

The physics of scattering of transverse phonons in superconductors, as observed in experiments using transversely polarized ultrasonic waves, is distinctly different from the longitudinal case. This interesting fact has been largely overlooked in textbooks in superconductivity until now, in spite of the fact that the physics has been well described and understood. In early experiments [46] the attenuation of transverse acoustic waves in very pure metals was found to drop precipitously just below T_c. This effect became known as the 'rapid fall'. It was somewhat erroneously considered to be due purely to Meissner screening of the fields associated with the transverse wave. It turns out, rather, that this difference offers an interesting possibility to test no less than a hallmark of the BCS physics, namely the coherence factors [47], cf. Eqs 4.14. With transverse waves these factors do not drop out of the transition rates as they did in the longitudinal case just discussed. An experimental test to reveal their possible presence is therefore a highly worthwhile enterprise.

The difference referred to, is due to the different electronic responses to the ionic displacements of longitudinal and transverse acoustic waves in a conducting medium. In the presence of a longitudinal wave the electron gas is driven by the electrostatic fields created by the 'piling up' and 'thinning out' of positive ionic charge as the wave propagates through the lattice. This essentially electrostatic response is very fast, limited only by the plasma frequency, and therefore occurs completely in phase with the ionic motion. The interaction between the primary ionic wave and the electron gas is of a scalar nature, and independent of the direction of both electron spin σ and momentum $\hbar k$. The response of the electron gas is often described as that of maintaining (approximate) *charge neutrality*.

The case of transverse waves is wholly different; there is no piling up or thinning out of charge in the transverse motion of the lattice. Charge neutrality, preexisting before injection of the acoustic wave, is preserved in the electronic response as well. In this case the transverse ionic current generates a time varying magnetic field, which in its turn drives an electronic current. The coupling between the ionic motion and the electron gas is therefore of electromagnetic nature and much slower than in the longitudinal case, involving the electronic skin depth δ. The presence of the induced magnetic field as an intermediary between ionic and electronic currents becomes extremely important when entering into the superconducting state. This necessarily leads to Meissner screening of electronic currents below T_c, an effect which is not present in the longitudinal case. Screening occurs in the entire range of the acoustic wave, in the bulk, contrary to how we previously encountered the Meissner effect, as a response occurring in a thin layer λ near the surface. As transverse wave propagation is easily generated at a single wavevector q at a time, this opens up for a determination of the superconducting penetration depth λ versus q in the bulk of the material.

As was mentioned earlier, the coherence factors now appear in a different way: The interaction between a free charge and a magnetic field occurs via the product $p \times A = \hbar k \times A$, where A is the vector potential. This term is independent of spin, but changes sign on replacing k by $-k$. When evaluating the integrals discussed above this property turns out to require the opposite sign in the coherence factor function C introduced in Eq. 4.68. Although the coherence factors dropped out against a density of states factor in the longitudinal case, this clearly cannot happen here, and the coherence factors remain in the problem. Furthermore, screening has to be taken care of, and finite frequency electrodynamics enters in an important way. We now outline the approach which has been followed by Fossheim [42, 47].

The attenuation coefficient α for ultrasonic waves propagating in a conducting medium can be expressed as the energy dissipated per second divided by the energy input per second, both calculated per unit area of the wavefront. In the presence of induced current densities and fields this leads to the expression

$$\alpha = \frac{1/2 Re(j^* E)}{1/2 \rho v_i^2 v_s} \tag{4.74}$$

Here we need an expression for the electronic current density j whose complex conjugate appears in Eq. 4.74, and the electric field E within the superconductor, which is by no means zero in this electrodynamic problem. Other factors in the above expression are the density of the material ρ, the ionic velocity v_i and the velocity of sound v_s. The quantity we have to express is primarily the field E, since the current density can always be written as $j = \sigma E \equiv (\sigma_1 + i\sigma_2)E$. Here we have defined the complex frequency dependent conductivity σ as consisting of a real part σ_1 and an imaginary part σ_2. Next we express the electric field by these conductivities. For this purpose we combine Maxwell's equations

$$\nabla \times \boldsymbol{E} = -\frac{\partial \boldsymbol{B}}{\partial t} \tag{4.75}$$

$$\nabla \times \boldsymbol{B} = \mu_0(\boldsymbol{j}_s + \boldsymbol{j}_i) = \mu_0(\sigma \boldsymbol{E} + \boldsymbol{j}_i). \tag{4.76}$$

Here we have introduced the magnetic induction $\boldsymbol{B}$ created by the ionic current density $\boldsymbol{j}_i$ of the transverse acoustic wave, and the associated self-consistent electric field $\boldsymbol{E}$ generated by the time-dependent magnetic induction. We will assume the waveform to be $\exp i(qx - \omega t)$ for all fields and currents in Eqs 4.75 and 4.76. To solve these equations for $\boldsymbol{E}$ we assume that the wave is traveling in the positive x-direction, generating the transverse ionic currents along the y-axis, and consequently the induction $\boldsymbol{B}$ is along the z-direction. In this plane wave scheme the spatial derivatives are nonzero only when taken with respect to x. We combine the two Maxwell equations by taking the curl of Eq. 4.75 and the time derivative of Eq. 4.76, and find

$$\nabla \times \frac{\partial \boldsymbol{B}}{\partial t} = \mu_0 \frac{\partial}{\partial t}(\sigma \boldsymbol{E} + \boldsymbol{j}_i) \tag{4.77}$$

which is to be combined with

$$\nabla \times \nabla \times \boldsymbol{E} = -\nabla \times \frac{\partial \boldsymbol{B}}{\partial t} \tag{4.78}$$

Recalling that for transverse waves there is no net charge accumulation, we have $\nabla \times \boldsymbol{E} = 0$ in the second of these equations. Using this to simplify Eq. 4.78, and after equating the right-hand side of Eq. 4.77 with the left-hand side of Eq. 4.78, we arrive at

$$\boldsymbol{E} = \frac{i\mu_0\omega}{q^2 - i\mu_0\omega(\sigma_1 + i\sigma_2)}\boldsymbol{j}_i \tag{4.79}$$

and eventually to

$$\boldsymbol{E} = \frac{-\sigma_1 + i\left(\dfrac{q^2}{\mu_0\omega} + \sigma_2\right)}{\left(\dfrac{q^2}{\mu_0\omega} + \sigma_2\right)^2 + \sigma_1^2}\boldsymbol{j}_i \tag{4.80}$$

This expression is fully valid in both the normal and the superconducting states. Referring to Eq. 4.74 we now normalize the attenuation coefficient in the superconducting state against that in the normal state, in both cases taking the expression for the electric field from Eq. 4.80. The procedure is now to carry out

$$\frac{\alpha_s^t}{\alpha_n^t} = \frac{Re(\boldsymbol{j}_s \times \boldsymbol{E}_s)}{Re(\boldsymbol{j}_n \times \boldsymbol{E}_n)} = \frac{\sigma_{1s}|\boldsymbol{E}_s|^2}{\sigma_{1n}|\boldsymbol{E}_n|^2} \qquad (4.81)$$

We have attached the suffixes s and n to the respective fields in the superconducting and normal states. To further develop Eq. 4.80 we introduce one simplifying aspect: the imaginary part of the normal state conductivity can be shown to be negligible in comparison with the real part at usual ultrasonic frequencies up to several hundred megahertz, hence σ_{2n} is omitted hereafter. Note, however, that σ_{2s} can by no means be neglected. After some algebra the final, normalized attenuation was found by Fossheim as [47]

$$\frac{\alpha_s^t}{\alpha_n^t} = \frac{\dfrac{\sigma_{1s}}{\sigma_{1n}}\left[(q\delta)^4 + 1\right]}{\left(\dfrac{\sigma_{1s}}{\sigma_{1n}}\right)^2 + \left[\dfrac{\sigma_{2s}}{\sigma_{1n}} + (q\delta)^2\right]^2} \qquad (4.82)$$

We leave it as an exercise to carry out the manipulations required to obtain this result from Eq. 4.81. In Eq. 4.82 we have introduced the parameter $\delta = (1/\mu_0\omega\sigma_{1n})^{1/2}$. Except for a factor $\sqrt{2}$ this is the usual normal state skin depth for electromagnetic fields. We note from the way the skin depth appears together with the imaginary conductivity in the denominator of Eq. 4.82, that normal state screening plays a role. The product of skin depth δ and wavevector q – essentially the ratio of skin depth to acoustic wavelength – measures the effectiveness of the normal state screening since the term $(q\delta)^2$ adds to the term σ_{2s}/σ_{1n} which represents superconducting screening. The latter contains the Meissner effect. Thus in a subtle way Eq. 4.82 describes a continuous crossover between normal state screening and superconducting screening. What is being screened, of course, are the induced currents j, and B-fields that provide the coupling between the transverse lattice waves and the electron gas. And, as these are screened out, so is the electric field E. Therefore the whole coupling mechanism breaks down in a narrow temperature range below T_c. This eliminates a dissipation channel, and the acoustic attenuation drops dramatically, as mentioned above, the so-called 'rapid fall' which was referred to before. It is a phenomenon full of essential superconductor physics, as we demonstrate in more detail, next.

Importantly, the coherence factors did not drop out of the problem, but are present in the conductivities. We proceed to discuss these aspects, making use of the transition rates again, i.e. Eq. 4.71, as was done originally by Mattis and Bardeen [48]. It is appropriate to associate the transition rate with the processes that determine σ_{1s} since this is the loss part of the complex conductivity. The resulting integral, still with the coherence factors intact, cannot be solved analytically, but an approximate analytic form was found by Cullen and Ferrell [49]

$$\frac{\sigma_{1s}}{\sigma_{1n}} = 1 + \frac{1}{2}\frac{\Delta}{kT}\ln\left(\frac{8\Delta}{e\hbar\omega}\right) - \left(8.42/\pi^2\right)\left(\frac{\Delta}{kT}\right)^2 \tag{4.83}$$

Mattis and Bardeen also gave σ_{2s}/σ_{1n} in integral form, but an analytic solution was only given under conditions which are outside of the accessible range in the experiments we are discussing here. The work of Fossheim [47] introduced instead a nonlocal modification of the London expression for σ_{2s}/σ_{1n}, as follows:

$$\frac{\sigma_{2s}}{\sigma_{1n}} = 2\left(\frac{1}{\mu_0\omega\sigma_{1n}}\right)\lambda_L^{-2}(0) \times \Delta T \times K(q\xi_0)$$

$$= 2\left(\frac{\delta}{\lambda_L(0)}\right)^2 \times \Delta T \times K(q\xi_0) \tag{4.84}$$

Here $\lambda_L(0)$ is the zero degree London penetration depth, $\Delta T = (T_c - T)$, and $K(q\xi_0)$ is the nonlocal Pippard kernel (see [47]). Since this term Eq. 4.84 appears not as a factor in the expression for α_s^t/α_n^t, but added to the term $(q\delta)^2$ this approximation is not as essential to have in the correct BCS form as is the case with σ_{1s}/σ_{1n}, which contains the dominating effect of the coherence factors. That aspect is essential very close to T_c as it turns out that the presence of the coherence factors in σ_{1s}/α_{1n} dominate the behavior of the attenuation α_s^t/α_n^t in the first few mK below T_c, below which it is overlapped and followed by the Meissner screening. And while at the transition the attenuation would have zero slope if the Meissner screening was the dominant effect, on the contrary, with the coherence factors present a sharp break occurs, with vertical slope. Perhaps no other phenomenon defines T_c as sharply as this in superconductivity.

The issue is therefore whether experiments can confirm the latter prediction. Experiments in indium [47] and in aluminium [42] shown in Figure 4.9 provide substantial verification of the presence of the coherence factors, showing the predicted sharp break on the sub-millikelvin scale. In fact, in Ref. [47], T_c was resolved with $\pm 2/10000$ K resolution! In the aluminium results, there appeared a correction factor in the parameter $\lambda_L(0)$ which was interpreted as an effect of electronic effective mass enhancement by about 1.5 due to electron–phonon interaction, also in accord with cyclotron resonance experiments. Again, evidence for the predicted first order transition [50] was not found. Evidently, even the already very high resolution of this experiment was not sufficient to see it.

Calculations of the attenuation of transverse waves in several metals were carried out by [51], covering a wide range of frequencies. At frequencies of the order of 1 GHz, depending on the metal, it was found that the presence of the $q\delta$-factor alters the sign of the initial slope of α_s^t/α_n^t from a sharp downward break to an equally sharp upward break. This latter effect resembles very closely

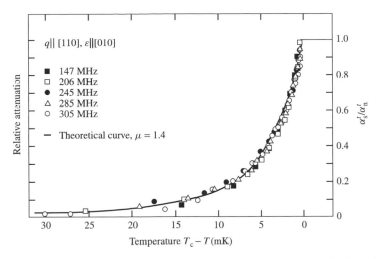

Figure 4.9 Attenuation of transversely polarized ultrasound in superconducting Al near T_c. Measurements by Fossheim *et al.* [42]. The continuous curve is a calculation based on theory by Fossheim [47]. μ is the mass enhancement factor mentioned in the text.

the behavior of the NMR relaxation rate at the superconducting transition. In essence, transverse ultrasonic experiments and NMR experiments contain the same information regarding the presence of BCS coherence factors, whether the ultrasonic frequency is high or low. It should be mentioned that the ultrasonic experiments referred to were carried out on ultra-pure 6N metals, where the electronic mean free path at low temperatures is of the order of millimetre length, making the product $q\ell \gg 1$, ℓ being the electronic mean free path. In this case the electronic attenuation in the normal state is temperature independent and proportional to q, and the rapid fall or rapid rise below T_c is a very strong and sharp effect.

4.6.5 Nuclear magnetic resonance relaxation below T_c

One of the earliest and most successful confirmations of the BCS theory was provided by the nuclear magnetic resonance measurements of the longitudinal relaxation rate $1/T_1$ in a superconductor by Hebel and Slichter [52]. The observed rise of the relaxation rate just below T_c, to a level above that in the normal state, confirmed a crucial aspect of the theory. A good idea of the expected behavior is obtained by use of the scattering rate already given in Eq. 4.71. By again introducing the derivative of the Fermi–Dirac function like before, and adding symmetric contributions from the ranges $E \geq \Delta$ and $E \leq -\Delta$ after allowing $\omega \to 0$, one finds the following integral expressing the relative NMR relaxation rate α_s/α_n in the superconducting state normalized against that in the normal state as

$$\frac{\alpha_s}{\alpha_n} = -2 \int_\Delta^\infty \frac{E^2 + \Delta^2}{E^2 - \Delta^2} \left(\frac{\partial f}{\partial E} \right) dE. \tag{4.85}$$

This integral diverges logarithmically. If ω had been allowed to retain a finite value, the divergence would be avoided, and the result would be of the order of $\ln(\Delta/\hbar\omega)$, as discussed by Tinkham [53], of the order of 10 for typical cases. This is far larger than what is measured. Various schemes have been discussed in the literature to reduce this discrepancy. The main ones have been to take account of anisotropy of the energy gap, and to assume nonuniform lifetimes of quasiparticles over the Fermi-surface. What has become quite clear after the discovery of high-T_c superconductors is that the rise in relaxation rate below T_c is tied to the symmetry of the wavefunction. The relaxation peak is present in s-wave superconductors, as was worked out by BCS, but not in d-wave superconductors like the high-T_c cuprates.

An example of a low-temperature measurement

Superconductivity research during the first 75 years since 1911, was always a domain of low temperature physics, and a large part still is, in spite of high-T_c superconductivity. Figure 4.10 shows an example of an experimental setup which led to the ultrasonic data near T_c quoted previously in this section [42]. The measurements in aluminium required careful temperature control and measurements down to 1.1 K. This is just within the temperature range accessible by pumping away the evaporating gas from liquid helium ^{4}He, whose boiling point is 4.2 K at STP. The cylindrical chamber shown in the figure is immersed in helium, which at these temperatures ($T < 2.2$ K) is a superfluid, in a cryogenic glass Dewar and filled with superfluid through the tube indicated in the figure as the pumping line.

 Inside the chamber shown here, the sample is mounted onto the lower end of a quartz crystal acting as an acoustic delay line, centrally located in solid copper block for temperature stabilization. The use of superfluid helium in addition, offers excellent conditions for temperature stabilization since it is well known to sustain zero temperature gradients. This is ideal for the present purpose which requires a temperature stability better than 1/1000th of a degree. Other details of the physical attachments are shown in the figure. Note in particular the pumping outlet. It has a narrow constriction, 0.65 mm in diameter. While this arrangement still allows efficient pumping of gas from the chamber due to the short length of the hole, and the low density of the gas, it reduces substantially the flow of superfluid He film out of the chamber. If the superfluid is allowed to flow up along the inner wall of the pumping line over its full diameter, a phenomenon which is well known to occur, it will reach high enough temperatures to evaporate, thereby reducing the pumping efficiency sufficiently to prevent the lowest temperatures to be reached. The

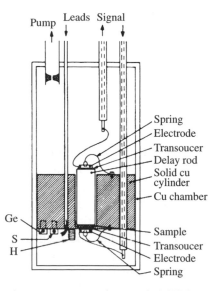

Figure 4.10   Experimental arrangement used to reach 1.1 K by pumping in liquid helium.

constriction reduced the film flow substantially, and the system worked exactly as planned, allowing 1.1 K to be reached. This as a good example of the type of know-how which is sometimes needed to perform good low-temperature measurements. The insert in the chamber is also equipped with a temperature sensor and heater for control of the temperature to submillikelvin level.

4.7   The Ginzburg–Landau theory

4.7.1   Some remarks on Landau theory

The Ginzburg–Landau theory [10] for superconductivity represents one of the most useful tools available for the theoretical description of superconductivity. It starts with a free energy expansion, completely in line with the general Landau theory [54] for condensed matter, with particular attention being paid to the term representing the gradient of the ordering quantity.

In Section 4.1 we have referred to magnetic analogies to superconductivity for an intuitively simplified discussion of the quantum mechanical wavefunction. We begin here similarly with a brief discussion of the Landau free energy for a ferromagnet, where the ordering quantity is the spin. When spins, or the corresponding magnetic moments μ_i, are summed up and averaged over the volume they occupy, the magnetization may be expressed as $M = \sum_i \mu_i / V$. M may take different orientations depending on symmetry, in order to minimize the total energy. Examples of such behaviour are commonly found in magnetic domains in ferromagnets, where neighbouring domains may have the magnetization pointing in orthogonally different directions due to anisotropy. The domain walls between these small volumes are a kind of topological defects where the spins gradually change direction form one domain to the neighboring one, a configuration called Bloch wall, that costs extra energy to produce. This energy appears in the free energy expansion as the square of the gradient of M. This arrangement is still energetically favorable due to the resulting reduced field energy outside the material.

Apart from this, the Landau expansion is permissible when assuming that we perform the calculation of the free energy relatively close to T_c, although not too close in case fluctuations might become important. The point here is that we can define a small parameter in this range, namely $m = M(T)/M(0)$, which is the ratio between the actual magnetization $M(T)$, and the saturation value $M(0)$ at $T = 0$. This quantity $m(T)$ may be used as an expansion parameter to whatever order is necessary. We have still one more aspect to take care of: There is always a background energy $F_0(r, T)$, which is independent of m. This can most easily be thought of as the entire free energy of the system at $m = 0$, for instance just above the magnetic ordering temperature T_c.

The Landau free energy density is then:

$$F(r, T) = F_0(r, T) + \alpha m^2 + \frac{1}{2}\beta m^4 + \frac{1}{4}c\,|\nabla m|^2 \tag{4.86}$$

Notice here that there are no terms of odd power in the m-expansion. Physically this is due to the obvious fact that the free energy in a ferromagnet must be independent of whether the spins point up or down. This constraint applies to most systems, whether magnetic or not, and is always caused by symmetry.

We will now for simplicity confine the analysis to the interior of one domain where M and m are constant, and disregard the remaining terms. Minimizing the remaining free energy, requires

$$\frac{\partial F}{\partial m} = 0 \tag{4.87}$$

We find

$$2\alpha m + 2\beta m^3 = m(\alpha + \beta m^2) = 0 \tag{4.88}$$

The solutions are:

$$m = 0 \quad \text{and} \quad m^2 = -\frac{\alpha}{\beta} \tag{4.89}$$

Generally, both α and β are themselves functions of temperature and are to be thought of here as the lowest order term in a temperature dependent expansion of the coefficients of Eq. 4.86. By studying the minimization conditions in more detail one finds that the expansion of α must have a zero'th order term equal to zero, so to lowest order

$$\alpha = \alpha_1(T - T_c) \tag{4.90}$$

where α_1 is a constant. For β we find that the zero-order term β_0 is finite, and we stop the expansion there. This adds up to

$$m^2 = \frac{\alpha_1}{\beta_0}(T_c - T)$$

$$m = \left(\frac{\alpha_1}{\beta_0}\right)^{\frac{1}{2}} (T_c - T)^{\frac{1}{2}} \qquad (4.91)$$

So, the magnetization grows from 0 at T_c, like the square root of $T_c - T$. Above T_c its value is $m = 0$ in accordance with the first solution in Eq. 4.89. Both aspects are illustrated in Figure 4.11. What happens physically is that above T_c spins are totally disordered, producing $\boldsymbol{m} = 0$. Below T_c, however, ordering increases gradually like the square root of temperature measured relative to T_c.

Because of the gradual associated ordering taking place in the magnetic spins when T is lowered below T_c, the quantity m is called the '*order parameter*' of the ferromagnetic transition. We also see that m is indeed suitable as a small quantity to expand in. The above gives an idea of the basic content of Landau theory, applicable to a wide range of systems where ordering takes place below a certain temperature. We have so far neglected the gradient term. This will often not be permissible, as we will see in superconductors.

4.7.2 Ginzburg–Landau theory for superconductors

The Landau approach works specially well in systems with long correlation length, called coherence length ξ in superconductors. We should measure this length on an atomic scale. Because ξ in low-T_c superconductors can be as long as thousands of interatomic distances (recall for instance Al, with $\xi(0) = 1600 \, \text{nm}$) superconductors offer a particularly suitable arena for the application of Landau theory, in that case called Ginzburg–Landau theory. As, on the other hand, ξ in high-T_c materials is only of the order of one, or a few, unit cell distances, one

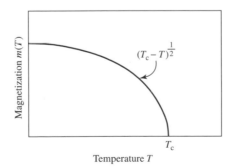

Figure 4.11 Magnetization $m(T)$ versus temperature according to standard Landau theory described in the text.

should rightly be concerned about the validity of Ginzburg–Landau theory in such systems. Experience has shown that for many purposes it works quite well even there, when discussing equilibrium properties. However, one has to exercise great caution. For instance, as we have shown in Section 2.6 one should expect very vigorous thermal fluctuation in the order parameter, the superconducting wavefunction ψ, on approaching T_c. Indeed this causes immense complications in high-T_c as compared to low-T_c superconductors. Leaving those problems aside for now we proceed to develop the Ginzburg–Landau theory of superconductivity. We emphasize the following: The superconducting wavefunction ψ does possess the property required for an order parameter: Its equilibrium value is zero above T_c and increases gradually below T_c when written as

$$\psi = n^{\frac{1}{2}}e^{i\varphi} = |\psi|e^{i\varphi} \tag{4.92}$$

where n is the density of Cooper pairs. Whether we refer to the BCS theory or use the language of the two-fluid model we know that the appearance of the factor n ensures that ψ has the required property. The phase ϑ is another quantity of great importance, and one that we shall dwell considerably on in later chapters.

The next important step is to write the gradient term in the free energy expansion in a quantum mechanically appropriate manner. We have to define the momentum operator in a magnetic field, writing

$$\boldsymbol{p} = (-i\hbar\nabla - 2e\boldsymbol{A}) \tag{4.93}$$

where the symbols have their usual meaning in quantum mechanics, and we have introduced the charge $-2e$ of a Cooper pair; and below we introduce its mass m.

We are now ready to sum up the terms in the Ginzburg–Landau free energy density for a superconductor:

$$F_s(\boldsymbol{r}, T) = F_n(\boldsymbol{r}, T) + \alpha\,|\psi|^2 + \frac{\beta}{2}\,|\psi|^4 + \frac{1}{2m}\,|(-i\hbar\nabla - 2e\boldsymbol{A})\psi|^2 + \frac{1}{2\mu_0}B^2 \tag{4.94}$$

We remark here that since all the terms in Eq. 4.94 represent energy densities, the total free energy of the system is

$$F_s(T) = \int_V d^3r\, F_s(\boldsymbol{r}, T) \tag{4.95}$$

For the ensuing calculations we will define a vector $\boldsymbol{G} \equiv (-i\hbar\nabla - 2e\boldsymbol{A})$, and we remind ourselves that the vector potential $\boldsymbol{A}$ is related to the magnetic induction $\boldsymbol{B}$ by $\boldsymbol{B} = \nabla \times \boldsymbol{A}$.

Next we carry out the minimization procedure for the free energy by making variations in ψ and ψ^*, i.e. by letting $\psi \to \psi + \delta\psi$, and $\psi^* \to \psi^* + \delta\psi^*$. The equilibrium condition $\delta F_s(T) = 0$ will then give us the Ginzburg–Landau equations, two equations of paramount importance in superconductivity. We can immediately write

$$\delta F_s(\boldsymbol{r}, T) = \alpha(\psi + \delta\psi)(\psi^* + \delta\psi^*) - \alpha\psi\psi^*$$
$$+ \frac{\beta}{2}\left[(\psi + \delta\psi)^2(\psi^* + \delta\psi^*)^2 - \psi^2\psi^{*2}\right]$$
$$+ \frac{1}{2m}\left[G(\psi + \delta\psi)G^*(\psi^* + \delta\psi^*) - G\psi G^*\psi^*\right] \quad (4.96)$$

with G as defined above. Next we simplify:

$$\delta F_s(\boldsymbol{r}, T) = \alpha(\psi\delta\psi^* + \psi^*\delta\psi) + \beta(\psi|\psi|^2\delta\psi^* + \psi^*|\psi|^2\delta\psi)$$
$$+ \frac{1}{2m}(G\psi G^*\delta\psi^* + G\delta\psi G^*\psi^*) \quad (4.97)$$

The last parenthesis is

$$G\psi G^*\delta\psi^* + G\delta\psi G^*\psi^* = (-i\hbar\nabla - 2e\boldsymbol{A})\psi(i\hbar\nabla - 2e\boldsymbol{A})\delta\psi^*$$
$$+ (-i\hbar\nabla - 2e\boldsymbol{A})\delta\psi(i\hbar\nabla - 2e\boldsymbol{A})\psi^* \quad (4.98)$$

Summing up:

$$\delta F_s(T) = \int_V \mathrm{d}^3\boldsymbol{r}\left\{(\alpha\psi\delta\psi^* + \beta\psi|\psi|^2\delta\psi^*\right.$$
$$\left. + \frac{1}{2m}(-i\hbar\nabla - 2e\boldsymbol{A})\psi(i\hbar\nabla - 2e\boldsymbol{A})\delta\psi^* + c.c.)\right\} \quad (4.99)$$

After a partial integration we obtain, using $\nabla\boldsymbol{A} = 0$

$$\delta F_s(T) = \int_V \mathrm{d}^3\boldsymbol{r}\left\{[(\alpha\psi + \beta|\psi|^2\psi + \frac{1}{2m}(-i\hbar\nabla - 2e\boldsymbol{A})^2\psi)]\delta\psi^* + c.c.\right\}$$
$$(4.100)$$

Requiring $\delta F_s(T) = 0$ demands that the functions in front of $\delta\psi^*$ and $\delta\psi$ are zero, which means that we have

GL-I: $\boxed{\alpha\psi + \beta|\psi|^2\psi + \dfrac{1}{2m}(-i\hbar\nabla - 2e\boldsymbol{A})^2\psi = 0}$ $\quad (4.101)$

This is the first of the two Ginzburg–Landau equations, hereafter referred to as GL-I. Next, as a preparation for the derivation of the second Ginzburg–Landau equation, GL-II, we introduce Maxwell's equation $\nabla \times \boldsymbol{H} = \boldsymbol{J}$ which we use in the form $\nabla \times \boldsymbol{B} = \mu_0\boldsymbol{J}$. We write it out as

$$\mu_0\boldsymbol{J} = \nabla \times \boldsymbol{B} = \nabla \times (\nabla \times \boldsymbol{A}) = \nabla(\nabla \times \boldsymbol{A}) - \nabla^2\boldsymbol{A} = -\nabla^2\boldsymbol{A} \quad (4.102)$$

in the London gauge where $\nabla \times \boldsymbol{A} = 0$. When we later encounter the term $-\frac{1}{\mu_0}\nabla^2\boldsymbol{A}$ we will recognize this as the supercurrent $\boldsymbol{J}$.

The procedure by which GL-II is obtained is to vary A in the free energy $F_s(r, t)$, i.e. we let $A \to A + \delta A$ and find the corresponding variation in $F_s(r, t)$. For this minimization of the free energy density $F_s(T)$ with respect to the vector potential A we need only retain the A-dependent parts of $F_s(r, t)$. We write this as $F_s(r, t, A)$. Next we take the variation

$$\delta F_s(r, t, A) = F_s(r, t, A + \delta A) - F_s(r, t, A) \tag{1.102}$$

In writing out the expression here we temporarily introduce the symbol p for $-i\hbar\nabla$ to simplify the mathematics. The proper operator is immediately reintroduced in the next step. We find

$$\delta F_s(r, t, A) = \frac{1}{2m}\left[(p - 2e(A + \delta A))\psi\right]\left[(p^* - 2e(A + \delta A))\psi^*\right]$$

$$- \frac{1}{2m}\left[(p - 2eA)\psi\right]\left[(p^* - 2eA)\psi^*\right]$$

$$+ \frac{1}{2\mu_0}\left[(\nabla \times (A + \delta A))^2 - (\nabla \times A)^2\right]$$

$$= -\frac{e}{m}\left[i\hbar\psi^*\nabla\psi + i\hbar\psi\nabla\psi^* + 4e\,|\psi|^2\,A\right]\delta A$$

$$+ \frac{1}{\mu_0}(\nabla \times \delta A)(\nabla \times A) \tag{4.104}$$

Now $\delta F_s(r, t, A)$ is to be integrated over the superconductor volume. We do the last term first:

$$\frac{1}{\mu_0}\int d^3r(\nabla \times \delta A)(\nabla \times A) = -\frac{1}{\mu_0}\int d^3r\nabla^2 A \times \delta A \tag{4.105}$$

Therefore, the entire $\delta F_s(T)$ becomes

$$\delta F_s(T) = \int d^3r\left[\frac{ie\hbar}{m}(\psi\nabla\psi^* - \psi^*\nabla\psi) + \frac{4e^2}{m}|\psi|^2\,A - \frac{1}{\mu_0}\nabla^2 A\right]\delta A = 0 \tag{4.106}$$

This requires the square bracketed term to be zero. Using the result from Eq. 4.102 the equation for the supercurrent is obtained:

$$J = -\frac{1}{\mu_0}\nabla^2 A = \frac{e}{m}\left[(i\hbar\psi^*\nabla\psi - i\hbar\psi\nabla\psi^*) + 4e\,|\psi|^2\,A\right] \tag{4.107}$$

or

$$\text{GL-II:} \quad \boxed{J = \frac{e}{m}\left[\psi^*(-i\hbar\nabla - 2eA)\psi + c.c.\right]} \tag{4.108}$$

This is the second Ginzburg–Landau equation.

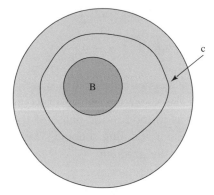

Figure 4.12 Flux quantization. The larger area is the superconductor. The small hatched area illustrates a normal domain inside the superconductor where a magnetic flux penetrates a cross section of the sample. The contour c encloses this area completely.

To be precise, in deriving the GL equations so far we have omitted a surface integral:

$$\int d\psi^*(-i\hbar\nabla - 2eA)\psi d\sigma + c.c. \tag{4.109}$$

This is correct to do if $(-i\hbar\nabla - 2eA)_\perp\psi = 0$, i.e. when the normal component of the operator as applied at the surface gives zero. On inserting this condition into GL-II we find that it expresses the condition that no current is entering through the surface, $J_\perp = 0$. This is physically sound when the surface is a boundary to vacuum or an insulating material. We could turn the question around and ask what boundary condition on ψ would in general make $J_\perp = 0$. To answer this question, we apply the operator $(-i\hbar\nabla - 2eA)_\perp$ in GL-II and require $\boldsymbol{J}_\perp = 0$, i.e.:

$$\boldsymbol{J}_\perp = \frac{e}{m}\left[\psi^*(-i\hbar\nabla\psi - 2eA)_\perp\psi + c.c.\right] = 0 \tag{4.110}$$

We continue the discussion of this problem in Chapter 5.

4.7.3 Flux quantization

Imagine that some region of a superconductor is in the normal state due to magnetic flux penetrating it, as illustrated in Figure 4.12. Thus, we may think of the superconductor as a multiply connected domain described by a surface of genus one. In this case, given the order parameter $\psi = |\psi|e^{i\varphi}$ the line integral of the gradient $\nabla\varphi$ around a closed contour encircling the normal domain has to be given by $2\pi N$, where $N \in \mathbf{Z}$. This follows since the order parameter must be singled valued at any given point, and $\int_1^2 d\mathbf{l}\nabla\varphi = \varphi_1 - \varphi_2$ independent of the path connecting the start-and endpoint. Thus we have

$$\oint d\mathbf{l} \cdot \nabla \varphi = 2\pi N \tag{4.111}$$

Now, in the presence of a magnetic field, this is only slightly modified. The magnetic field couples to the superconducting order parameter via the vector potential $\mathbf{A}$ in such a way as to modify the phase of the superconducting order parameter by the usual minimal coupling scheme. The gauge invariant phase of the order parameter now has the gradient

$$\nabla \varphi_{GI}(\mathbf{r}) = \nabla \varphi(\mathbf{r}) - \frac{2e}{\hbar} \mathbf{A} \tag{4.112}$$

where we have used the fact that the charge of the Cooper pair is $2e$. Such a modification is necessary in order to preserve gauge-invariance of the supercurrent, a physical observable. We have $2e/\hbar = 2\pi/(h/2e) = 2\pi/\Phi_0$, where the $\Phi_0 = h/2e = 2.0679 \times 10^{-15} \, T \, m^2$. It is this gauge-invariant order parameter which now is subjected to the constraint

$$\oint d\mathbf{l} \nabla \varphi_{GI} = 2\pi N \tag{4.113}$$

as the requirement of singled-valuedness must still hold. Hence, we obtain

$$\frac{2\pi}{\Phi_0} \oint d\mathbf{l} \times \mathbf{A} = 2\pi N \tag{4.114}$$

By using this in conjunction with Stokes' theorem we have

$$\oint_C d\mathbf{l} \times \mathbf{A} = \iint_S d\mathbf{S} \nabla \times \mathbf{A} = \iint_S d\mathbf{S} \mathbf{B} \equiv \Phi = N\Phi_0 \tag{4.115}$$

where $\mathbf{B} = \nabla \times \mathbf{A}$ and Φ is the magnetic flux through the surface S enclosed by the contour C. This flux is thus seen to be quantized in units of Φ_0. Note that the only place where any information about the superconducting state enters into this is in the statement that the charge of the superconducting condensate is $2e$. Had we had a normal metal, we could apply precisely the same reasoning to the phases of the single-electron wavefunction which couples to a vector potential with the charge e instead of $2e$. In that case, we would again get flux-quantization, but with twice the periodicity in magnetic field.

5

Weak Links and Josephson Effects

5.1 Weak links, pair tunnelling, and Josephson effects

5.1.1 Introductory remarks

When two superconductors are brought into contact in such a way that the critical current in the contact region is much lower than that of the individual constituents, the contact is called a 'weak link'. Before contact is established the two superconducting constituents have independent wavefunctions, and therefore arbitrary and independent phases φ. Each superconductor wavefunction Ψ is characterized by its amplitude $|\psi|$ and phase φ:

$$\Psi = |\psi|e^{i\varphi} \tag{5.1}$$

After establishment of the weak link, coherence is established across the barrier, with a phase difference $\Delta\varphi$ causing interference between the previously independent wavefunctions. When a Cooper-pair tunnels from the left side to the right of the junction, say, it probes successively the phases of both wavefunctions, the 'left' one with phase φ_l and the 'right' one with phase φ_r. More precisely, the system can be described as having one wavefunction as a whole. A typical realization of a weak link is a SIS tunnel junction, consisting of two superconducting films, separated by a very thin oxide layer, typically $1-2$ nm thick. For low-T_c junctions the most commonly used superconductors are Nb and Pb. Niobium has turned out to be the best one overall. The critical current density of these junctions may be in the range $10^3 - 10^4$ A/cm^2, far below the typical critical current density of bulk superconductors. Transport across the barrier is a tunnel process, and this explains the low current. Tunnelling of Cooper-pairs naturally occurs with a probability far below that which could be expected for single particles.

Superconductivity: Physics and Applications   Kristian Fossheim and Asle Sudbø
© 2004 John Wiley & Sons, Ltd   ISBN 0-470-84452-3

Josephson predicted in 1962 [8] that such a junction would have entirely new and unexpected properties: It should be able to sustain a supercurrent without application of a voltage; and if driven by an external current source to exceed its critical current, radiation of high frequency electromagnetic waves would appear. These predictions were soon confirmed experimentally, and moved superconductivity into a new era.

A weak link system can be established in several ways (Figure 5.1). Instead of a thin oxide layer which was mentioned, other materials may be used, for instance a normal metal, corresponding to an SNS junction, in which case the metal layer can be much thicker. Another well established method is to create the whole system from a single superconductor, by splitting for example a continuous film into two regions, leaving just a narrow constriction between, of typical dimensions like the coherence length ξ. This latter construction is referred to as a Dayem bridge. In this case the critical current density is the same in the bridge and the bulk, but the overall critical current of the device is much lower than in the two parts. This is sufficient to make it a weak link.

A recently developed method, specially suited for high-T_c superconductors, takes advantage of their very short, nanometre size coherence length ξ. In these superconductors a very narrow interruption of superconducting properties can be achieved by depositing a high-T_c film across a naturally occurring grain boundary in a substrate like SrTiO$_3$. The grain boundary forces the superconductor to develop a chain of defects along the length of the grain boundary. Above the grain boundary, therefore, there is a weak link between the two parts of the superconductor film. Such contacts are commonly referred to as grain boundary junctions or bi-crystal junctions. So-called break junctions and step junctions are variations on this method, where the topology of the substrate plays a role in creating the weak link.

The observation already made, that the phase difference $\Delta\varphi$ between the two sides of the junction provides the driving force for the transport of Cooper-pairs, is something we already could have seen from the Ginzburg–Landau equations. The two driving fields for a supercurrent were found to be the vector potential A

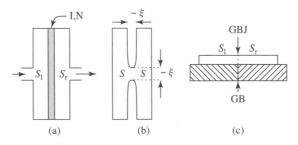

Figure 5.1   Different types of weak links, (a) between superconductors S_l and S_r with normal (N) or insulating (I) barrier between; (b) narrow constriction; (c) grain boundary junction GBJ in a superconductor deposited in top of a substrate with a grain boundary (GB).

and the superconductor phase gradient $\nabla\varphi$. In the case of a Josephson junction between two homogeneous superconductors the phase gradient becomes discretized, i.e $\nabla\varphi_r \rightarrow (\varphi_l - \varphi_r)/d$ where d is the barrier thickness. So, already based on the GL-equations the physics community should not have been too surprised in 1962 by the prediction of a phase difference driven, voltage independent supercurrent across a thin barrier between two superconductors. Once the idea has been put forward one realizes the effect should be expected, provided Cooper-pair tunnelling can take place.

5.1.2   DC Josephson effect: the Feynman approach

The basic equation used in analysing the Josephson effect is the time dependent Schrödinger equation

$$i\hbar\frac{\partial\Psi}{\partial t} = E\Psi \tag{5.2}$$

In a superconductor $\Psi = |\Psi|e^{i\varphi} = \rho_s^{\frac{1}{2}}e^{i\varphi}$, φ being the macroscopic superconducting phase and ρ_s the density of Cooper-pairs. However, when a coupling with overlap exists between the wavefunction to the left, Ψ_l, and to the right, Ψ_r, a term must be added to account for the 'leakage' of wavefunction so that Ψ_l changes with time at a rate which is proportional to the amount of leakage from Ψ_r into the left side. And similarly, a symmetric equation must exist for the rate of change of Ψ_r. This results in the following modification of Eq. 5.2.

$$i\hbar\frac{\partial\Psi_l}{\partial t} = E_l\Psi_l + K\Psi_r \tag{5.3}$$

$$i\hbar\frac{\partial\Psi_r}{\partial t} = E_r\Psi_r + K\Psi_l \tag{5.4}$$

Here E_l and E_r are the ground state energies of the unperturbed system when $K = 0$, i.e. with no transfer of charge. We do not need to know E_l and E_r, but remark that the difference $E_l - E_r = -2eV$ is fixed by the potential difference V. We will take $E_l = E_r$ when $K = 0$, i.e. for a thick barrier. If we further choose the zero of energy to be midway between E_l and E_r, we can share the potential energy difference between the two states symmetrically [55] by writing:

$$i\hbar\frac{\partial\Psi_l}{\partial t} = eV\Psi_l + K\Psi_r \tag{5.5}$$

$$i\hbar\frac{\partial\Psi_r}{\partial t} = -eV\Psi_r + K\Psi_l \tag{5.6}$$

The functions Ψ_l and Ψ_r may be written

$$\Psi_l = \rho_l^{\frac{1}{2}} e^{i\varphi_l}; \quad \Psi_r = \rho_r^{\frac{1}{2}} e^{i\varphi_r} \tag{5.7}$$

i.e. each function is assumed to have a well-defined macroscopic phase, constant in space, and a well defined Cooper-pair density. This first limitation will have to be lifted later, but for the moment we can limit the discussion to a contact area that is very small, so that no spatial variation of φ is possible. We can now insert the wave functions with timedependent ρ_l and ρ_r into the differential equations (Eqs 5.5 and 5.6) and equate real and imaginary parts separately, a procedure which immediately leads to the following four relations, with $\gamma \equiv \varphi_r - \varphi_l$:

$$\frac{\partial \varphi_r}{\partial t} = -\frac{K}{\hbar} \left(\frac{\rho_l}{\rho_r} \right)^{\frac{1}{2}} \cos \gamma + \frac{eV}{\hbar} \tag{5.8}$$

$$\frac{\partial \varphi_l}{\partial t} = -\frac{K}{\hbar} \left(\frac{\rho_r}{\rho_l} \right)^{\frac{1}{2}} \cos \gamma - \frac{eV}{\hbar} \tag{5.9}$$

$$\frac{\partial \rho_r}{\partial t} = -\frac{2K}{\hbar} (\rho_l \rho_r)^{\frac{1}{2}} \sin \gamma; \quad \frac{\partial \rho_l}{\partial t} = \frac{2K}{\hbar} (\rho_l \rho_r)^{\frac{1}{2}} \sin \gamma \tag{5.10}$$

We note that variations of phase φ is due to the combined effects of the phase difference γ across the contact, and the potential difference V. Variations in pair density, on the other hand, are solely due to phase difference γ.

The supercurrent across the contact is straightfowardly calculated by use of

$$J = -2e \frac{\partial \rho_l}{\partial t}; \quad \text{or} \quad J = -2e \frac{\partial \rho_r}{\partial t} \tag{5.11}$$

The time derivatives are already given above, so that we have:

$$J = \frac{4Ke}{\hbar} (\rho_l \rho_r)^{\frac{1}{2}} \sin \gamma \tag{5.12}$$

which we often rewrite as

$$J = J_0 \sin \gamma \tag{5.13}$$

a major result in the Josephson theory. It tells us that a supercurrent is driven across the thin barrier separating two superconductors simply by the superconducting phase difference across the barrier.

The predictions from the theory outlined above have been extensively confirmed experimentally. An example of an IV-characteristic showing the DC Josephson current is seen in Figure 5.2. The data are from an $S_1 IS_2$ structure,

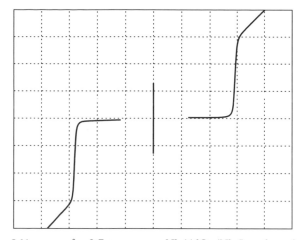

Figure 5.2 The I-V curve of a 0.7-μm square Nb/AlO$_x$/Nb Josephson junction. The vertical scale is 20 μA/div and the horizontal scale is 1 mV/div.[56]

with S$_1$ = Nb and S$_2$ = Pb. It shows the Josephson DC current at finite voltage which was discussed previously. Note that these measurements show the I-V curve for both positive and negative bias.

The fact that a voltage is not needed to drive a supercurrent is something we already know well from the Meissner effect. In Chapter 6 we will find that the vector potential provides the driving mechanism. The Josephson effects established the phase of the wavefunction as a driving mechanism on equal footing with the vector potential, or magnetic field. The superconducting phase thereby took on a more firm and real existence.

Observations of the corresponding physical effects followed, first few in numbers, then in abundance in the years after Josephson's prediction. Superconductivity

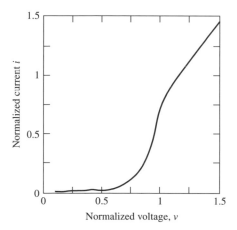

Figure 5.3 Quasiparticle current in intrinsic junction in Bi$_2$Sr$_2$CaCu$_2$O$_8$ crystal [57].

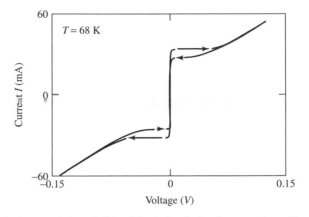

Figure 5.4 I-V characteristic of $(Pb_{0.2}Bi_{0.8})_2Sr_2CaCu_2O_8$ crystal at $T = 68\,K$ showing intrinsic Josephson effect. After Kleiner *et al.* [58].

broadened its territory enormously by these discoveries, as mentioned before. In particular a range of unprecedented possibilities opened up in the area of electronic applications.

An unusual realization of the Josephson effect was observed in a single crystal high-T_c superconductor $Bi_2Sr_2Cu_2Cu_3O_{10}$. This strongly anisotropic, layered structure consists of alternating superconducting CuO_2-layers separated by non-superconducting ones. The material can behave like a stack of SIN- and SIS-structures. Hence, currents flowing along the c-axis normal to the layers have been observed, corresponding to either quasiparticle current or supercurrent, as seen in Figures 5.3 and 5.4 [57, 58].

5.2   AC Josephson effect

Josephson's discovery included, however, another effect, one which even in hindsight may be seen as a surprising phenomenon an applied voltage across an junction would bring about an oscillatory variation of phase difference. We can find this fact from analysis of the time dependence of the phase difference by taking the difference between Eq. 5.8 and Eq. 5.9. This gives the Josephson frequency ω_J, when $\rho_l = \rho_r$, by

$$\frac{\partial \gamma}{\partial t} = \frac{\partial}{\partial t}(\varphi_r - \varphi_l) = \omega_J = \frac{2e}{\hbar}V \qquad (5.14)$$

which is in an observable range, since $V = 1\,\mu V$ leads to $f_J \equiv \omega_J/2\pi = 483.6\,MHz$. We are speaking here of electromagnetic phenomena related to the movement of charge at elevated frequency: The Josephson contact emits electromagnetic radiation at the frequency ω_J.

Equation 5.14 for the Josephson frequency can be integrated to give the time dependence of the phase difference in zero magnetic field:

$$\gamma = \gamma_0 + \frac{2e}{\hbar}Vt = \gamma_0 + \omega_J t \qquad (5.15)$$

Clearly then, as the Josephson current depends on $\sin \gamma$, it will now be modified to

$$J = J_0 \sin(\gamma_0 + \omega_J t) \qquad (5.16)$$

It is this oscillatory Cooper-pair current which generates the microwave electromagnetic radiation. In the opposite case, when microwave radiation is incident on the junction, steps appear at regular intervals $V_n = n(\frac{\hbar}{2e})\omega_J$ on the I-V curve. These steps, caused by absorption of n quanta of electromagnetic radiation at the Josephson frequency, are called Shapiro-steps after their discoverer. Figure 5.5 illustrates the I-V characteristic during illumination by electromagnetic radiation at $\omega = \omega_J$.

5.2.1   Alternative derivation of the AC Josephson effect

The simplified derivation of the AC Josephson effect did not explicitly bring out the role of the vector potential A, related to the B-field by $\nabla \times A = B$. We therefore rederive the AC Josephson current in a manner which makes the role of A explicit. In so doing we have to use a gauge invariant form of the phase difference across the barrier. The necessity of doing this is seen in the

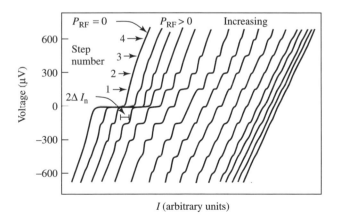

Figure 5.5   Microwave-induced Shapiro steps in I-V curves of a Nb point contact Josephson junction at 4.2 K at different microwave power levels irradiating the junction at 72 GHz. I-V curves are shifted for clarity [59].

GL-equations where the correct quantity to consider is

$$\nabla\varphi - \frac{2e}{\hbar}A \tag{5.17}$$

To bring out the phase difference between the two sides we integrate from left to right and get:

$$\gamma_{l\rightarrow r} = \int_1^r \left(\nabla\varphi - \frac{2e}{\hbar}A\right)dl = \int_1^r \left(\nabla\varphi - \frac{2\pi}{\Phi_0}A\right)dl \tag{5.18}$$

where it is understood that the integration starts on the left-hand side of the junction (see Figure 5.1a), at a point which is deeper than the penetration of magnetic field, and ends similarly on the right side. We have also inserted Φ_0 for $h/2e$. Taking the vector potential $A_z(x)$ to be parallel to the direction of integration z, normal to the contact, corresponds to placing the B-field parallel to the contact plane, and pointing along the y-axis, with $B_y(x) = \partial A_z/\partial x$.

On integrating we obtain

$$\gamma_{l\rightarrow r} = (\varphi_r - \varphi_l) - \frac{2\pi}{\Phi_0}\int_1^r A_z\,dz \tag{5.19}$$

The phases φ_r and φ_l may be taken here to be fixed phases which exist in the absence of any AC current. Even though φ may not be well defined in the junctions, we may write the integral of the gradient as the difference between the endpoints and neglect a $2\pi \times n$ contribution. When we take the time derivative of this last relation we find, using $E_z = -\frac{\partial A_z}{\partial t}$,

$$\frac{\partial\gamma}{\partial t} = -2\pi\,\Phi_0^{-1}\int_l^r \frac{\partial A_z}{\partial t}dz = 2\pi\,\Phi_0^{-1}\int E_z\,dz = 2\pi\,\Phi_0^{-1}V \tag{5.20}$$

$$\frac{\partial\gamma}{\partial t} = \frac{2e}{\hbar}V = \omega_J \quad \text{as before.} \tag{5.21}$$

This derivation brings out explicitly the point that the AC Cooper-pair super-current across the contact is driven by a time-dependent vector potential, which again is created by a static electric field by the voltage V.

A relevant question is now whether this current, being a Cooper-pair super-current, is lossless. A DC super-current can run forever as long as it is truly DC. But here, we are discussing time-dependent currents, i.e. accelerated mass and charge, and the process can only be sustained as long as we feed energy into the circuit. This will be the case in any super-current charge transport.

In addition to this loss, observation shows that energy is radiated from the contact as a microwave electromagnetic field. Thus the Josephson effects are quite complex phenomena. The Shapiro steps (see Figure 5.5) in the I-V diagram

resulting from feeding microwave power into the contact is another manifestation of the energy exchange involving Cooper-pair super-current. More precisely,
transfer of a Cooper-pair across the barrier costs an energy $2eV = \hbar\omega_J$, exactly
the photon energy. Some of this energy is radiated, as mentioned.

5.3 Josephson current in a magnetic field

One of the truly remarkable and useful aspects of the Josephson contact comes
about when its sensitivity to magnetic field is considered. Let the planar junction
be placed in the xy-plane, with the magnetic field parallel to the y-axis. The
GL-equations tells us that the DC supercurrent will be established according to

$$J_s = \frac{\rho e}{m}(\hbar\nabla\varphi - 2eA) \tag{5.22}$$

This current is the equivalent of the Meissner-effect according to London theory,
but with the addition of a $\nabla\varphi$ driving term. Eq. 5.22 is easily rewritten as

$$\nabla\varphi = \frac{2e}{\hbar}\left(\frac{m}{2e^2\rho}J_s + A\right) \tag{5.23}$$

Now choose an integration contour as shown in Figure 5.6, where the film
thickness direction is along the z-axis, and the B-field along the $+$ y-axis.
We define the sum of length parameters $\lambda_L + \lambda_R + \delta = d_J$. The length of the
junction is L.

Next, we perform an integration of Eq. 5.23 along the contours C_L and C_R
in the superconductors, resulting in

$$\varphi_{R_a}(x) - \varphi_{R_b}(x + \delta x) = \frac{2e}{\hbar}\int_{C_R}\left(A + \frac{m}{2e^2\rho}J_s\right)\delta l \tag{5.24}$$

$$\varphi_{L_b}(x + \delta x) - \varphi_{L_a}(x) = \frac{2e}{\hbar}\int_{C_L}\left(A + \frac{m}{2e^2\rho}J_s\right)\delta l \tag{5.25}$$

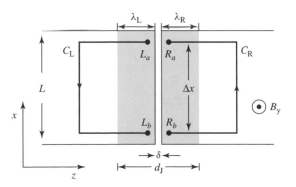

Figure 5.6 A good choice of integration path.

Now we exploit the following possibility: (i) The integration parallel to the junction may be performed so deep in the superconductor that screening current density is zero on both sides; (ii) that part of the integration which is done within the penetration depth of field and screening currents may be performed at right angles to that current. Figure 5.6 illustrates these two points. Forming now the phase difference $\varphi(x + \Delta x) - \varphi(x)$ this can be achieved by manipulation of Eqs 5.24 and 5.25:

$$\varphi(x + \delta x) - \varphi(x) \equiv \left[\varphi_{L_b}(x + \delta x) - \varphi_{R_b}(x + \delta x)\right] - \left[\varphi_{L_a}(x) - \varphi_{R_a}(x)\right] \tag{5.26}$$

$$\varphi(x + \delta x) - \varphi(x) = \frac{2e}{\hbar}\left[\int_{C_L} \mathbf{A} \cdot d\mathbf{l} + \int_{C_R} \mathbf{A} \cdot d\mathbf{l}\right] \approx \frac{2e}{\hbar} \oint \mathbf{A} \cdot d\mathbf{l} \tag{5.27}$$

when we have taken $\lambda_L + \lambda_R \gg \delta$, which is a good approximation in most cases. Also the contribution from the two gaps will tend to be of opposite sign and equal size. Integration across the thickness δ can therefore be neglected. The result is that the phase difference across the barrier depends on x, as we get

$$\varphi(x + \delta x) - \varphi(x) = \frac{2e}{\hbar} \oint \mathbf{A} \cdot d\mathbf{l}$$

$$= \frac{2e}{\hbar} B_y(\lambda_R + \lambda_L + \delta)\delta x$$

$$= 2\pi \Phi_0^{-1} B_y d_J \delta x \tag{5.28}$$

which in integrated form becomes

$$\varphi(x) = 2\pi \Phi_0^{-1} B_y d_J x + \varphi_0 \tag{5.29}$$

The shielding supercurrent density can now be written down using the standard Josephson relation Eq. 5.13.

$$J = J_0 \sin(2\pi \Phi_0^{-1} B_y d_J x + \varphi_0)$$

$$= J_0 \sin\left(\frac{2\pi}{\lambda_\varphi}x + \varphi_0\right) \tag{5.30}$$

$$\text{with} \quad \lambda_\varphi = \frac{\Phi_0}{B_y d_J}$$

The interesting result is that the Josephson current across the barrier is a sine-function, in other words varying in sign and amplitude along the direction perpendicular to the magnetic field, as illustrated in Figure 5.7.

Looking at the whole pattern of super-currents in a Josephson junction in the presence of a magnetic field one recognizes the similarity with vortices

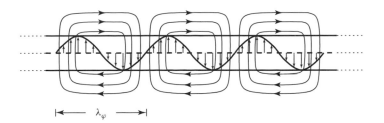

Figure 5.7 A periodic pattern of supercurrent circulation in a Josephson junction.

penetrating a type II superconductor. In addition to the current crossing the barrier, charge conservation requires currents to flow along the contact in the direction transverse to tunnelling currents. This establishes a periodic pattern of supercurrent circulation, with the period λ_φ defined previously (see Figure 5.7). The associated vortex patterns are called Josephson vortices. We note here that although similar to the usual vortices in type II superconductors, they are of a different nature since they have no normal core. Consequently, no upper critical field H_{c2} is associated with the Josephson vortices.

From Figure 5.7 one can immediately conclude (neglecting end effects) that a rectangular junction will obey a simple rule that the current is zero every time the relation

$$L = n\lambda_\varphi = n\frac{\Phi_0}{B_y d_J} \tag{5.31}$$

is satisfied, or at values of the magnetic field which obey:

$$B_y = n\frac{\Phi_0}{L d_J}; \quad n = 1, 2, 3\ldots \tag{5.32}$$

in other words every time an integral number of flux quanta is located in the junction. In case $\varphi_0 = 0$ maximum current will appear at every field corresponding to the condition

$$L = (n + \tfrac{1}{2})\lambda_\varphi \tag{5.33}$$

or

$$B_y = (n + \tfrac{1}{2})\frac{\Phi_0}{L d_J}; \quad n = 0, 1, 2\ldots \tag{5.34}$$

The total current is found by integration of Eq. 5.30. Assuming we have a rectangular contact of area $S = LW$ the current is found by integrating over the area S:

$$I = \int_S J_0 \sin\varphi(x)\mathrm{d}S = \Re e \int_S i J_0 e^{i\left(\frac{2\pi}{\lambda_\varphi}x - \varphi_0\right)}\,\mathrm{d}S \tag{5.35}$$

where J_0 is assumed to be constant, and $\mathrm{d}S = W\,\mathrm{d}L$

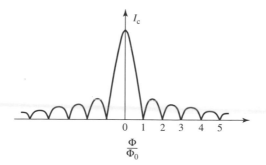

Figure 5.8 DC Josephson effect in magnetic field. Maximum zero-voltage current as a function of the magnetic flux.

The integration gives the following Josephson diffraction current:

$$I_{max} = I_0(0) \left| \frac{\sin(\pi \Phi_B / \Phi_0)}{\pi \Phi_B / \Phi_0} \right| \tag{5.36}$$

Here we have removed a factor $\sin \varphi_0$ whose maximum is 1. Figure 5.8 shows a graph of $I(B)$.

Based on these results one can conclude that the principle for a sensitive detector of magnetic field has been established due to the sensitivity of the current in a Josephson junction to magnetic field. But the possible device described above is easily surpassed by a different design, which is far more sensitive. This is the SQUID, which we discuss next.

5.4   The SQUID principle

The SQUID (an acronym for superconducting quantum interference device), employs a design with two current paths in a loop-like structure. Current is fed into the loop on one side and collected on the other, in principle as sketched in Figure 5.9. In reality, the structure is usually made of flat layers of thin films, in a superconductor-insulating-superconductor sequence. Figure 5.9 can be thought of as a cross-section through such a structure. A complete contour is drawn inside the two superconductor branches, deep enough that we can regard the current as vanishingly small. In other words we assume: $J_s = \frac{\hbar \rho e}{m}(\nabla \varphi - \frac{2eA}{\hbar}) = 0$ which requires $\nabla \varphi = \frac{2e}{\hbar} A$ along the dashed contour C.

We integrate this relation along C, and add also to this the integration across both junctions to make a complete closed contour, to obtain

$$(\varphi_r(a) - \varphi_l(a)) - (\varphi_r(b) - \varphi_l(b)) = \frac{2e}{\hbar} \oint A \cdot dl \tag{5.37a}$$

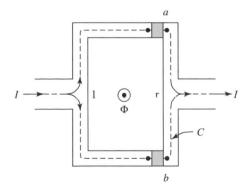

Figure 5.9   A double Josephson junction structure.

or

$$\gamma_a - \gamma_b = \frac{2e}{\hbar} \oint A \cdot dl = \frac{2\pi \, \Phi}{\Phi_0} \qquad (5.37b)$$

where $\gamma_a = \varphi_r(a) - \varphi_l(a);\quad \gamma_b = \varphi_r(b) - \varphi_l(b)$.

The total lumped current I can now be written as a sum of the currents across junction a and junction b:

$$I = I_0 \sin(\varphi_r(a) - \varphi_l(a)) + I_0 \sin(\varphi_r(b) - \varphi_l(b)) \qquad (5.38a)$$

$$I = I_0 \sin \gamma_a + I_0 \sin \gamma_b \qquad (5.38b)$$

resulting in

$$I = 2I_0 \cos\left(\frac{\pi \, \Phi}{\Phi_0}\right) \sin\left(\gamma_b + \frac{\pi \, \Phi}{\Phi_0}\right) \qquad (5.38c)$$

by use of standard trigonometric relations for the addition of sine functions.

The maximum current which the device can carry without dissipation is

$$I_{max} = 2I_0 |\cos(\pi \, \Phi / \Phi_0)| \qquad (5.39)$$

With this device a flux change $\Delta \Phi_B \ll \Phi_0$ can be detected. And since the area where the flux is felt by the SQUID (i.e. the whole loop) is much larger than the Josephson contact discussed previously, this shows that the SQUID can be many orders of magnitude more sensitive than a single contact in magnetic field measurements. This result is the basis for most applications of the SQUID. We shall return to that issue in Chapter 11. Calculation based on Eq. 5.39 for the resulting current through a two-point contact is illustrated in Figure 5.10.

The flux Φ introduced in the analysis above consists of two parts; the external flux Φ_e due to the applied magnetic field, and the flux $L I_{scr}$ due to the screening currents I_{scr} in the loop with inductance L. In other words

$$\Phi = \Phi_e - L I_{scr} \qquad (5.40)$$

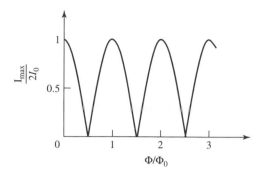

Figure 5.10 Maximum current which is possible to pass through a SQUID, as a function of the flux through it, measured in units of Φ_0.

Clearly, the maximum current I_{max} referred to previously is influenced by the presence of flux generated by screening currents, and the simple result given in Eq. 5.39 is strictly valid only when LI_{scr} is negligible. In Chapter 11 we discuss the equations of motion for the DC SQUID in greater detail, and give references to extensive literature.

A new and interesting development in SQUID physics and technology has taken place by the discovery of d-wave pairing in high-T_c cuprate superconductors. We recommend reading about the outstanding work described by Mannhart in the a topical contribution to this book, in Chapter 13.

5.5 The Ferrell–Prange equation

What happens to screening currents in a Josephson junction exposed to an external magnetic field? We now address this question, referring to Figure 5.11, a sketch of the situation we discuss.

We recognize the main structure as that of a Josephson junction of length L in the x-direction and with an external field applied along the y-axis. In the presence of the applied B-field screening currents are generated. In the superconductor these penetrate to a usual depth λ. However, where the screening current crosses the insulator barrier, the density of Cooper-pairs is much lower than in the superconductor bulk or film. Hence the screening length is much larger in the junction. This length is usually called the Josephson penetration depth λ_J. The previously defined lengths d_J and δ are also indicated in the figure. Our task is now to express λ_J in terms of other known parameters.

We can now repeat the procedure previously adopted, i.e. integrating the relation $\nabla\varphi = \frac{2e}{\hbar}A$ along a closed contour crossing the junction at distances dx like before.

$$\gamma(x + dx) - \gamma(x) = \frac{2e}{\hbar} \oint A \cdot dl \tag{5.41}$$

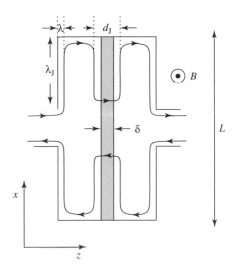

Figure 5.11   Illustration of geometries relevant to derivation of the Ferrel–Prange equation.

The right-hand side becomes

$$\frac{2e}{\hbar} \oint A \cdot dl = \frac{2e}{\hbar} d\Phi = 2\pi \frac{d\Phi}{\Phi_0} \tag{5.42}$$

where $d\Phi$ is the flux enclosed within the contour, between the contour crossings at x and $x + dx$. The integration on the left side of Eq. 5.41 at x and $x + dx$ leads to

$$d\gamma = \gamma(x + dx) - \gamma(x) = 2\pi \frac{d\Phi}{\Phi_0} \tag{5.43}$$

in the same notation and by the same argument as before. We can write this as

$$\frac{d\gamma}{dx} = \frac{2\pi}{\Phi_0} \frac{d\Phi}{dx} \tag{5.44}$$

We can now introduce the magnetic field instcad of the flux Φ by observing that

$$B = \frac{d\Phi}{d_J dx} \tag{5.45}$$

Introducing this into Eq. 5.44 gives us

$$\frac{d\gamma}{dx} = \frac{2\pi d_J}{\Phi_0} \frac{d\Phi}{d_J dx} = \frac{2\pi d_J}{\Phi_0} B$$

$$B = \frac{\Phi_0}{2\pi d_J} \frac{d\gamma}{dx} \tag{5.46}$$

From Maxwell's equations we get the supercurrent density $J_s = \nabla \times H = \mu_0 \nabla \times B = \mu_0 \frac{dB}{dx}$ along the z-axis. When this is inserted into Eq. 5.46 together with $J = J_0 \sin \gamma$ we find:

$$\frac{d^2\gamma}{dx^2} = \frac{2\pi J_0 \mu_0 d_J}{\Phi_0} \sin \gamma \qquad (5.47)$$

which is usually written

$$\frac{d^2\gamma}{dx^2} = \frac{1}{\lambda_J^2} \sin \gamma \qquad (5.48)$$

This is the Ferrell–Prange equation [60] which predicts how the screening field penetrates into and parallel to the Josephson junction, with

$$\lambda_J = \left(\frac{\Phi_0}{2\pi d_J \mu_0 J_0} \right)^{\frac{1}{2}} \qquad (5.49)$$

When the phase difference γ is very small $\sin \gamma$ can be approximated by γ. In this case we get

$$\frac{d^2\gamma}{dx^2} = \frac{1}{\lambda_J^2} \gamma \qquad (5.50)$$

which is easily solved as

$$\gamma(x) = \gamma(0) e^{-x/\lambda_J} \qquad (5.51)$$

Using this in the previously found relation for B, Eq. 5.46, we obtain

$$B(x) = B(0) e^{-x/\lambda_J} \qquad (5.52)$$

The Josephson penetration depth can be estimated on the basis of typical values for the parameters entering in Eq. 5.49. With $J_0 \sim 10^2 \, \text{A/cm}^2$ and $d_J \sim 10^{-5}$ cm, one finds $\lambda_J \sim 0.1$ mm. The result is that λ_J is much larger than λ, as we argued from the outset.

5.6   The critical field H_{c1} of a Josephson junction

What field could we guess is required to force Josephson vortices into a Josephson junction using an external field? For this question we already have a relevant reference: The argument leading to an expression for H_{c1} in bulk superconductors. In the bulk we expect H_{c1} to be determined by distributing one quantum of flux Φ_0 over the characteristic area for penetration of flux in the presence of the Meissner effect, i.e. $\pi\lambda^2$. This leads to the approximate expression $H_{c1} \approx \Phi_0/\pi\lambda^2$ in bulk superconductors.

In the case of a Josephson junction there are two characteristic lengths, one along the junction, λ_J, and one transverse to it, d_J, as introduced before. We should therefore expect H_{c1} in the junction to be approximately given by

$$H_{c1} \approx \frac{\Phi_0}{d_J\lambda_J} \tag{5.53}$$

which is indeed correct to within a numerical factor of order unity. The precise expression turns out to be

$$H_{c1} = \frac{2}{\pi^2}\frac{\Phi_0}{d_J\lambda_J}. \tag{5.54}$$

5.7 Josephson vortex dynamics

We have only discussed static vortices above. A lot of interesting physics is connected with dynamic phenomena, i.e. Josephson vortices in motion. Because of the fact that these do not have a normal core, as mentioned, they can move extremely rapidly and with minute losses within the junction. Vortex dynamics is described by the Sine–Gordon equation. In deriving the Josephson penetration depth λ_J via the Ferrell–Prange equation we came close to this problem. One only needs to add a dynamic term to the equation, namely the acceleration term $\frac{\partial^2\gamma}{\partial t^2}$ to change the validity from that of a static to a dynamic description, resulting in

$$\frac{\partial^2\gamma}{\partial x^2} + \frac{\partial^2\gamma}{\partial t^2} = \frac{1}{\lambda_J^2}\sin\gamma \tag{5.55}$$

The solution of this equation is solitary-like, i.e. representing the Josephson vortex as a solitary wave in a non-linear medium. Figure 5.12a shows the location of the vortex in the junction, and Figure 5.12b shows a representation of the solitary wave in more detail. This solitary wave moves along the junction with a speed close to that of the speed of light due to its low dissipation in the absence of a normal core.

Moreover, high frequency radiation is generated from the junction as the solitons reach the surface in rapid succession. The solitary waves can be driven by a current source or a voltage source applied to the junction.

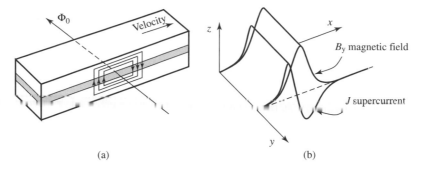

(a)   (b)

Figure 5.12   The Josephson vortex as a solitary wave. After Pedersen [61].

A practical problem arises in the use of these junctions as high-frequency radiation sources due to the great mismatch of impedance between the junction and the surrounding vacuum. This mismatch is caused by the difference in wave velocities. In the case of solitary wave analysis it is necessary to consider an extended or long junction, with a length $L \gg \lambda_J$. In a long junction one includes the effects of the self field in addition to the externally applied magnetic field. This corresponds to taking into account the screening currents in the SQUID, a possibility we already pointed out. For a general introduction to soliton physics in Josephson junctions, see reference [61].

5.8   Josephson plasma in cuprate high-T_c superconductors

An interesting development related to the Josephson effects was the prediction [62] and observation by numerous groups, among them Matsuda and coworkers [63], of plasma effects in $Bi_2Sr_2CaCu_2O_{8+\delta}$. Neighbouring superconducting CuO_2 layers are coupled by the Josephson effect through the block layers between the CuO_2 layers, and the Josephson plasma is generated by the coupling between the Josephson current flowing along the c-axis and the electromagnetic field, leading to current oscillations at frequencies well below the superconducting gap frequency, in the GHz to THz range. Longitudinal and transverse plasma waves propagating along the c-axis and in the ab-plane are of quite different nature. These phenomena have already been developed to an advanced stage, and it has become possible to make devices that emit electromagnetic waves in the THz range. Another, related phenomenon is the interaction between the plasma and vortices. This leads to strong dependencies of the plasma frequencies on the state of the vortex matter created by an external magnetic field. Furthermore, the static interaction between pancake vortices and Josephson vortices has led to new insight into so-called crossing lattices of vortex matter.

6

London Approximation to Ginzburg–Landau Theory ($|\psi|$ constant)

6.1 The London equation and the penetration depth λ_L

6.1.1 Early electrodynamics and the London hypothesis

Figure 6.1 describes an essential property of Meissner screening: The interior of a superconductor is shielded against a static magnetic field, a phenomenon we already discussed briefly in Chapter 1. We will now discuss the physics of this effect and its history. Already years before the discovery of Meissner screening, speculations had been made by Onnes among others, that the current in a superconductor might involve surface currents. The discovery made by Meissner and Ochsenfeld [3] in 1933 proved that this was correct, in the case of diamagnetic response to an external magnetic field. Logically the magnetic field could be expelled from the interior only by setting up a surface current. But the current could not exist only *in* the surface. It would have to penetrate to a certain depth called λ. If this were not so, a finite current should exist in a layer of zero thickness, which would imply an infinite density of free charge. Since there is of the order of one free charge pr metal atom, the penetration depth must be finite, and its length must be controlled by the free electron number density n_e. The more electrons are present per volume, the more effective is the screening, and the shorter must λ be. The prediction for an *electrodynamic* screening length λ had already been made by de Haas Lorentz [64] in 1925. In modern SI units it reads

$$\lambda = (m_q/\mu_0 n_s q^2)^{\frac{1}{2}} \tag{6.1}$$

where q is the charge of the carriers in the superconducting state, m_q their mass, and n_s their number density. This is the same formula that we use for screening

Superconductivity: Physics and Applications   Kristian Fossheim and Asle Sudbø
© 2004 John Wiley & Sons, Ltd   ISBN 0-470-84452-3

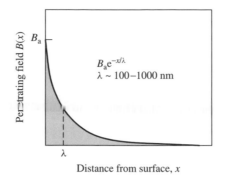

Figure 6.1   Meissner screening.

of a *static* magnetic field today. The prediction would give a penetration depth of about 50 nm in typical metals, a value which later has turned out to be in good agreement with observation. The derivation of the result for λ goes as follows: Near the surface of a superconductor an electric field can exist, accelerating the superconducting electrons frictionlessly according to

$$m_q \dot{v} = -q E \tag{6.2}$$

where, v is the carrier velocity and E the electric field. Time derivative is indicated by the dot. Since current density can be written as $J = -n_s q v$ this gives an expression for E:

$$E = -\frac{m_q \dot{v}}{q} = \frac{\mu_0 m_q \dot{J}}{\mu_0 n_s q^2} = \mu_0 \lambda^2 \dot{J} \tag{6.3}$$

Taking the curl on both sides we obtain

$$\nabla \times E - \mu_0 \lambda^2 \nabla \times \dot{J} = 0 \tag{6.4}$$

Using $\nabla \times E = -\dot{B}$ we find

$$\mu_0 \lambda^2 \nabla \times \dot{J} + \dot{B} = 0 \tag{6.5}$$

Applying Maxwell's equation $\nabla \times B = \mu_0 J$ leads to

$$\lambda^2 \nabla \times \nabla \times \dot{B} + \dot{B} = 0 \tag{6.6}$$

The first term is equal to $-\nabla^2 \dot{B}$, hence

$$\nabla^2 \dot{B} = \frac{1}{\lambda^2} \dot{B} \tag{6.7}$$

The solution is an exponential decay of $\dot{B}$ from the surface towards the interior, over the characteristic distance λ. Note that this is proposed as the electrodynamics of a superconductor. It is fundamentally different from the skin depth of a metal since we assumed frictionless motion of charge. Heinz and Fritz London [65] observed that on performing the time integration of Eq. 6.7 one is left with an equation between a static magnetic field B and a stationary current J, with an additional unknown integration constant. Setting the constant of integration equal to zero, they obtained

$$B = -\mu_0 \lambda_L^2 \nabla \times J \qquad (6.8)$$

Here we have renamed the parameter λ as λ_L. Equation 6.8 is the famous (second) London equation, which correctly describes the Meissner effect mathematically. Furthermore, on introducing the vector potential A via $\nabla \times A = B$, we obtain the profound relationship between the vector potential and the supercurrent

$$J = -\frac{1}{\mu_0 \lambda_L^2} A = -\Lambda A \qquad (6.9)$$

Notice that this equation is analogous to Ohm's law: $J = \sigma E$. Hence, in a superconductor exposed to a static magnetic field it is the vector potential A that drives a stationary supercurrent J. The factor Λ is a response function, analogous to the conductivity σ in a normal metal.

Although this equation seemed to express the Meissner effect correctly, and as such was a major achievement, its justification on microscopic grounds remained unknown since the London treatment is purely phenomenological. To justify the result from first principles remained a challenge for more than 20 years, until the famous BCS paper of 1957 [7]. In any case, the phenomenological basis for the important length parameter λ_L, the London penetration depth, had now been established and a prediction made for the screening length of a *static B*-field. It is mathematically the same parameter as was originally found by de Haas Lorentz, a fact which has gone largely unnoticed in the scientific literature.

For completeness we repeat the simple steps leading to the Meissner effect as a consequence of the finite screening length λ_L for static magnetic fields. Our starting point is now the London Eq. 6.8, which we rewrite as

$$B + \lambda_L^2 \nabla \times \nabla \times B = 0 \qquad (6.10)$$

The simplest situation is the 1-dimensional case with the field applied parallel to the z-axis along the surface of a superconductor of long length along y, and with the x-axis measuring the distance from the surface into the superconductor, as already depicted in Figure 6.1.

Equation 6.10 may be reformulated into

$$B = \lambda_L^2 \nabla^2 B \qquad (6.11)$$

with the solution in the 1-dimensional case

$$B(x) = B_a e^{-\frac{x}{\lambda_L}} \tag{6.12}$$

B_a is the externally applied induction field related to the applied field H_a by $B_a = \mu_0 H_a$. The solution of the London equation shows the significance of the parameter λ_L as the screening length over which the applied field is reduced to B_a/e. The length λ_L is mainly controlled by the superfluid density n_s. BCS theory showed later that due to pairing of electrons the mass m_q and charge q are twice that of the normal electron gas, and the full value of n_s is half of the density n in the normal state. We could introduce factors of 2 in the mass, charge and particle density, but the end result is the same for λ_L. These factors cancel out, and the zero degree value of λ_L acquires the same value by either choice. In the case of pure metals the main variations in n_s from one metal to another come from different valency. The typical values of the calculated λ_L lie in the range 40–50 nm in metallic superconductors such as Al, Sn and Pb. Because there is an effective mass involved, as well as different band structures, deviations from the ideal values are to be expected. In doped superconductors on the other hand, n_s is of course determined by the doping level rather than by band structure effects.

The solution (Eq. 6.12) precisely describes the Meissner effect: when an external static magnetic field is applied to a superconductor, supercurrents arise spontaneously in the surface layer to a characteristic depth λ_L so as to create an opposing field in the sample, exactly cancelling the applied field inside. It depends on the shape of the superconducting sample how the currents run, and whether the screening is complete or not (Figure 6.2). A partial deeper penetration of field will often occur, depending on the demagnetization factor of the sample with respect to the direction of the applied field. We refer back to Section 1.7 for a brief discussion of this point.

Experimental values for the relative penetration depth λ were determined in Ref. [66] which reported on the temperature dependence of the quantity λ/λ_0 where λ_0 is the value near 0 K. It showed a steep descent of λ/λ_0 below and

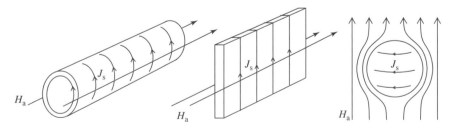

Figure 6.2   Some examples of sample shapes and screening currents in an external magnetic field.

away from the transition temperature T_c. This corresponded well with the two-fluid model of Gorter and Casimir [67], which would give $\left(\frac{\lambda}{\lambda_0}\right)^2 = \frac{1}{1-[T/T_c]^4}$. These authors had introduced the idea that the electron gas in the superconducting state consists of a uniform mixture of two electron fluids, one being the normal component and having properties like in the normal state, the other one being the superfluid component observed only below T_c, and being responsible for the superconducting properties. In the two-fluid model they argued that a fraction w of superconducting electrons would condense into an ordered state, and vary from zero at $T = T_c$ to unity at $T = 0$ according to $1 - (T/T_c)^4$. When this was put into the expression for λ the experimentally observed behavior would result. Figure 6.3 shows data by Schawlow [68] confirming that this form even agrees well with BCS theory.

After World War II, Pippard [69] performed measurement of the surface impedance of metals at radio-frequencies up to 1200 MHz. In the normal state the penetration is the familiar classical skin depth δ, which, upon passing below T_c becomes λ. The wave-vector dependence had not yet been sorted out; the possibility remained therefore that the measured λ could be different from the λ_L which had been predicted for the DC case. Even so, the data seemed in harmony with the predictions of the two-fluid model.

Further measurements of λ in the presence of a static magnetic field led Pippard to propose the existence of a new length parameter, the superconducting coherence length ξ, which would cause the observed entropy change in pure tin at the superconducting transition to be distributed to a depth as much as 20 times the penetration depth λ.

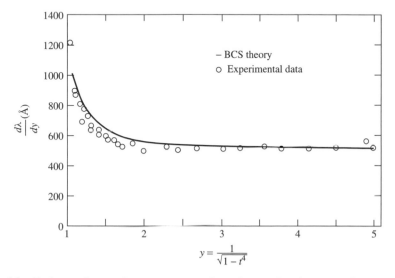

Figure 6.3 Early results on the temperature dependence of λ in a type I superconductor [68].

A decisive step was taken when Pippard showed that, contrary to the prediction of the London theory, λ depended very sensitively on impurity scattering. With the addition of only 3% indium to tin λ changed by a factor of 2, while at the same time the changes in T_c and H_c were insignificant. In the London theory λ depended only on m/n, offering no hint at an explanation. Figure 6.4 shows data demonstrating the effect of the mean free path l on penetration depth λ.

Pippard realized that the key point was a breakdown of the underlying assumption of local response in the London theory. We shall return to Pippard's analysis in Section 7.3.

6.1.2   Derivation of the London equation from the free energy

The London equation applies in the presence of a magnetic field, giving rise to both a kinetic energy density F_k and a magnetic energy density F_m. The superconductor is already at a temperature $T < T_c$ when the London equation takes effect, so the total energy includes also the condensation energy density F_s. The total energy density F for the problem is found by integrating the three densities over the volume V:

$$F = \int_V F_s(r)\mathrm{d}^3r + \int_V F_k(r)\mathrm{d}^3r + \int_V F_m(r)\,\mathrm{d}^3r \tag{6.13}$$

We may regard the condensation energy density as unaffected by the application of a magnetic field. Then $\int F_s(r)\,\mathrm{d}^3r = F_s$ may be regarded as a constant

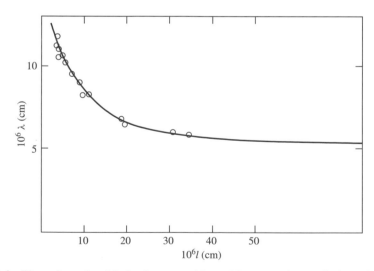

Figure 6.4   Pippard non-local behaviour as evidenced by mean free path dependent penetration depth. After J. Waldram, thesis 1961 [70] and Lynton, 1962 [71].

background term. The kinetic energy is

$$F_k = \int F_k(r) d^3r = \int \frac{1}{2} m v_s^2(r) n_s d^3r \tag{6.14}$$

$$= \frac{\mu_0}{2} \int \lambda_L^2 J^2(r) \, d^3r \tag{6.15}$$

which is obtained on writing $J = -n_s e v_s$. The magnetic energy is

$$F_m = \int F_m(r) \, d^3r = \int \frac{1}{2\mu_0} b^2(r) \, d^3r \tag{6.16}$$

In Eq. 6.16 we have introduced the vector b allowing it to represent local induction on a scale smaller than λ_L. For later treatment of the physics of the flux lines this allows a distinction between induction on scales larger than λ_L represented by B and on scales smaller than λ_L. Hereafter we write λ for λ_L.

After combining all three terms, introducing b, and using the Maxwell equation $\mu_0 J = \nabla \times b$ we find

$$F = F_s + \frac{1}{2\mu_0} \int \left[\lambda^2 (\nabla \times b(r))^2 + b^2(r) \right] d^3r \tag{6.17}$$

We want to vary $b(r)$ an amount $\delta b(r)$ and find the corresponding δF, i.e. we replace $b(r)$ with $b(r) + \delta b(r)$ and take the difference in F before and after the operation. The resulting δF to lowest order in δb is found to be

$$\delta F = \frac{1}{2\mu_0} \int \left[2b\delta b + 2\lambda^2 (\nabla \times b)(\nabla \times \delta b) \right] d^3r \tag{6.18}$$

To obtain the condition under which the free energy is minimized we set $\delta F = 0$. We find that this requires

$$b(r) + \lambda^2 \nabla \times \nabla \times b(r) = 0 \tag{6.19}$$

Here we have recovered the London equation. The physical meaning of the London equation is therefore to express the minimization condition for the free energy of the superconductor in the presence of a magnetic field. This is a result worth contemplating. It takes some of the mystery out of the London equation. There is a good, understandable reason for this equation, which was originally postulated without proof.

6.2   The energy of a single flux line

Upon subjecting a type II superconductor to a field higher than H_{c1} the Meissner screening breaks down, and individual flux quanta penetrate the superconductor. An important problem with regard to the physics of the vortex state, referred to as the *mixed state* between B_{c1} and B_{c2}, is the determination of the energy contents of a flux line. We will calculate this quantity as the energy per unit of length. We begin by deriving an expression for $b(r)$, and then integrate the energy density associated with the field and currents in a flux line. The aim of the calculation will be to find, in the London approximation, the sum of kinetic and magnetic energies whose densities were already given in Eqs 6.15 and 6.16, respectively. The background energy F_s is of no interest in this context. We find

$$F = \frac{1}{2\mu_0} \int_{r > \xi} \left[b^2(r) + \lambda^2 \left(\nabla \times b(r) \right)^2 \right] \mathrm{d}^3 r \qquad (6.20)$$

The limit of integration is a reminder that we cannot include the vortex core, of radius ξ, in this derivation because the London equation does not apply when n_s varies in space. The justification for this approximation is that $\xi \ll \lambda$ in most type II superconductors, notably in all high-T_c materials. The energy associated with the vortex core is therefore quite small. The energy F is also often referred to as the line tension. It tells us how much energy is required to create extra length of flux line. Flux lines are not like rubber bands, there is no elastic energy associated with stretching a straight flux line. However, it does require extra length to bend a flux line in a superconductor, either uniformly or locally, and the energy per unit length required to achieve this elongation is the line tension.

A practical way of approaching this problem was discovered by the Russian physicist Abrikosov[1] (1957) [4, 5]. We replace the singularity inside $r < \xi$ when we let $\xi \to 0$ by a 2D delta-function $\delta_2(r)$. Let us first discuss that aspect. The London equation now takes the form:

$$b + \lambda^2 \nabla \times \nabla \times b = \Phi_0 \delta_2(r) \qquad (6.21)$$

Notice that higher order terms in δb in Eq. 6.18 are neglected. The deltafunction represents a solution where $b(r)$ is not smooth.

In Eq. 6.21 Φ_0 is a vector along the direction of the induction normal to the surface, where Φ_0 is the flux associated with an area of radius $|r| \sim \lambda$.

---

[1]It was not immediately obvious that the approach of Abrikosov was correct. It is known that Landau rejected it when Abrikosov first presented it to him. This delayed the publication of Abrikosov's theory by years. The consequences of Abrikosov's idea were far-reaching, however, leading to the prediction of a flux line lattice, a truly remarkable result which was later to be confirmed experimentally.

Integrating Eq. 6.21 immediately gives

$$\int \left[\boldsymbol{b} + \lambda^2 \nabla \times \nabla \times \boldsymbol{b} \right] \mathrm{d}\boldsymbol{\sigma} = \int \boldsymbol{\Phi}_0 \times \delta_2(\boldsymbol{r}) \, \mathrm{d}\boldsymbol{\sigma} = \Phi_0 \qquad (6.22)$$

where $\mathrm{d}\boldsymbol{\sigma}$ is a surface element, and we integrate over the plane perpendicular to the flux line. The last relation is obtained since $\boldsymbol{\Phi}_0$ and $\mathrm{d}\boldsymbol{\sigma}$ are parallel, and by the properties of the 2D delta-function. Next we carry out

$$\lambda^2 \int [\nabla \times (\nabla \times \boldsymbol{b})] \mathrm{d}\boldsymbol{\sigma} = \lambda^2 \oint_{r=R} (\nabla \times \boldsymbol{b}) \mathrm{d}\boldsymbol{l} = \lambda^2 \oint_{r=R} \mu_0 \boldsymbol{J} \mathrm{d}\boldsymbol{l} \approx 0 \quad (6.23)$$

The last form approximates zero because we take the line integral at such large distance, $R \gg \lambda$, that the current associated with the flux line vanishes. From Eq. 6.22 we are now left with a relation which determines the total flux associated with a single flux line:

$$\int \boldsymbol{b} \mathrm{d}\boldsymbol{\sigma} = \Phi_0 \qquad (6.24)$$

We have proved that Φ_0 is the flux associated with the local field $\boldsymbol{b}$ of a single flux line when ξ can be regarded as vanishingly small compared to λ. This corresponds to insisting that $\delta \, |\psi| = 0$, which is the overall approximation made in the present chapter.

We can now go back to solving (6.22) for the region $\xi < r \ll \lambda$. We first rewrite it approximately as

$$\int_{\xi < r \ll \lambda} \boldsymbol{b} \mathrm{d}\boldsymbol{\sigma} + \lambda^2 \oint_{\xi < r \ll \lambda} (\nabla \times \boldsymbol{b}) \mathrm{d}\boldsymbol{l} = \Phi_0 \qquad (6.25)$$

In this case the integral $\int \boldsymbol{b} \cdot \mathrm{d}\boldsymbol{\sigma} \approx b(0)\pi r^2$ can be neglected since only a fraction $(r^2/\lambda^2)\Phi_0$ of Φ_0 is inside the radius r. By the foregoing argument 6.25 reduces to

$$\lambda^2 \oint_{\xi < r \ll \lambda} (\nabla \times \boldsymbol{b}) \mathrm{d}\boldsymbol{l} = \lambda^2 |\nabla \times \boldsymbol{b}| 2\pi r = \Phi_0 \qquad (6.26)$$

Hence, in the range $\xi < r \ll \lambda$ we find the approximate expression

$$|\nabla \times \boldsymbol{b}| = \frac{\Phi_0}{2\pi \lambda^2 r} \qquad (6.27)$$

Since $\boldsymbol{b} \| z$ we have $|\nabla \times \boldsymbol{b}| = -\mathrm{d}b/\mathrm{d}r$. Combining this with Eq. 6.27 gives

$$-\mathrm{d}b = \frac{\Phi_0 \mathrm{d}r}{2\pi \lambda^2 r} \qquad (6.28)$$

which, upon integration results in

$$-\int_b^{b_0} db = \frac{\Phi_0}{2\pi\lambda^2} \int_r^\lambda \frac{dr}{r} = \frac{\Phi_0}{2\pi\lambda^2} \ln\left(\frac{\lambda}{r}\right) \tag{6.29}$$

The solution for b is

$$b = \frac{\Phi_0}{2\pi\lambda^2} \ln\left(\frac{\lambda}{r}\right) + b_0 \tag{6.30}$$

where the constant b_0 gives an indication of how much of Φ_0 lies outside of $r = \lambda$. A rigorous analysis leads to the following result:

$$b = \frac{\Phi_0}{2\pi\lambda^2} K_0\left(\frac{r}{\lambda}\right) \tag{6.31}$$

where K_0 is a zero'th order Bessel function [72]. Limiting forms of K_0 may be used to obtain

$$\text{For } r \ll \lambda \qquad b \approx \frac{\Phi_0}{2\pi\lambda^2} \ln\left(\frac{\lambda}{r}\right) \tag{6.32}$$

$$\text{For } r \gg \lambda \qquad b \approx \frac{\Phi_0}{2\pi\lambda^2} \left(\frac{\pi\lambda}{2r}\right)^{\frac{1}{2}} e^{-\frac{r}{\lambda}} \tag{6.33}$$

Our goal was to find the energy per unit length, or line-tension, of the vortex. Recall that

$$F = \frac{1}{2\mu_0} \int_{r>\xi} \left[b^2 + \lambda^2 (\nabla \times b)^2\right] d^3r \tag{6.34}$$

Here, we have already argued that the first term may be neglected. So we have, to a good approximation

$$F = \frac{1}{2\mu_0}\lambda^2 \int_{r>\xi} (\nabla \times b)^2 \, d^3r \tag{6.35}$$

Upon partial integration this becomes

$$F = \frac{1}{2\mu_0}\lambda^2 \oint_{r=\xi} d\boldsymbol{l}\, b \, |\nabla \times b| \tag{6.36}$$

$$F = \frac{1}{2\mu_0}\lambda^2 2\pi\xi b(\xi) \, |\nabla \times b(\xi)| \tag{6.37}$$

Inserting here previously found expressions for $b(\xi)$ and $|\nabla \times b(\xi)|$ we obtain

$$F = \frac{1}{2\mu_0} \lambda^2 2\pi \xi \frac{\Phi_0}{2\pi \lambda^2} \ln \left(\frac{\lambda}{\xi}\right) \frac{\Phi_0^2}{2\pi \lambda^2 \xi} = \frac{\Phi_0}{4\pi \mu_0 \lambda^2} \ln \left(\frac{\lambda}{\xi}\right) \qquad (6.38)$$

This expression can be rewritten, by elimination of λ^2, as

$$F = \frac{\pi^3 \mu_0}{6} H_c^2 \xi^2 \ln \left(\frac{\lambda}{\xi}\right) \qquad (6.39)$$

We could improve this result by adding the condensation energy of the core $(\mu_0/2) H_c^2 \xi^2$. This energy is far less than F derived so far. The precise total energy including condensation energy has been shown to be

$$F = \frac{\Phi_0^2}{4\pi \lambda^2 \mu_0} \left(\ln \frac{\lambda}{\xi} + \varepsilon\right); \quad \varepsilon \approx 0.1 \qquad (6.40)$$

where the small correction term ε comes from the condensation energy.

6.2.1 Energy of a flux line: alternative derivation. An exercise

In the following we take a more formal approach to the same problem as above. We consider a cubic sample with dimensions $(2L)^3$, with a straight flux line f running through the middle in the z-direction (see Figure 6.5). We introduce the local field as h and observe that

$$h_x = h_y = 0 \qquad (6.41)$$

$$\frac{\partial h_z}{\partial z} = 0 \qquad (6.42)$$

The energy of the flux line in the London picture is:

$$F = \int dV \left(\frac{\mu_0 h^2}{2} + \frac{\lambda^2 \mu_0}{2} |\nabla \times h|^2\right) \qquad (6.43)$$

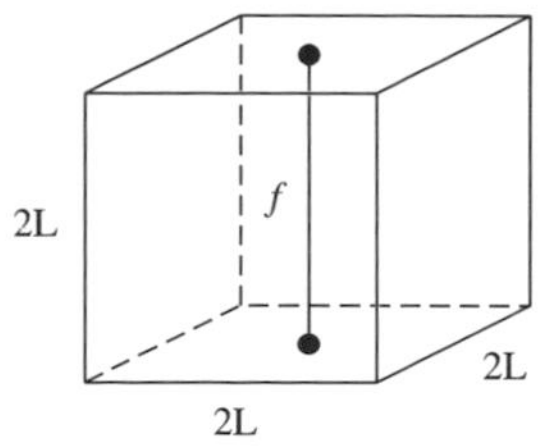

Figure 6.5 Flux line.

We have, further

$$\nabla \times h = \left(\frac{\partial h_z}{\partial y}, -\frac{\partial h_z}{\partial x}, 0 \right) \tag{6.44}$$

and

$$|\nabla \times h|^2 = \left(\frac{\partial h_z}{\partial y} \right)^2 + \left(\frac{\partial h_z}{\partial x} \right)^2 \tag{6.45}$$

With Eq. 6.45 inserted into Eq. 6.43 we get

$$F = \frac{\mu_0}{2} \int dV \left(h^2 + \lambda^2 \left[\left(\frac{\partial h_z}{\partial y} \right)^2 + \left(\frac{\partial h_z}{\partial x} \right)^2 \right] \right) \tag{6.46}$$

We Fourier transform h in the x- and y-directions. Since h points in the z-direction, we drop vector-signs on it.

$$h = \frac{1}{4L^2} \int d^2k e^{-i k \cdot r} B(k) \tag{6.47}$$

where $B(k) = \int d^2r e^{i k \cdot r} h(r)$

Next, we look at London's equation in the Abrikosov modification

$$h + \lambda^2 \nabla \times (\nabla \times h) = \frac{\Phi_0}{\mu_0} \delta(r - r_f) \tag{6.48}$$

We define the origin of the system to be on the flux line $r_f \equiv 0$. From Eqs 6.41 and 6.42 we get

$$\nabla \times (\nabla \times h) = \left(0, 0, -\frac{\partial^2 h_z}{\partial x^2} - \frac{\partial^2 h_z}{dy^2} \right) \tag{6.49}$$

which again has only z-component. By applications of Eqs 6.47 and 6.49 we can write Eq. 6.48 as

$$\frac{1}{4L^2} \int e^{-i k r} B(k) \left(1 + \lambda^2 (k_x^2 + k_y^2) \right) d^2k = \frac{1}{4L^2} \frac{\Phi_0}{\mu_0} \int d^2k e^{-i k r} \int d^2r e^{i k r} \delta(r) \tag{6.50}$$

By equating Fourier coefficients we arrive at

$$B(k) = \frac{\Phi_0}{\mu_0} \frac{1}{1 + \lambda^2 (k_x^2 + k_y^2)} \int d^2r e^{i k r} \delta(r) \tag{6.51}$$

By inserting Eq. 6.47 into Eq. 6.46 we get

$$F = \frac{\mu_0}{2} \frac{1}{(2L)^2} \int d^2r \, dz \, d^2k \, d^2k' e^{-i\mathbf{k}\cdot\mathbf{r}} e^{-i\mathbf{k}'\mathbf{r}} B(\mathbf{k}) B(\mathbf{k}') \tag{6.52}$$

Integration over $\mathbf{r}$ gives

$$F = \frac{\mu_0}{2} \frac{1}{(2L)^2} \int dz \, d^2k \, d^2k' \delta(\mathbf{k} + \mathbf{k}') B(\mathbf{k}) B(\mathbf{k}')(1 - \lambda^2(k_x k'_x + k_y k'_y)) \tag{6.53}$$

Now we integrate over $\mathbf{k}'$

$$F = \frac{\mu_0}{2} \frac{1}{(2L)^2} \int dz \, d^2k B(\mathbf{k}) B(-\mathbf{k})(1 + \lambda^2(k_x^2 + k_y^2)) \tag{6.54}$$

We put Eq. 6.51 into $B(\mathbf{k})$

$$F = \frac{\Phi_0}{2\mu_0} \frac{1}{(2L)^2} \int dz \, d^2k \frac{1}{1 + \lambda^2(k_x^2 + k_y^2)} \int d^2r e^{i\mathbf{k}\mathbf{r}} \delta(\mathbf{r}) \tag{6.55}$$

The integral $\int d^2r' e^{-i\mathbf{k}\mathbf{r}'} \delta(\mathbf{r}') = (2L)^2$ is now taken, and Eq. 6.55 becomes

$$F = \frac{\Phi_0^2}{2\mu_0} \int dz \, d^2k \frac{1}{1 + \lambda^2(k_x^2 + k_y^2)} \tag{6.56}$$

We convert the integral over k_x and k_y to one over θ and $|k|$

$$F = \frac{\Phi_0^2}{2\mu_0} \int dz \, dk \, d\theta \frac{k}{1 + \lambda^2 |k|^2} = \frac{\Phi_0^2}{2\mu_0} \int dz 2\pi \int_0^\infty \frac{k}{1 + \lambda^2 k^2} dk \tag{6.57}$$

This integral diverges, but with a cutoff $k \equiv 1/\xi$ it becomes finite, and we have:

$$\frac{F}{2L} = \frac{\Phi_0^2}{2\pi 2\mu_0} \frac{1}{2\lambda^2} \ln \left| \frac{\lambda^2}{\xi^2} + 1 \right| \tag{6.58}$$

For $\xi \ll \lambda$ we get

$$\hat{F} = \frac{F}{2L} = \frac{\Phi_0^2}{4\pi \mu_0 \lambda^2} \ln \frac{\lambda}{\xi} \tag{6.59}$$

6.3   Interacting flux lines: the energy of an arbitrary flux line lattice

So far we dealt with a single flux line. In real systems we normally have a large amount of such objects. We will now formalize the calculation of the total energy for an arbitrary collection of flux lines. We use again the symbol b for the locally varying induction B, so that B is given by $B = \langle b \rangle$ averaged over an area much larger than that of a single flux line.

Let us first note that, like before, the energy associated with currents and fields in the London limit is

$$F = \frac{1}{2\mu_0} \int d^3r \left[b(r)b(r) + \lambda^2 (\nabla \times b)(\nabla \times b) \right] \tag{6.60}$$

Next, we do the Fourier transform of the two terms, one by one. Using

$$b(r) = \int \frac{d^3k}{(2\pi)^3} b(k) e^{ikr} \tag{6.61}$$

we find

$$b(r) \times b(r) = \int \frac{d^3k}{(2\pi)^3} \int \frac{d^3k'}{(2\pi)^3} b(k) e^{ikr} b(k') e^{ik'r} \tag{6.62}$$

Next, do

$$\nabla \times b(r) = \nabla \times \int \frac{d^3k}{(2\pi)^3} b(k) e^{ikr} \tag{6.63}$$

$$= \int \frac{d^3k}{(2\pi)^3} ik \times b(k) e^{ikr} \tag{6.64}$$

We want

$$(\nabla \times b)(\nabla \times b) = \int \frac{d^3k}{(2\pi)^3} \int \frac{d^3k'}{(2\pi)^3} e^{i(k+k')r} (k \times b(k))(k' \times b(k'))(-1) \tag{6.65}$$

Using the property of the delta function

$$\int d^3r e^{i(k+k')r} = 2\pi^3 \delta(k + k') \tag{6.66}$$

the energy integral now becomes

$$F = \frac{1}{2\mu_0} \int \frac{d^3k}{(2\pi)^3} \int \frac{d^3k'}{(2\pi)^3} \left\{ b(k)b(k') - \lambda^2 (k \times b(k))(k' \times b(k')) \right\}$$
$$\times (2\pi)^3 \delta(k + k')$$
$$= \frac{1}{2\mu_0} \int \frac{d^3k}{(2\pi)^3} \left\{ b(k)b(-k) + \lambda^2 (k \times b(k))(k \times b(-k)) \right\} \tag{6.67}$$

Finally

$$F = \frac{1}{2\mu_0} \int \frac{d^3k}{(2\pi)^3} b(k)b(-k) \left[1 + \lambda^2 k^2\right] \tag{6.68}$$

Here we need to insert $b(k)$ and $b(-k)$, which we find by Fourier-transforming the London–Abrikosov equation:

$$b + \lambda^2 \nabla \times (\nabla \times b) = \Phi_0 \delta_2(r - r_i) \tag{6.69}$$

which can be rewritten, with $\vec{\nabla} \cdot \vec{b} = 0$,

$$b - \lambda^2 \nabla^2 b = \Phi_0 \delta_2(r - r_i) \tag{6.70}$$

This is the equation to be solved for b in the presence of a single flux line, as is indicated by the right hand side of the equation: One flux quantum, and one delta-function. We will now proceed to generalize the problem to that of an arbitrary collection of flux lines, each of arbitrary form. This is an ambitious goal, but now we have the tools to do it. For such a collection of flux lines the right hand side has to sum up the contributions from all flux lines, and an integral must be taken along each line i to account for its contribution to the field at a point r. We now write

$$b + \lambda^2 \nabla \times (\nabla \times b) = \Phi_0 \sum_i \int dr_i \delta(r - r_i) \tag{6.71}$$

By standard manipulation of the left-hand side Eq. 6.71 becomes

$$b - \lambda^2 \nabla^2 b = \Phi_0 \sum_i \int dr_i \delta(r - r_i) \tag{6.72}$$

We need the Fourier transforms of b and $\nabla^2 b$:

$$b(r) = \int \frac{d^3k}{(2\pi)^3} e^{ikr} b(k)$$

$$\nabla^2 b = \int \frac{d^3k}{(2\pi)^3} b(k) \nabla^2 e^{ikr}$$

$$= \int \frac{d^3k}{(2\pi)^3} b(k)(-k^2) e^{ikr} \tag{6.73}$$

Inserting these in Eq. 6.72 we find

$$\int \frac{d^3k}{(2\pi)^3} e^{ikr} \left\{ b(k) + \lambda^2 k^2 b(k) \right\} = \Phi_0 \int \frac{d^3k}{(2\pi)^3} \sum_i \int dr_i e^{ik(r-r_i)} \tag{6.74}$$

On comparing integrands, we find

$$b(k) = \Phi_0 \sum_i \int dr_i \frac{e^{-ikr_i}}{1 + \lambda^2 k^2} \tag{6.75}$$

Next, on inserting $b(k)$ in Eq. 6.69 we get

$$F = \frac{1}{\mu_0} \int \frac{d^3k}{(2\pi)^3} b(k)b(-k)(1 + \lambda^2 k^2) \tag{6.76}$$

$$= \frac{\Phi_0^2}{2\mu_0} \int \frac{d^3k}{(2\pi)^3} \sum_{ij} \iint dr_i dr_j \frac{e^{ik(r_i - r_j)}}{(1 + \lambda^2 k^2)} \tag{6.77}$$

We write the result by means of an effective potential $V(r_i - r_j)$ as follows

$$F = \frac{\Phi_0^2}{2\mu_0} \sum_{ij} \oint \oint dr_i dr_j V(r_i - r_j) \tag{6.78}$$

where the interaction potential between two vortex segments is given by

$$V(r_i - r_j) = \int \frac{d^3k}{(2\pi)^3} \frac{e^{ik(r_i - r_j)}}{(1 + \lambda^2 k^2)}$$

$$= \frac{1}{4\pi\lambda^2|r_i - r_j|} \exp(-|r_i - r_j|/\lambda) \tag{6.79}$$

In Eq. 6.78, we are supposed to sum over all pairs of vortex segments dr_i and then integrate over all vortex segments along each and every vortex line. This therefore includes self interaction involving individual vortex lines whose various vortex segments interact with one another, as well as vortex segments on distinct vortex lines. Note that when $i = j$ (self-energy term), we should really prescribe a core cutoff $r = \sqrt{x^2 + y^2 + z^2} \rightarrow \sqrt{x^2 + y^2 + z^2 + \xi^2}$ in order to mimic the core-attraction term that cancels the formal divergence in $V(r)$ at small distances. The symbol $\oint$ means that the line-integrations along vortex lines are supposed to be carried out either over vortex lines that completely penetrate the sample, or along closed vortex loops of the system. This follows from the fact that vortex lines cannot start or stop inside the superconducting sample. Note also that the interaction vanishes when $dr_i \perp dr_j$, exhibiting explicitly the vectorial nature of the vortex interaction. An immediate consequence of this is that when parallel vortex lines are pushed close to each other, there is a natural tendency for them to bend and twist in a response to this, since this will lower the interaction energy of the system. This bending comes at the cost of increasing the self energy of the vortex lines, since there is also a line tension in the problem. Nonetheless, we conclude that in an ensemble of interacting vortex lines, approximating them as rigid objects may not be quite adequate.

The above results apply to the spatially isotropic case, i.e. the case when the penetration length and the coherence length do not depend on which direction in the sample they are measured along. An important situation which differs from this is the case of uniaxial anisotropy, which must be used for the case when we have a system of superconducting layers stacked on top of each other with a relatively weak Josephon coupling between the layers. This is a necessary modification for describing the vortex system in for instance high-temperature superconductors. It is a somewhat laborious, although straightforward, task to go through the above steps again including uniaxial anisotropy, and we will limit ourselves to giving a brief outline of the derivation of the results here. Enough details are provided for the reader to fill in the missing pieces, the exposition follows that of Sudbø and Brandt [73, 74]. The physical picture of the vortex system is the same as for the isotropic case, however the result differs in important details. We denote as the $\hat{c}$-axis the direction of the uniaxial anisotropy, while the plane perpendicular to this symmetry axis is called the ab-plane. In this case, we introduce two penetration lengths λ_{ab} and λ_c, which are the penetration depths for currents in the ab-plane and along the $\hat{c}$-axis, respectively. Using these definitions, the analogue of Eq. 6.20 reads

$$F = \frac{1}{2\mu_0} \int \left[\mathbf{b}^2 + (\nabla \times \mathbf{b})\Lambda(\nabla \times \mathbf{b}) \right] \tag{6.80}$$

where Λ is a tensor with components $\Lambda_{\alpha\beta} = \Lambda_1 \delta_{\alpha\beta} + \Lambda_2 c_\alpha c_\beta$, where c_α denotes the Cartesian components of the $\hat{c}$ unit vector along the crystalline α-axis, $\alpha \in (x, y, z)$. Here, we have defined $\Lambda_1 = \lambda_{ab}^2$, while $\Lambda_2 = \lambda_c^2 - \lambda_{ab}^2$. The analog og Eq. (6.21) is then given by

$$\mathbf{b} + \nabla \times [\Lambda(\nabla \mathbf{b})] = \Phi_0 \sum_i \oint d\mathbf{r}_i \delta(\mathbf{r} - \mathbf{r}_i) \tag{6.81}$$

Following the steps for the isotropic case by Fourier transforming Eq. 6.81, solving for $\mathbf{b}$ and inserting the result in Eq. 6.80, we arrive at an expression very much like Eq. 6.78. The solution for the $\mathbf{b}$-field in this case reads

$$b_\alpha(\mathbf{r}) = \Phi_0 \sum_i \oint dr_i^\beta V_{\alpha\beta}(\mathbf{r} - \mathbf{r}_i)$$

$$V_{\alpha\beta}(\mathbf{r}) = \int \frac{d^3k}{(2\pi)^3} V_{\alpha\beta}(\mathbf{k}) \exp(i\vec{k} \cdot \vec{r})$$

$$V_{\alpha\beta}(\mathbf{k}) = \frac{1}{1 + \Lambda_1 k^2} \left(\delta_{\alpha\beta} - \frac{\Lambda_2 q_\alpha q_\beta}{1 + \Lambda_1 k^2 + \Lambda_2 q^2} \right] \tag{6.82}$$

where we have introduced the auxiliary vector $q = k \times \hat{c}$. Inserting this solution into the expression Eq. 6.80, we arrive at

$$F = \frac{\Phi_0^2}{2\mu_0} \sum_{ij} \oint \oint \mathrm{d}r_{\alpha i}\, \mathrm{d}r_{\beta j}\, V_{\alpha\beta}(r_i - r_j) \tag{6.83}$$

Here, $\mathrm{d}r_{\alpha i}$ denotes the Cartesian α-component of a vortex segment. Note how the interaction now has become tensorial, rather than vectorial, as it was in Eq. 6.78.

The hard part of the calculation is to compute the necessary integrals required for expressing $V_{\alpha\beta}$ in real-space, see the second of the expressions in Eq. 6.82. This is done by noting that we may write V as $V_{\alpha\beta}(r) = V_1(r)\delta_{\alpha\beta} + V_2^{\alpha\beta}(r)$, where V_1 is the only term that survives in the isotropic limit $\Lambda_2 = 0$, and is obtained from the first term in $V_{\alpha\beta}(k)$. V_1 is essentially given by Eq. 6.79, with $\lambda = \lambda_{ab}$. To compute $V_2^{\alpha\beta}(r)$, we consider first the auxiliary integral

$$I_0 = \frac{\Lambda_2}{(2\pi)^3} \int \mathrm{d}^3 k \frac{e^{ikr}}{(1 + \Lambda_1^2 k^2)[1 + (\Lambda_1 + \Lambda_2)k_\perp^2 + \Lambda_1 k_z^2]} \tag{6.84}$$

On performing the azimuthal and $k_\perp$ integrations, we get

$$I_0 = \frac{2}{(2\pi)^2} \int_0^\infty \mathrm{d}k_z \cos(k_z) \frac{K_0(z_1) - K_0(z_2)}{1 + \Lambda_1 k_z^2} \equiv I_0^1 - I_0^2 \tag{6.85}$$

Here, we have introduced $z_i = \alpha_i \sqrt{1 + \Lambda_1 k_z^2}$, $\alpha_i = c_i \sqrt{x^2 + y^2}$, $c_1 = 1/\sqrt{\Lambda_1 + \Lambda_2}$, $c_2 = 1/\sqrt{\Lambda_1}$, and K_0 is a modified Bessel function of zeroth order. Doing the necessary integrations over these Bessel functions does not appear to be straightforward, but fortunately is not necessary either. To see this, note that in terms of I_0, we have

$$V_2^{\alpha\beta}(r) = \left[\left(\partial_x^2 + \partial_y^2\right)\delta_{\alpha\beta} - \partial_x \partial_y\right] I_0 \tag{6.86}$$

where $\partial_x = \partial/\partial x$ etc. Now, introduce the auxiliary quantity

$$\Gamma_i \equiv \frac{\partial I_0^i}{\partial \alpha_i} \tag{6.87}$$

In terms of this we have, using the chain rule for differentiation

$$\partial_\alpha \partial_\beta I_0^i = \frac{\partial \Gamma_1}{\partial \alpha_1} \partial_\alpha \alpha_1 \partial_\beta \alpha_1 + \Gamma_1 \partial_\alpha \partial_\beta \alpha_1$$
$$- \frac{\partial \Gamma_2}{\partial \alpha_1} \partial_\alpha \alpha_2 \partial_\beta \alpha_2 - \Gamma_2 \partial_\alpha \partial_\beta \alpha_2 \tag{6.88}$$

Therefore, we do not need to compute I_0^i, but only $\partial I_0^i/\partial\alpha_i$. After a bit of lengthy, but straightforward algebra, this is found to be given by

$$\Gamma_i = \frac{\partial I_0^i}{\partial\alpha_i} = -\sqrt{\frac{\pi\tilde{z}_i}{2\Lambda_1}}\frac{1}{\alpha_i}K_{1/2}(\tilde{z}_i) \qquad (6.89)$$

where $K_{1/2}(x) = \sqrt{\pi/2x}\exp(-x)$ is a modified Bessel function of order $1/2$, and $\tilde{z}_i = \sqrt{(\alpha_i^2\Lambda_1 + z^2)/\Lambda_1}$.

Performing the necessary derivatives on Γ_i and α_i, and collecting all of the above, we find that $V_{\alpha\beta}(r)$ is given by

$$V_{\alpha\beta}(r) = V_1(r)\delta_{\alpha\beta} + V_2^{\alpha\beta}(r); \; (\alpha,\beta) \in (x,y,z)$$

$$V_1(r) = \frac{1}{4\pi\lambda_{ab}^2 r}\exp(-\tilde{r})$$

$$V_2^{\alpha\beta}(r) = \frac{1}{4\pi\lambda_{ab}^2\rho}\left[G_1(r)\delta_{\alpha\beta} + G_2(r)\frac{x_\alpha x_\beta}{\rho^2}\right]; \; (\alpha,\beta) \in (x,y) \quad (6.90)$$

We have $V_2^{zz}(r) = V_2^{xz}(r) = V_2^{yz}(r)$, when the average vortex direction is along the $\hat{c}$-axis. The functions G_i are given by

$$G_i(r) = a_i\exp(-\tilde{r}) - b_i\exp(-\tilde{\rho})$$

where $\tilde{r} = r/\lambda_{ab}$, $\tilde{\rho} = (\rho^2 + \Gamma^2 z^2)^{1/2}/\lambda_c$, $a_1 = 1 - a_2$, $b_1 = 1 - b_2$, $a_2 = 2 + \rho^2/\lambda_{ab}^2\tilde{r}$, $b_2 = 2 + \rho^2/\lambda_c^2\tilde{\rho}$, $r^2 = \rho^2 + z^2$, and $\rho^2 = x^2 + y^2$, Note how this reduces to Eq. 6.78 when $\lambda_{ab} = \lambda_c$, since this implies that $a_2 = b_2$, $a_1 = b_1$, and $\tilde{\rho} = \tilde{r}$. This in turn means that $G_1 = G_2$ such that $V_2 = 0$, and hence $V_{\alpha\beta}$ reduces to a vectorial interaction, as in the isotropic case.

Again, when flux lines with finite core radii are in close contact, Eq. 6.90 should be supplemented by a scalar core attraction, which we for simplicity may mimic by using a cutoff in the potential $V_{\alpha\beta}$, achieved by replacing r by $\sqrt{r^2 + \xi_{ab}^2}$, where ξ_{ab} is the ab-plane vortex-core radius.

The above assumes that the flux lines on average are directed along the z-axis, i.e. that the external magnetic field is directed along the z-axis. In fact, it is possible to generalize the above even further, to the case where the external magnetic field forms an arbitrary angle θ with the z-axis. The generalizations were first obtained by Sardella in 1992 and only published much later [75], and discovered independently by Nguyen [76]. The coordinate z-axis is oriented along the external magnetic field, while the direction of the crystal $\hat{c}$-axis forms an angle θ with the z-axis. The (x, y)-plane is perpendicular to the z-axis. Deriving the analogue of Eq. 6.90 for the tilted case follows the same path as above (the details may be found in Appendix A of Ref. [76]). The result is a new potential given by

$$V_{\alpha\beta}(r) = \frac{1}{4\pi\lambda_{ab}^2 (\mathbf{r} \times \hat{c})^2} \left[G_1(\mathbf{r})(\delta_{\alpha\beta} - c_\alpha c_\beta) - G_2(\mathbf{r}) \frac{(\mathbf{r} \times \hat{c})_\alpha (\mathbf{r} \times \hat{c})_\beta}{(\mathbf{r} \times \hat{c})^2} \right] \quad (6.91)$$

where we have defined the two auxiliary functions

$$G_1(\mathbf{r}) = e^{-r/\lambda_{ab}} - e^{-\tilde{r}/\lambda_c} \quad (6.92)$$

and

$$G_2(\mathbf{r}) = \left[2 + \frac{(\mathbf{r} \times \hat{c})^2}{\lambda_{ab} r} \right] e^{-r/\lambda_{ab}} - \left[2 + \frac{(\mathbf{r} \times \hat{c})^2}{\lambda_c \tilde{r}} \right] e^{-\tilde{r}/\lambda_c} \quad (6.93)$$

Here, we have introduced $\tilde{r} = \sqrt{(\mathbf{r} \times \hat{c})^2 + (\lambda_{ab}/\lambda_c)^2 (\mathbf{r} \times \hat{c})^2}$. Moreover, c_α is the projection of $\hat{c}$ onto the αth (x, y, z) axis.

This is as far as we can go without specifying the flux line geometry and configuration. These results can be put to good use, as we shall see several examples of now.

6.4 Self energy of a single straight flux line in the London approximation

The result just arrived at includes all possible interactions in the present limit. It cannot be reduced further on analytic form without specifying some geometric constraint. A normal one would be to assume a perfectly regular lattice consisting of ordered, straight flux lines. We shall reduce it to a single, straight flux line. This is a problem we already solved, but here we follow a more general approach. The quantities $d\mathbf{r}_j$ and $d\mathbf{r}_i$ are now to be found at two different positions $\mathbf{r}_j$ and $\mathbf{r}_i$ *on the same flux line*. Our form for F can now be used to sum up the self-energy due to the interaction between these pieces of the flux line (see Figure 6.6). Let us reduce the sum for F in steps:

$$F = \frac{\Phi_0^2}{2\mu_0} \sum_{ij} d\mathbf{r}_i \, d\mathbf{r}_j V(\mathbf{r}_i - \mathbf{r}_j) \qquad \text{General expression} \quad (6.94)$$

$$\rightarrow \frac{\Phi_0^2}{2\mu_0} \int d\mathbf{r}_i \, d\mathbf{r}_j V(\mathbf{r}_i - \mathbf{r}_j) \qquad \text{One flux line} \sum_{ij} = 1 \quad (6.95)$$

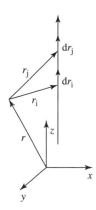

Figure 6.6   Coordinates used in the calculation of the self energy of a single flux line.

$$\rightarrow \frac{\Phi_0^2}{2\mu_0} \int dz_i \, dz_j \, V(z_i - z_j) \qquad \text{One straight flux line along the } z \text{-axis}$$

(6.96)

$$\rightarrow \frac{\Phi_0^2}{2\mu_0} L_z \int dz \, V(z) \qquad \begin{array}{l} \text{after substitution } z_i - z_j = z \\ \text{and integration over one vari-} \\ \text{able.} \end{array} \qquad (6.97)$$

We can now write

$$\begin{aligned}
\frac{F}{L_z} &= \frac{\Phi_0^2}{2\mu_0} \int \frac{d^2k}{(2\pi)^3} \frac{1}{1 + \lambda^2 k^2} \int dz^{ik_z z} \\
&= \frac{\Phi_0^2}{2\mu_0} \int \frac{d^3k}{(2\pi)^3} \frac{1}{1 + \lambda^2 k^2} \times 2\pi \delta(k_z) \\
&= \frac{\Phi_0^2}{2\mu_0} \int \frac{dk_x}{2\pi} \int \frac{dk_y}{2\pi} \frac{1}{1 + \lambda^2 (k_x^2 + k_y^2)}
\end{aligned} \qquad (6.98)$$

To make this a finite energy we must introduce a cut-off so that $r > \xi$. This corresponds to a cut-off $k_\perp < 1/\xi$. The integration over k_x and k_y is simplified by transforming to polar coordinates.

$$\begin{aligned}
\frac{F}{L_z} &= \frac{\Phi_0^2}{2\mu_0} \frac{1}{2\pi} \int_0^{\frac{1}{\xi}} dk \frac{2\pi k}{1 + \lambda^2 k^2} \\
&= \frac{\Phi_0^2}{2\mu_0} \frac{1}{(2\pi)^2} \int_0^{\frac{\lambda^2}{\xi^2}} \frac{du}{2\lambda^2 k} \frac{2\pi k}{1 + u} \qquad \text{where} \quad u = \lambda^2 k^2
\end{aligned} \qquad (6.99)$$

The final result is

$$\frac{F}{L_z} = \frac{\Phi_0^2}{4\pi\mu_0\lambda^2}\ln\kappa \quad \kappa = \frac{\lambda}{\xi} \tag{6.100}$$

The problem is now solved for the free energy per length of a straight flux line in the London approximation for the isotropic case.

6.5   Interaction between two parallel flux lines

We discuss next a simple application of the expression for the interaction energy between flux lines, specified to two parallel, straight flux lines along the z-axis, with flux line no 1 at $r_1 = (x_1, y_1, z_1)$, and no 2 at $r_2 = (x_2, y_2, z_2)$, as shown in Figure 6.7. This is an important example, as it can be viewed as the beginning stage of studying the forces at work in the flux line lattice. We already derived the general energy expression we have to start from; which we now specify for two parallel flux lines:

$$U_{12} = \frac{\Phi_0^2}{\mu_0} \iint dr_1 dr_2 V(r_1 - r_2) \tag{6.101}$$

$$V(r_1 - r_2) = \int \frac{d^3k}{(2\pi)^3} \frac{e^{ik(r_1-r_2)}}{1 + \lambda^2 k^2} \tag{6.102}$$

$V(r_1 - r_2)$ is the potential energy between two arbitrary line elements dr_1 and dr_2, and is valid whatever configuration the two flux lines assume.

Carrying out the integral (Eq. 6.102) gives

$$V(r_1 - r_2) = \frac{1}{2\pi\lambda^2} \frac{e^{-r/\lambda}}{r} = \frac{1}{2\pi\lambda^2} \frac{e^{-|r_1-r_2|/\lambda}}{|r_1 - r_2|} \tag{6.103}$$

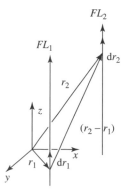

Figure 6.7   Coordinates used in calculating the interaction energy between two parallel, straight flux lines.

This is essentially the well-known Yukawa potential. Since the vortices are parallel to the z-axis, we have $d\mathbf{r}_1 = dz_1\hat{k}$; $d\mathbf{r}_2 = dz_2\hat{k}$ where $\hat{k}$ is a unit vector along the positive z-axis. We introduce for the distance $|\mathbf{r}_1 - \mathbf{r}_2|$:

$$|\mathbf{r}_1 - \mathbf{r}_2| = \left[(x_1 - x_2)^2 + (y_1 - y_2)^2 + (z_1 - z_2)^2\right]^{\frac{1}{2}}$$

$$= \left[\rho^2 + (z_1 - z_2)^2\right]^{\frac{1}{2}} \equiv \left[\rho^2 + z_2'^2\right]^{\frac{1}{2}} \tag{6.104}$$

where $z_2' = z_2 - z_1$, and $\rho^2 = (x_1 + x_2)^2 + (y_1 - y_2)^2$. Inserting this expression into the integrand of Eq. 6.101 results in

$$U_{12} = \frac{\Phi_0^2}{\mu_0} \int_{-\infty}^{\infty} dz_1 \int_{-\infty}^{\infty} dz_2' \times \frac{1}{2\pi\lambda^2} \frac{e^{-(\rho^2 + z_2'^2)^{\frac{1}{2}}/\lambda}}{(\rho^2 + z_2'^2)}$$

$$= \frac{\Phi_0^2}{2\pi\mu_0\lambda^2} \int_{-\infty}^{\infty} dz_1 \int_{-\infty}^{\infty} dz_2' \frac{e^{-(\rho^2 + z'^2)^{\frac{1}{2}}/\lambda}}{(\rho^2 + z_2'^2)} \tag{6.105}$$

The integral over dz_1 gives the length of vortex line no 1. We divide it over to the left-hand side and get the energy pr unit length U_{12}. The integral over z_2 gives

$$U_{12} = \frac{\Phi_0^2}{2\pi\mu_0\lambda^2} K_0\left(\frac{\rho}{\lambda}\right) \tag{6.106}$$

where ρ is the distance between the flux lines in the (x,y) plane, and $K_0(\frac{\rho}{\lambda})$ is a modified Bessel function. We find that the result can be written as the product of the flux Φ_0 times the field $h(\rho)$ at distance ρ.

$$U_{12} = \Phi_0 h(\rho); \quad h(\rho) = \frac{1}{\mu_0} b(\rho) \tag{6.107}$$

The total energy of two interacting, parallel flux lines is therefore:

$$U_{12}^{\text{tot}} = 2F + \Phi_0 h(\rho) \tag{6.108}$$

We can now find the force between the two filaments in the x-direction:

$$f_{2,x} = -\frac{\partial U_{12}}{\partial x_2} - \frac{\partial(2F)}{\partial x_2} = -\Phi_0 \frac{\partial h(\rho)}{\partial \rho} \tag{6.109}$$

The interaction force is repulsive. If we introduce the current $\boldsymbol{j_y}$, by Maxwell's equation; $\boldsymbol{j}_y = -\hat{j}\frac{\partial}{\partial\rho}h$, where $\hat{j}$ is the unit vector along y, we find

$$f_{2,x} = j_y\Phi_0 \tag{6.110}$$

We notice in particular that the force in the x-direction on a flux line is zero only when the superfluid current (and velocity) at that point is $j_y = 0$. This corresponds well with the Lorentz force on a charge e, $\boldsymbol{F} = e\boldsymbol{v} \times \boldsymbol{B}$. The whole situation of two interacting vortices can be easily illustrated by letting the currents and fields of the two overlap. One finds repulsive interaction when the vortices are parallel, and attractive when they are anti-parallel as is also implied by the mathematical analysis. Interestingly, when two flux lines of opposite rotation are made to coincide, they annihilate, since the sum of all currents must everywhere be zero. Clearly, the situation of anti-parallel flux lines is an extremely unstable one. See also Chapter 8, Figure 8.1.

6.6   Interaction between two flux lines at angle α

In high-T_c materials the consensus is that a flux line liquid exists above a certain flux line lattice (FLL) melting temperature T_m. A situation, which has been predicted to exist between T_m and $T_c(H)$ is that of an entangled flux line fluid. The behaviour of this fluid will be quite dependent on whether the flux lines are stuck in their entanglement, or are able to disentangle and straighten out. The answer to this question should be sought in the flux-line–flux-line interaction energy. In the case of parallel flux lines there is distinct repulsion, as we have seen in Section 6.5. But in a liquid state we can envisage flux lines which are softly bending and interwoven between each other, as shown in Figure 6.8. The flux lines at positions of closest spatial approach interact at some angle α, which is different from one place to another. One interesting question which arises here

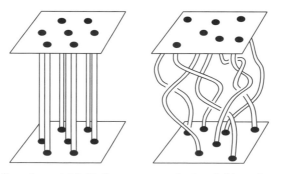

Figure 6.8 Flux lines in a rigid Abrikosov vortex lattice (left), and a putative entangled flux-line system in the molten phase (right).

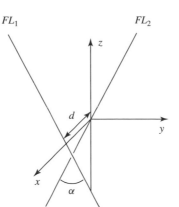

Figure 6.9 Coordinates for calculation of interaction between two flux lines at an angle α. Both flux lines are located in the (y, z)-plane, one displaced a distance d along the x-axis from the other. Hence the distance of closest approach is d. The angle they are splayed is denoted by α, in the present case the splaying is symmetric around the z-axis.

is the following: What is the energy cost of moving two nearly intersecting flux lines across each other to disentangle? This situation may be closely modeled by letting two flux lines be placed at a crossing angle α, like in Figure 6.9.

To calculate the interaction energy we can again revert to the general formalism employed before. The difference between this case and the one treated in Section 6.5, is that the interacting line elements are at an angle α, and at a distance d, which means that the line elements $d\boldsymbol{r}_1$ and $d\boldsymbol{r}_2$ are at such angle, wherever their location is on line 1 and line 2. The integral to work out is now:

$$U_{1,2}(\alpha) = \int_{FL_1} d\boldsymbol{r}_1 \int_{FL_2} d\boldsymbol{r}_2 V(\boldsymbol{r}_1 - \boldsymbol{r}_2) \qquad (6.111)$$

The integration over the variables in parallel position was carried out in Section 6.5. Taking advantage of that integration result, and inserting the $\cos \alpha$ for the vector product of $d\boldsymbol{r}_1 \times d\boldsymbol{r}_2$, where $|d\boldsymbol{r}_2| = dy/\sin\alpha$, and after dividing by the length of line 1 on both sides, we get

$$U_{1,2}(\alpha, d) = \frac{\Phi_0^2}{8\pi^2\lambda^2} \int_{-\infty}^{\infty} \frac{dy}{\sin\alpha} \cos\alpha K_0\left(\frac{(d^2+y^2)^{\frac{1}{2}}}{\lambda}\right) \qquad (6.112)$$

where K_0 is again the same Bessel function encountered in Section 6.5. We find

$$U_{1,2}(\alpha, d) = \frac{\Phi_0^2}{8\pi\lambda^2} \cot\alpha\, e^{-|d|/\lambda} \qquad (6.113)$$

In fact, if we consider the fact that the systems also may be uniaxially anisotropic, then we may parametrize this in the London model by introducing two penetration lengths, namely λ_{ab} which is the penetration length in the (a, b)-plane $((x, y)$-plane) for a magnetic field oriented along the c-axis (z-axis), and λ_c which is the penetration length along the (a, b)-plane for a screening current along the c-axis. Eq. 6.112 then generalizes to

$$U_{1,2}(\alpha, d) = \frac{\Phi_0^2}{4\pi\lambda^2}\left[\cot(\alpha/2)e^{-|d|/\lambda_{ab}} - \tan(\alpha/2)e^{-|d|/\lambda_c}\right] \qquad (6.114)$$

as was shown by Sudbø and Brandt [77]. This is seen to reduce to Eq. 6.113 when $\lambda_{ab} = \lambda_c$. Note how the angle at which flux lines do not interact (electromagnetically) in fact increases when the anisotropy ratio λ_c/λ_{ab} increases. For the isotropic case, the interaction changes sign at $\alpha = \pi/2$. When $\lambda_c/\lambda_{ab} \to \infty$, the interaction never changes sign, but approaches zero as $\alpha \to \pi$. The *neutrality angle* where the electromagnetic interaction between two rigid flux lines vanishes therefore increases with anisotropy.

6.7   General flux-line lattice elastic matrix in the London approximation

As a final application of Eqs 6.83 and 6.91, let us use them to derive an expression for the general elastic fluctuation matrix of a flux-line lattice in an arbitrarily tilted external magnetic field, valid beyond the continuum approximation for the flux-line system. This means that we explicitly take into account the underlying Abrikosov lattice structure when deriving the fluctuation Hamiltonian for flux lines around their ground state configuration. The continuum approximation is a special case of the result we will derive, and as a byproduct we will obtain expression for the flux-line *liquid* elastic moduli in this limit when the external magnetic field is tilted an arbitrary angle with respect to the crystal $\hat{c}$-axis. The exposition here closely follows that of Sardella [78] and Nguyen and Sudbø [76]. The starting point is Eq. 6.83

$$F = \frac{\Phi_0^2}{2\mu_0}\sum_{ij}\oint\oint d\mathbf{r}_{\alpha i}d\mathbf{r}_{\beta j}V_{\alpha\beta}(\mathbf{r}_i - \mathbf{r}_j) \qquad (6.115)$$

We now rewrite this by being slightly more explicit in writing the line integrals that are involved. The line integrals are along directed flux lines, the average direction of the flux lines is the z-axis, and hence we have $d\mathbf{r}_{\alpha i} = dz(d\mathbf{r}_{\alpha i}/dz)$. We now insert this into Eq. 6.115 and obtain

$$F = \frac{\Phi_0^2}{2\mu_0}\sum_{ij}\int dz\int dz'\frac{d\mathbf{r}_{\alpha i}}{dz}V_{\alpha\beta}(\mathbf{r}_i(z) - \mathbf{r}_j(z'))\frac{d\mathbf{r}_{\beta j}}{dz'} \qquad (6.116)$$

where the information on the shape of the flux lines is encoded in $d\mathbf{r}_{\beta j}/dz$. We now imagine that the ground state of this system is the Abrikosov flux-line lattice, and expand the free energy in fluctuations around this ground state, keeping terms up to quadratic order in the displacements $\mathbf{s}_j(z)$ around the ground state, where we have

$$\mathbf{r}_i(z) = \mathbf{R}_j + \mathbf{s}_j(z) \tag{6.117}$$

where $\mathbf{R}_j$ denotes the positions of flux lines in the ground state Abrikosov vortex lattice, and $\mathbf{s}_j(z)$ denotes the fluctuations around these positions. The first thing to notice is that $d\mathbf{r}_{\beta j}(z)/dz = 1$ if $\beta = z$, and $d\mathbf{r}_{\beta j}(z)/dz = d\mathbf{s}_j(z)/dz$ if $\beta = (x, y)$. Let us now expand the tensor $V_{\alpha\beta}(\mathbf{r}_i(z) - \mathbf{r}_j(z'))$ up to quadratic order in the displacements $\mathbf{s}_j(z)$, obtaining

$$
\begin{aligned}
V_{\alpha\beta}(\mathbf{r}_i(z) - \mathbf{r}_j(z')) = {} & V_{\alpha\beta}(\mathbf{R}_i - \mathbf{R}_j) \\
& + [\mathbf{s}_i(z) - \mathbf{s}_j(z')]\nabla V_{\alpha\beta}(\mathbf{r})_{\mathbf{r}=\mathbf{R}_i-\mathbf{R}_j} \\
& + \frac{1}{2}[\mathbf{s}_i(z) - \mathbf{s}_j(z')]_n[\mathbf{s}_i(z) - \mathbf{s}_j(z')]_m \\
& \times \nabla_n \nabla_m V_{\alpha\beta}(\mathbf{r})_{\mathbf{r}=\mathbf{R}_i-\mathbf{R}_j} + \cdots
\end{aligned}
\tag{6.118}
$$

Inserting this back into the expression for the energy, Eq. 6.116, we obtain, upon collecting terms up to quadratic order and noticing that terms that are linear by necessity must vanish since we are expanding around a minimum, we have for the elastic energy $\Delta F = F - F_0$, where F_0 is the total energy of the ground state Abrikosov flux-line lattice (NB We must keep terms that are linear in displacements when expanding the tensor $V_{\alpha\beta}(\mathbf{r}_i(z) - \mathbf{r}_j(z'))$!)

$$
\begin{aligned}
\Delta F = {} & \frac{\Phi_0^2}{2\mu_0} \sum_{ij} \int dz \int dz' \left(\frac{ds_{\alpha i}}{dz} \frac{ds_{\beta j}}{dz'} V_{\alpha\beta}(\mathbf{R}_i - \mathbf{R}) \right. \\
& + \frac{1}{2}[\mathbf{s}_i(z) - \mathbf{s}_j(z')]_n[\mathbf{s}_i(z) - \mathbf{s}_j(z')]_m \nabla_n \nabla_m V_{\alpha\beta}(\mathbf{r})_{\mathbf{r}=\mathbf{R}_i-\mathbf{R}_j} \\
& \left. + 2\frac{ds_{\alpha i}}{dz}[\mathbf{s}_i(z) - \mathbf{s}_j(z')]_\beta \nabla_\beta V_{z\alpha}(\mathbf{r})_{\mathbf{r}=\mathbf{R}_i-\mathbf{R}_j} \right)
\end{aligned}
\tag{6.119}
$$

where $(\alpha, \beta) \in (x, y)$. The next step is to express this in terms of the Fourier-modes of the displacement vectors $\mathbf{s}_i(z)$. Now, because we are assuming an underlying Abrikosov lattice which gives rise to a Brillouin zone, the wave-vectors of the displacement vectors are only defined up to an arbitrary reciprocal lattice vector $\mathbf{Q}$. Hence, when we Fourier-expand the displacement vectors $\mathbf{s}_j(z)$, we express the Fourier transform through a sum over reciprocal lattice vectors

Q of the Abrikosov vortex lattice and an integral over the wavevector **k** that runs over the *first* Brillouin zone. This means that we can write

$$\Delta F = \frac{1}{2} \sum_{\mathbf{k}} s_\alpha(-\mathbf{k}) \Phi_{\alpha\beta}(\mathbf{k}) s_\beta(\mathbf{k}) \tag{6.120}$$

which defines the elastic matrix $\Phi_{\alpha\beta}(\mathbf{k})$. In the above expression, **k** runs over the first Brillouin zone. Such an expression is arrived at by partial integrations in Eq. 6.119, and then introducing the Fourier transforms of $s_j(z)$ as follows

$$s_i(z) = \int \frac{d^3k}{(2\pi)^3} s_\beta(\mathbf{k}) \, e^{i\mathbf{k}\mathbf{R}_i} \tag{6.121}$$

Moreover, because of the underlying Abrikosov flux-line lattice we must take into account that the wavevectors **k** are only defined up to reciprocal lattice vectors of the Abrikosov lattice, whence

$$\sum_i e^{i\mathbf{k}\mathbf{R}_i} = (2\pi)^2 \frac{B}{\Phi_0} \sum_{\mathbf{Q}} \delta_2(\mathbf{k} - \mathbf{Q}) e^{ik_z z} \tag{6.122}$$

where $\delta_2(\mathbf{k} - \mathbf{Q})$ is a two-dimensional Dirac δ-function. Inserting all of this into Eq. 6.119, we find Eq. 6.120 with

$$\Phi_{\alpha\beta}(\mathbf{k}) = \frac{B^2}{2\mu_0} \sum_{\mathbf{Q}} \left[f_{\alpha\beta}(\mathbf{K}) - f_{\alpha\beta}(\mathbf{Q}) \right] \tag{6.123}$$

where we have introduced the function

$$f_{\alpha\beta}(\mathbf{k}) = k_z^2 \tilde{V}_{\alpha\beta}(\mathbf{k}) + k_\alpha k_\beta \tilde{V}_{zz}(\mathbf{k}) - k_z k_\beta \tilde{V}_{z\alpha}(\mathbf{k}) - k_z k_\alpha \tilde{V}_{z\beta}(\mathbf{k}) \tag{6.124}$$

and where $\tilde{V}_{\alpha\beta}(\mathbf{k})$ is the Fourier transform of Eq. 6.91, namely

$$\tilde{V}_{\alpha\beta}(\mathbf{k}) = \frac{1}{1 + \Lambda_1 k^2} \left[\delta_{\alpha\beta} - \frac{\Lambda_2 q_\alpha q_\beta}{1 + \Lambda_1 k^2 + \Lambda_2 q^2} \right] \tag{6.125}$$

In Eq. 6.123, we introduced $\mathbf{K} = \mathbf{Q} + \mathbf{k}$ and moreover symmetrized the tensor $f_{\alpha\beta}(\mathbf{k})$. From Eq. 6.119, as it stands, we would have obtained

$$f_{\alpha\beta}(\mathbf{k}) = k_z^2 \tilde{V}_{\alpha\beta}(\mathbf{k}) + k_\alpha k_\beta \tilde{V}_{zz}(\mathbf{k}) - 2k_z k_\beta \tilde{V}_{z\alpha}(\mathbf{k}) \tag{6.126}$$

However, for numerical purposes, when computing elastic modes of the flux-line lattice, it is often convenient to use the symmetrized form since it guarantees positive definite eigenvalues of the elastic matrix. The expression Eq. 6.123 is very useful, among other things it directly lends itself to a computation of the elastic shear moduli of the Abrikosov flux-line lattice, of which there are several when the magnetic field is tilted away from the $\hat{c}$-axis.

The continuum limit of the elastic description of the Abrikosov flux-line lattice is obtained by omitting all terms in the sum over $\mathbf{Q}$ except $\mathbf{Q} = 0$. Since the underlying lattice is not taken into account, this is basically an elastic description of a flux-line liquid, so the shear moduli are not accessible in this approximation. The bulk and tilt moduli are, however. Let us see how we can extract them from what we have derived above. In the flux-line liquid limit, taking only the $\mathbf{Q} = 0$ terms and using that $f_{\alpha\beta}(\mathbf{k} = 0) = 0$, from Eqs 6.120 and 6.123 we simply have the elastic free energy given by

$$\Delta F = \frac{B^2}{2\mu_0} \sum_{\mathbf{k}} s_\alpha(-\mathbf{k}) f_{\alpha\beta}(\mathbf{k}) s_\beta(\mathbf{k}) \tag{6.127}$$

When the external magnetic field is tilted away from the $\hat{c}$-axis by an angle θ we have *four* distinct elastic moduli of the flux-line liquid, such that the elastic description on symmetry grounds is given by the expression

$$\Delta F = \frac{1}{2} \sum_{\mathbf{k}} \left[c_{11}(\mathbf{k})(\mathbf{ks})^2 + c_{44}^\perp(\mathbf{k})(k_z s_x)^2 + c_{44}^\parallel(\mathbf{k})(k_z s_y)^2 \right.$$
$$\left. + 2c_{14}(\mathbf{k})(k_z s_x)(\mathbf{ks}) \right] \tag{6.128}$$

The elastic moduli that enter her can be obtained by comparing Eq. 6.128 with Eq. 6.127, and they are found to be given by Refs [73, 76, 78]

$$c_{44}^\perp(\mathbf{k}) = \frac{B^2}{4\mu_0} \frac{1 + [\lambda_{ab}^2 + (\lambda_c^2 - \lambda_{ab}^2)\sin^2\theta]k^2}{[1 + \lambda_{ab}^2 k^2][1 + \lambda_{ab}^2 k^2 + (\lambda_c^2 - \lambda_{ab}^2)q^2]}$$

$$c_{44}^\parallel(\mathbf{k}) = \frac{B^2}{4\mu_0} \frac{1}{1 + \lambda_{ab}^2 k^2 + (\lambda_c^2 - \lambda_{ab}^2)q^2}$$

$$c_{11}(\mathbf{k}) = \frac{B^2}{4\mu_0} \frac{1 + [\lambda_{ab}^2 + (\lambda_c^2 - \lambda_{ab}^2)\cos^2\theta]k^2}{[1 + \lambda_{ab}^2 k^2][1 + \lambda_{ab}^2 k^2 + (\lambda_c^2 - \lambda_{ab}^2)q^2]}$$

$$c_{14}(\mathbf{k}) = \frac{B^2}{4\mu_0} \frac{(\lambda_c^2 - \lambda_{ab}^2)\sin\theta\cos\theta k^2}{[1 + \lambda_{ab}^2 k^2][1 + \lambda_{ab}^2 k^2 + (\lambda_c^2 - \lambda_{ab}^2)q^2]} \tag{6.129}$$

Here, the set of moduli given by $c_{44}^\perp(\mathbf{k})$ and $c_{44}^\parallel(\mathbf{k})$ are known as tilting moduli. When the flux lines on average are directed away from the symmetry $\hat{c}$-axis, there appear two tilting moduli in the problem. The reason is that when the system is layered and the z-axis is tilted an angle θ away from the $\hat{c}$-axis, we have a hard and a soft tilting mode depending on whether the tilting away from the z-axis is parallel to the crystal ab-plane, or out of the crystal ab-plane.

(Similar effects are present in the shear moduli of the flux-line lattice, which do not concern us here since we are considering the flux-line limit). Notice that for $\theta = 0$, we have $c_{44}^{\perp}(\mathbf{k}) = c_{44}^{\parallel}(\mathbf{k})$. The third of the four elastic moduli $c_{11}(\mathbf{k})$ is the bulk modulus, or compressional modulus, of the flux line liquid. The fourth of the elastic moduli, namely $c_{14}(\mathbf{k})$ does not have a counterpart in an isotropic superconductor, nor in an anisotropic superconductor when the magnetic field is parallel to the $\hat{c}$-axis. The new elastic modulus appearing is a mixed tilt- and bulk-modulus, suggesting that in a tilted magnetic field, the elastic tilting and compressional modes of the flux-line liquid interact with each other. These mixed elastic moduli vanish in the limit when $\mathbf{k} \to 0$. (For the shear moduli, we have a similar situation. We get a hard and a soft shear mode, as well as a mixed shear-tilt mode).

Note how all of the other elastic moduli, bulk as well as tilting moduli, take the value B^2/μ_0 when $\mathbf{k} = 0$. Mass anisotropy is not seen at all in this so-called local limit. Moreover, the elastic properties of the flux-line liquid are oblivious to the value of λ (or κ) in the local limit. At finite $\mathbf{k}$-vectors, the flux-line liquid softens considerably as the degree to which a superconductor is type II, increases. But the effect of mass anisotropy and the hardness of the superconductor (degree of type II) only influence the elastic properties of the flux-line lattice provided a non-local elastic theory is used. We shall come back to this point in Chapter 8, when we consider the Lindemann criterion for melting of the flux-line lattice.

7

Applications of Ginzburg–Landau Theory ($|\psi|$ spatially varying)

7.1 The temperature-dependent order parameter $|\psi(T)|$

As a first application of the Ginzburg–Landau (GL)-equations we derive the temperature dependence of the order parameter ψ. This can be done under the simplest possible conditions, with $B = 0$, and under homogeneous conditions $\nabla \psi = 0$. The free energy now reduces to

$$F(T) = F_0(T) + \alpha \, |\psi|^2 + \frac{\beta}{2} \, |\psi|^4 + \cdots \qquad (7.1)$$

Taking the derivative leads to the equilibrium condition

$$\frac{\delta F}{\delta \, |\psi|} = 0 \Rightarrow (\alpha + \beta \, |\psi|^2) \, |\psi| = 0 \qquad (7.2)$$

One solution is that $|\psi| = 0$, which is correct for the superconducting wavefunction above T_c. The other alternative is $\alpha + \beta|\psi|^2 = 0$. This gives

$$|\psi|^2 = - \left(\frac{\alpha}{\beta} \right) \qquad (7.3)$$

Since $\alpha = \alpha_1(T - T_c)$ where α_1 is a constant ,we find to a useful approximation, like in the magnetic case

$$|\psi|^2 = \frac{\alpha_1}{\beta}(T_c - T)$$

Superconductivity: Physics and Applications   Kristian Fossheim and Asle Sudbø
© 2004 John Wiley & Sons, Ltd   ISBN 0-470-84452-3

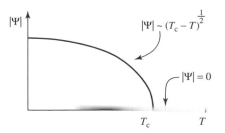

Figure 7.1   Temperature dependence of the order parameter $|\psi|$.

or

$$|\psi| = \left(\frac{\alpha_1}{\beta}\right)^{\frac{1}{2}} (T_c - T)^{\frac{1}{2}} \tag{7.4}$$

as sketched in Figure 7.1.

Since the interpretation of $|\psi|^2$ is that it represents the superfluid density n_s, we have

$$n_s \propto (T_c - T) \tag{7.5}$$

Some important remarks on the limitation of GL-theory are in order here. The last two results are both typical mean-field-like, a consequence of the tacit assumption of the GL-approach that fluctuations of ψ are to be neglected. In real cases, the temperature exponent 1/2 is good in the low-T_c materials, but not expected to hold in high-T_c materials. The difference comes from the properties we discussed in Section 2.6, that the coherence length is quite long in low-T_c and quite short in high-T_c materials, making fluctuations important in the latter case.

According to the results above, the superfluid density n_s – corresponding to the density of Cooper-pairs – vanishes at T_c. Again, this mean-field like prediction is correct for long ξ, low T_c materials. In these the Cooper-pairs do start to condense at T_c, and the superfluid density n_s and wavefunction ψ build up as the temperature is lowered below T_c. The lower the temperature, the more Cooper-pairs are created. However, again the situation does not prevail in high-T_c materials. It turns out that in those materials the wavefunction amplitude $|\psi|$ is *not* zero above T_c. We return to these profound problems in Chapter 10.

7.2   The coherence length ξ

The coherence length ξ is a temperature-dependent as well as a material-dependent quantity. We will now demonstrate how it may be derived from

the GL-theory, even without a magnetic field present. In this case the GL-I
equation immediately becomes

$$\alpha(T)\psi + \beta |\psi|^2 \psi + \frac{1}{2m}(-i\hbar\nabla)^2\psi = 0. \tag{7.6}$$

This is a differential equation for spatial variations of the order parameter. In
geometries where 1D calculation may be used for simplicity, it reads

$$-\frac{\hbar^2}{2m}\frac{d^2\psi}{dx^2} + \alpha(T)\psi + \beta |\psi|^2 \psi = 0. \tag{7.7}$$

On dividing through by the parameter α we observe that an operator defined
as $\frac{\hbar^2}{2m\alpha}\frac{d^2}{dx^2}$ in the first term of Eq. 7.7 must be dimensionless. Therefore, the
quantity ξ defined by

$$\xi \equiv \left(\frac{\hbar^2}{2m\alpha(T)}\right)^{\frac{1}{2}} \tag{7.8}$$

must have the dimension of a length. Such a finding is significant, and suggests
that this is to be understood as a characteristic length of the system. Indeed, this
is the temperature-dependent GL coherence length ξ. Its temperature dependence
is predicted to have the classical exponent $\nu = 1/2$, according to

$$\xi(T) = \xi_0 \begin{cases} (T/T_c - 1)^{-\frac{1}{2}}; & T > T_c \\ (1 - T/T_c)^{-\frac{1}{2}}; & T < T_c \end{cases} \tag{7.9}$$

where the length ξ_0 is $\hbar/(2m\alpha_1 T_c)^{\frac{1}{2}}$. The length $\xi(T)$ is of fundamental importance in superconductivity, representing the coherence length of the wavefunc-
tion. We may regard this as the shortest distance over which the wavefunction
is allowed to vary without generating pair breaking kinetic energy. We notice
in particular the important aspect that $\xi(T)$ diverges towards T_c, from below as
well as from above. This means that when the wavefunction is perturbed by for
instance a magnetic field, or by normal inclusions in the superconductor, the
length $\xi(T)$ is the healing length over which ψ recovers its full value. Clearly,
this situation also exists when flux lines or vortices penetrate a type II superconductor, where, as we know, $\psi \to 0$ at the centre. $\xi(T)$ is then the effective
radius within which the Cooper-pair density is suppressed, a fact which allows
the calculation of the energy cost of breaking the Cooper-pairs in a radius ξ
around the vortex centre. A similar role is played by $\xi(T)$ at a superconducting to normal interface in the laminar structure of type I superconductors. Here
ξ defines the wall thickness, quite analogous to domain walls in systems like
ferromagnets, ferroelectrics, etc.

    All results of the GL-theory are purely phenomenological. The coefficients of
the energy expansion, α_1, β and c are unknown. However, they can be determined

by relating them to measurable quantities, like ξ, λ and H_c. The consistency of data taken by independent methods may then be checked, with the prospect of asserting the overall picture, as well as imposing a quantitative control on the expressed relations between experimentally determined quantities.

The penetration depth λ is the other length parameter of particular importance in superconductivity. How can it be expressed by use of the GL-equations? Since λ is a measure of the depth of supercurrent response to a magnetic field, this suggests looking at the second GL-equation.

In situations where the Cooper-pair density remains unperturbed, like when a magnetic field penetrates to depth λ in the Meissner state, we can simply neglect the gradient of ψ in GL-II. What remains is

$$J = \frac{e}{m}\psi^*(-2eA)\psi + c.c. \tag{7.10}$$

$$= -\frac{4e^2|\psi|^2}{m}A \tag{7.11}$$

From the London equation we have $J = -(1/\mu_0\lambda^2)A$. This gives the relation $\frac{4e^2|\psi|^2}{m} = \frac{1}{\mu_0\lambda^2}$. Here we must have $|\psi|^2 = n_s$ for consistency of these independent derivations. The result clearly is in accord with the BCS-picture since it leads to the conclusion that

$$\lambda^2 = \frac{m}{(2e)^2\mu_0|\psi|^2} \tag{7.12}$$

This offers a somewhat more precise definition of the length λ, where $2e$ corresponds to the magnitude of the charge of a Cooper-pair, and m is the corresponding mass.

By now we have obtained results for ξ and λ, and can also express the quantity $\kappa = \lambda/\xi$, which is of fundamental importance in superconductivity.

7.2.1   Relations between λ, ξ and H_c

So far we have used the GL free energy expansion and the GL-equations to gain insight into the characteristic lengths of the superconductor. Next, we look for relations between these and the important thermodynamic quantity H_c. We consider again a homogeneous superconductor in the absence of a magnetic field. The free energy is

$$F_s(T) = F_n(T) + \alpha|\psi|^2 + \frac{\beta}{2}|\psi|^4 + \cdots \tag{7.13}$$

Using $|\psi|^2 = -\frac{\alpha}{\beta}$ we find

$$F_s(T) = F_n(T) - \frac{\alpha^2}{2\beta} \qquad (7.14)$$

This must agree with the difference in Gibbs's energy between the normal and superconducting states derived in Chapter 1, hence

$$\frac{\alpha^2}{2\beta} = \frac{\mu_0}{2} H_c^2 \qquad (7.15)$$

We can also find expressions for α and β from the preceding relations. Using

$$\mu_0 H_c^2 = -\alpha |\psi|^2; \quad |\psi|^2 = -\frac{\alpha}{\beta}; \quad \text{and} \quad \frac{1}{\lambda^2} = \frac{-4e^2 \mu_0^2 H_c^2}{\alpha m} \qquad (7.16)$$

we find

$$\alpha = \frac{-4e^2 \mu_0^2 H_c^2 \lambda^2}{m} \qquad (7.17)$$

and

$$\beta = \frac{\alpha^2}{\mu_0 H_c^2} = \frac{(4e^2)^2 \mu_0^3 H_c^2 \lambda^4}{m^2} \qquad (7.18)$$

By combining the equations containing H_c, ξ and λ we obtain the interesting result that

$$H_c \lambda \xi = \frac{\Phi_0}{2\pi \sqrt{2\mu_0}} \qquad (7.19)$$

Since the right-hand side is a constant, we see that on varying the temperature, the three quantities H_c, λ and ξ are predicted to vary in such a way that their product remains fixed. Equally interesting and useful is the combination

$$\lambda(T)\xi(T) = \frac{\Phi_0}{2\pi \sqrt{2\mu_0} H_c(T)} \qquad (7.20)$$

7.3 Two types of superconductors

In complete analogy with Ohm's law, $J = \sigma E$, which we pointed out is formally equivalent to the London equation, Pippard [69] adopted the corresponding non-local version of Ohm's law from Reuther and Sondheimer [79] for the so-called anomalous skin-effect,

$$J = \frac{3\sigma}{4\pi l} \int \frac{r(r \cdot E)e^{-r/l}}{r^4} d^3r \tag{7.21}$$

Non-locality enters the problem when the response to a field can only be determined correctly by integrating over a volume of the size of l^3 (3D case), where l is comparable to or longer than the distance δ, the depth over which the E-field varies. Consequently, in the London equation ξ would have to be introduced whenever the coherence length ξ is greater than λ, since λ is the penetration distance of B whose vector potential drives the current. From experience in studies of the anomalous skin-effect in pure, normal metals Pippard saw the analogy for the superconductor in the simplest case to be

$$J = -\frac{\xi(l)}{\xi_0} \Lambda A \tag{7.22}$$

where ξ_0 is the value of $\xi(l)$ in the limit of large l, i.e. for the pure case, and $\Lambda = (\mu_0 \lambda_L^2)^{-1}$. Clearly, *this equation predicts a reduced response when $\xi(l)$ is reduced by impurities.* Because experimentally λ is found to increase with decreasing l, $\xi(l)$ must decrease as l decreases, as can be seen from Eq. 7.20. Eq. 7.22 can be viewed as one which applies when $\xi \ll \lambda$, which we may call the London limit. It is also a limiting form of the general Pippard non-local relation, which he constructed on the basis of the combined insight from Eq. 7.22 and the Reuther–Sondheimer integral. The whole chain of arguments led Pippard to propose for superconductors, in complete analogy with the non-local response to an electric field in a normal metal

$$J = -\frac{3\Lambda}{4\pi\xi_0} \int \frac{r(r \cdot A)e^{-r/\xi}}{r^4} d^3r \tag{7.23}$$

where $\xi(l)^{-1} = \xi_0^{-1} + l^{-1}$.

Limiting cases which can be worked out are:

$$\lambda = (\xi_0/\xi)\lambda_L \qquad \text{for} \quad \xi \ll \lambda \qquad \text{(London limit)}$$

$$\lambda = \left[\frac{\sqrt{3}}{2\pi}\xi_0\lambda_L^2\right]^{\frac{1}{3}} \qquad \text{for} \quad \xi \gg \lambda \qquad \text{(Pippard limit)}$$

where we used the symbol λ_L for the original London value of λ. These expressions have been verified experimentally.

It is remarkable that the microscopic BCS theory [7] later confirmed the essential correctness of Pippard's bold adaptation of non-local electrodynamics from normal metals to superconductors. The BCS theory also gave an expression for the ξ_0-parameter: $\xi_0 = 0.18\hbar v_F/kT_c$, where v_F is the Fermi velocity, and $k = 1.38 \times 10^{-23}\, J/K$ is Boltzmann's constant. Essentially the same answer is obtained using the uncertainty principle. In that case the idea is to relate the coherence length to the energy gap.

The introduction of the coherence length lead to the extremely important distinction between two types of superconductors, i.e. those with $\xi > \lambda$ versus those with $\xi < \lambda$, called type I and type II superconductors, respectively.

A precise expression for this distinction is derived later in this chapter. At this point a semiquantitative, less stringent argument, proves the main point (see Figure 7.2). In both the parts of Figure 7.2 we are faced with a situation where a change takes place between a point at the origin, $r = 0$, where the normal state with $B = B_c$ ends, followed by a region where superconductivity recovers over a distance ξ, while screening extends over the distance λ. We refer first to the left figure, where the situation is sketched from the NS-interface into the superconducting region. Here $B(r)$ falls off from $r = 0$ according to $B(r) = B_c e^{-r/\lambda}$. Next, the superconducting wavefunction is suppressed over a much wider range, i.e. over the coherence length ξ. The energy argument now goes as follows: To expel the magnetic field from the volume of the superconductor costs the energy $\frac{1}{2\mu_0}B_c^2$ per volume. However, a volume equal to the product $A\lambda$ of laminar interface area A and penetration depth λ, is *not* screened against B_a, amounting to a total energy

$$E_1 = \lambda A \frac{1}{2\mu_0} B_c^2 \tag{7.24}$$

saved in setting up the laminar structure, compared to what it would cost to have a totally sharp boundary between normal and superconducting volumes, i.e. if $\lambda = 0$. On the other hand, the density of Cooper-pairs n_s is suppressed since the proximity to the normal interface forces the superconducting wave-function ψ

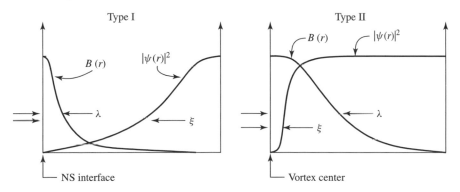

Figure 7.2 Variations of magnetic field B and wavefunction ψ at the normal-superconducting interface.

to go to zero at the normal boundary where $B = B_c$. The distance over which ψ is suppressed from full equilibrium value ψ_0 to zero, is precisely the length ξ that we introduced in Eq. 7.8. In the volume defined by the total laminar wall area times ξ, superconductivity is essentially destroyed. This costs an amount of energy corresponding to breaking the majority of Cooper-pairs in this volume,

$$E_2 = \xi A \frac{1}{2\mu_0} B_c^2.$$ (7.25)

So the net energy ΔE gained by setting up the laminar walls is given by the difference $\Delta E = E_1 - E_2$. For the situation to be stable ΔE must be positive:

$$\Delta E = \xi A \frac{1}{2\mu_0} B_c^2 - \lambda A \frac{1}{2\mu_0} B_c^2 = \frac{1}{2\mu_0} A B_c^2 (\xi - \lambda) > 0$$ (7.26)

We can regard this condition as the defining one for type I superconductivity, and note that it corresponds to the simple statement that $\xi > \lambda$, not very different from the exact condition: $\kappa < 1/\sqrt{2}$ to be shown later. The fact that the interface energy is positive in type I, makes the system want to establish as small an interface area as possible. It does so by setting up the laminar structure shown in Figure 7.3. We note from the picture that the lamina are not flat, but show a meandering structure as long as the field is directed normal to the flat sample surface.

Experiments have shown that the laminar structure can be straightened out and stabilized by a magnetic field tilted in the plane of the laminar direction [80, 81], as sketched in Figure 7.4.

With a flat geometry and a tilted field, as shown in Figure 7.4, the laminar structure acquires a thickness period where normal lamina thickness d_n and superconducting lamina thickness d_s add up to a total thickness $d = d_s + d_n$. In the normal lamina B is everywhere B_c, and in the superconducting lamina we take $B = 0$. Furthermore, if we assume that the lamina extend across a length L of a plate-like rectangular slab structure, there is a simple flux-conserving relationship between the laminar geometry, the thermodynamic field B_c and the applied field B_a. Conservation of total flux Φ_a applied to the specimen requires that

$$\Phi_a = B_a L d = 0 \times L d_s + B_c L d_n$$

$$\Rightarrow d_n = \frac{\Phi_a}{B_c L} = \frac{B_a}{B_c} d$$ (7.27)

Here we have disregarded the small tilt of $\boldsymbol{B}_a$ with respect to the surface normal. Since $d_n + d_s = d$ we find also

$$\frac{B_a}{B_c} d + d_s = d \Rightarrow d_s = \left(1 - \frac{B_a}{B_c}\right) d$$ (7.28)

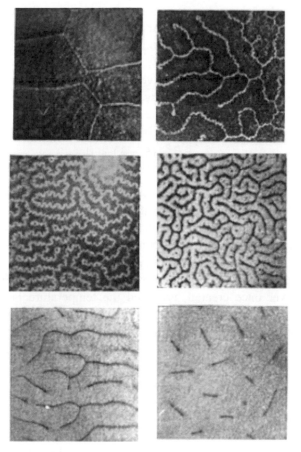

Figure 7.3 Meandering laminar structure with alternating normal and superconducting regions, typical for type I superconductors with the magnetic field applied normal to a flat slab. From [82]. Reproduced with permission of the Royal Society.

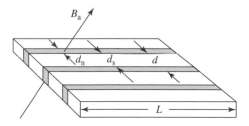

Figure 7.4 Sketch of laminar structure in type I superconductor, straightened by tilting the applied magnetic field B_a [80].

This shows how the laminar thicknesses d_n and d_s vary with applied field and temperature through the dependences of d_s and d_n on B_a and $B_c(T)$. This result applies as long as the total d is a constant. Original work in this area was done by Landau [83, 84].

Next we discuss the second alternative, $\kappa > 1$. The energy argument we used before, pertaining to the interface energy, applies again, but due to the condition $\xi < \lambda$ the energy ΔE required to create the interface is now negative:

$$\frac{1}{2\mu_0} B_c^2 A(\xi - \lambda) < 0 \tag{7.29}$$

So, to within a factor $1/\sqrt{2}$ as before, $\lambda > \xi$ is the condition for type II super-conductivity. Therefore now, the *larger* the interface area can be made, the more energy is gained. Can this go on without limit? Clearly no, because, as we have seen, flux is quantized in elementary units $\Phi_0 = h/2e$.

This is how far the division of flux *can* and *will* go. These flux quanta are established by circulating superfluid currents, vortices of Cooper-pair flow that will go on forever once created, provided the temperature is held below T_c. Because of the smallness of the flux quantum, there will be a huge number of these in a finite field. A field of $1\,T$ in a sample with $1\,cm^2$ cross-section will cause the sample to be penetrated by more than 10^{11} vortices! We shall later study the physics of these in greater detail.

As a consequence of $\nabla B = 0$ we can immediately conclude that any such quantized vortex, once created, must extend continuously all the way through a homogeneous superconductor or form closed loops. The physics of vortices constitutes the central issue in the science of superconductivity in type II super-conductors. We return to this subject in several later chapters of this book.

7.4   The structure of the vortex core

In Chapter 6 we derived, as a byproduct of finding the line energy, the radial dependence of the local field from a vortex:

$$b = \frac{\Phi_0}{2\pi\lambda^2} \ln\left(\frac{\lambda}{r}\right) + b_0 \tag{7.30}$$

This was obtained as an approximation when integrating curl $b \times dl$ for $\xi < r \ll \lambda$. The rigorous result is

$$b(r) = \frac{\Phi_0}{2\pi\lambda^2} K_0\left(\frac{r}{\lambda}\right) \tag{7.31}$$

where $K_0(\frac{r}{\lambda})$ is a zero'th order Bessel function [72]. This form flattens the dependence on r within a radius ξ, the 'hard core'. For convenience we repeat the limiting forms:

$$b(r) = \frac{\Phi_0}{2\pi\lambda^2} \ln\left(\frac{\lambda}{r}\right); \quad r \ll \lambda \tag{7.32}$$

$$b(r) = \frac{\Phi_0}{2\pi\lambda^2} \left(\frac{\pi\lambda}{2r}\right)^{\frac{1}{2}} e^{r/\lambda}; \quad r \gg \lambda \tag{7.33}$$

The structure of the vortex, therefore, is an inner volume of radius ξ where the field locally is high enough to destroy superconductivity by breaking most of the Cooper-pairs, i.e. n_s approaches zero. This means also that the wavefunction ψ must likewise approach zero. It is equal to zero only at the centre. How can we prove that n_s is so suppressed? Physically the upper critical field where the superconductor crosses over to the normal state, can be estimated by

$$B_{c2} \approx \frac{\Phi_0}{\pi\xi^2} \tag{7.34}$$

This implies that superconductivity is destroyed when the cores start to overlap. The reason why this leads to a normal state, is that at B_{c2} cores with $n_s \approx 0$, overlap. This argument is also bears on ψ, since $|\psi|^2 = n_s$. Figure 7.5 shows a sketch of a cut through the vortices, represented by n_s and $b(x)$. Figure 7.6

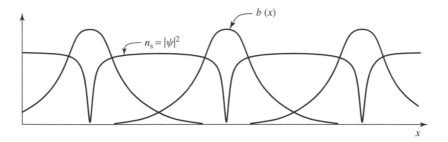

Figure 7.5 Cut through the centre of neighbouring vortices, showing $b(x)$ and $|\psi|^2(x)$.

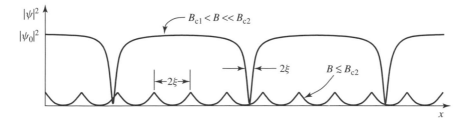

Figure 7.6 Cut through neighboring vortices showing $|\psi|^2(x)$ close to B_{c1}, and B_{c2}, respectively.

is a sketch of $|\psi|^2$ corresponding to two different values of B, cutting through vortices.

The properties of the vortex have been carefully investigated by scanning tunnelling spectroscopy. We refer to the topical contribution by Fischer in Chapter 13 for a detailed look into this subject, with reference to both low-T_c and high-T_c superconductors.

7.5   The length ξ and the upper critical field B_{c2}

7.5.1   Isotropic systems

Near the upper critical fields, as was pointed out above, the overlapping of cylindrical volumes of radius $\approx \xi$ suppresses the wavefunction everywhere to low amplitude, with some cusp-like increase at the circumferences of the $\pi \xi^2$ area. In Figure 7.6 we showed the situation both for $B_{c1} < B \ll B_{c2}$ and for $B \le B_{c2}$.

We observe that at magnetic fields close to B_{c2}, ψ and $|\psi|^2$ are everywhere very small. This allows a simplification of the GL free energy by dropping higher order terms in ψ. What remains then, is

$$F = F_0 + \frac{1}{2}\alpha\,|\psi|^2 + \frac{1}{2m}\left|(-i\hbar\nabla - 2e A)^2 \psi\right|^2 \qquad (7.35)$$

Minimization gives

$$\frac{\partial F}{\partial \psi} = \alpha\psi + \frac{1}{m}(-i\hbar\nabla - 2e A)^2 \psi = 0 \qquad (7.36)$$

or

$$\frac{1}{2m}(-i\hbar\nabla - 2e A)^2 \psi = -\alpha\psi \qquad (7.37)$$

This is formally a Schrödinger equation with α as the energy eigenvalue of the kinetic energy operator to the left, for a particle of mass m and charge $-2e$ in a field B.

This is a well-known problem in quantum mechanics, with solutions for the energy states:

$$\frac{1}{2m}m v_z^2 + \left(n + \frac{1}{2}\right)\hbar\omega_c = E_{n,v_z} \qquad (7.38)$$

where v_z is the velocity of the particle parallel to the field, and is unaffected by it, while the term $(n + 1/2)\hbar\omega_c$ comes from the circular (cyclotron) motion around the field in the plane normal to it. The complete motion is helical. In the following the velocity v_z may be set equal to zero; and we look for the lowest energy level $n = 0$. It corresponds to

$$-\alpha = \frac{1}{2}\hbar\frac{2eB}{m} = \hbar\omega_c \tag{7.39}$$

(Recall that $-\alpha$ is positive when T is below T_c, as is the case here.) The field that corresponds to the solution just given is B_{c2}, so

$$-\alpha = \frac{\hbar e B_{c2}}{m} \tag{7.40}$$

Finally, on combining with the relation $\alpha^2/2\beta = \frac{1}{2\mu_0}B_c^2$ one obtains

$$B_{c2} = \kappa\sqrt{2}B_c \tag{7.41}$$

where $\kappa = \lambda/\xi$. Next, we make use of a relation derived earlier: $-\alpha = \frac{\hbar^2}{2m\xi^2}$. Equating the two expression for α we find

$$B_{c2} = \frac{\Phi_0}{2\pi\xi^2} \tag{7.42}$$

The relation $B_{c2} = \kappa\sqrt{2}B_c$ holds an important message: When $\kappa > \frac{1}{\sqrt{2}}$ we have $B_{c2} > B_c$. $\kappa > \frac{1}{\sqrt{2}}$ is the exact condition for the appearance of a vortex phase in type II superconductors. When $\kappa < \frac{1}{\sqrt{2}}$ we get $B_{c2} < B_c$. In that case, as we lower the field from above, we meet the B_c field before B_{c2}. Here the Meissner effect sets in, and the mixed phase will not appear. We have type I behaviour. The importance of the value of $\kappa = \lambda/\xi$ is apparent. We have two types of superconductors:

$\kappa < \frac{1}{\sqrt{2}}$:   type I with laminar field distribution at $B_a > B_c$ when $n \neq 0$.

$\kappa > \frac{1}{\sqrt{2}}$:   type II with vortex state at $B_{c1} < B < B_{c2}$

The laminar structure of field penetration in type I is amply documented. It occurs when an inhomogeneous field surrounds a sample with demagnetization factor $n \neq 0$. The field may then exceed B_c in some areas, and some flux is allowed to penetrate the sample in order to lower the total energy. On penetration, the laminar structure appears as the device by which the sample avoids being exposed to a field $B > B_c$ anywhere. At this point it may be interesting to ask for an exact expression for B_{c1} too. A rigorous argument, which avoids use of thermodynamics, and only relies on geometric consideration, disregarding any finite core effects, is the following. B_{c1} is defined as the field at which the first vortex enters the superconductor. That being so, let one flux quantum be in the superconductor at the field B_{c1}. Let us for this purpose work in the London approximation where the normal core radius is set equal to zero. The flux quantum Φ_0 is now allowed to occupy the whole specimen, the field being distributed according to the function $e^{-r/\lambda}$. The critical field B_{c1} at which one flux quantum has entered can then be computed by weighting each area $dA = 2\pi r\, dr$ by the distribution function $e^{-r/\lambda}$. The field is the total flux divided by effective area as follows:

$$B_{c1} = \frac{\Phi_0}{\displaystyle\int_0^\infty e^{-\frac{r}{\lambda}} 2\pi r\, dr} = \frac{\Phi_0}{2\pi\lambda^2} \tag{7.43}$$

which is the exact answer in the London approximation with $|\psi|$ constant.

7.5.2 B_{c2}, ξ, and λ in anisotropic superconductors

So far we have discussed superconductivity in isotropic systems. However, for layered, highly anisotropic high-T_c compounds there is a need to develop methods to treat such cases. This is achieved by the Lawrence–Doniach (LD) theory [85, 86].

In a simple model picture, let us assume that the superconductor can be described by a state of alternating superconducting and insulating layers, of type SISISI, where S = superconductor and I = insulator, or SNSNSN, where N is a normal layer. In such systems, if the contact between S-layers is poor, the main charge transport between S-layers may take place by tunnelling, both for normal carriers and for Cooper-pairs. Among high-T_c materials the Bi-based compounds are good examples of such behaviour. In such a layered structure, let the x, y axes lie in the ab-plane, and the z-axis along the normal to the ab-plane. We will assume for simplicity that a- and b-axes are equivalent, corresponding to a tetragonal symmetry. In the high-T_c cuprates the orthorhombic structure is usually realized, but the distinction between a and b is far less significant than the difference between c and a, b. An immediate consequence is that the effective mass along a, b, called m_{ab} may be quite different than m_c. In this situation the inverse mass appears as a tensor quantity in transport problems and in summing up energy terms. The free energy for a stack of superconducting planes with restricted contact between layers can be written down as

$$F = \sum_n \int d^2r \left[\alpha\, |\psi_n|^2 + \frac{1}{2}\beta\, |\psi_n|^4 \right.$$

$$\left. + \frac{\hbar^2}{2m_{ab}} \left(\left| \frac{\partial\psi_n}{\partial x} \right|^2 + \left| \frac{\partial\psi_n}{\partial y} \right|^2 \right) + \frac{\hbar^2}{2m_c} \left| \frac{\partial\psi_n}{\partial z} \right|^2 \right] \tag{7.44}$$

The summation is over all planes, and integration is in each plane, in two dimensions. The distinction between the *ab*- and *c*-directions is most manifest in the coherence length, which is often much shorter in the *c*-direction than in the *ab*-plane, even shorter than the S-S-plane distance *s* in high-T_c cuprates. Here one may choose to discretize the last derivative in extreme cases of layeredness. This term would take the form

$$\frac{\hbar^2}{2m_c s^2} |\psi_n - \psi_{n-1}|^2 \tag{7.45}$$

and by the summation, each layer *n* will be found to interact with one nearest neighbour on each side. On writing $\psi_n = |\psi_n| e^{i\varphi_n}$, and $\psi_{n-1} = |\psi_n| e^{i\varphi_{n-1}}$ assuming identical layers, one finds by insertion in Eq. 7.44 that the energy for two adjacent planes is

$$\frac{\hbar^2}{m_c s^2} |\psi_n|^2 \, (1 - \cos(\varphi_n - \varphi_{n-1})) \tag{7.46}$$

This expression bears a striking resemblance to the Josephson coupling energy[1], as it ought to since the physics is the same when charge transport between layers takes place by tunnelling. On the other hand, in case we stick with the continuum version of derivation even along *z*, within some distance from T_c, due to the divergence of $\xi(T)$ we can instead use the continuum GL-equation found before, but with the modification that inverse mass is now a tensor expressed as $\overleftrightarrow{m}^{-1}$; $i = a$, b, c. The only change from the isotropic case, is that we have to write the product of the kinetic operator and the inverse mass tensor in a new form. As the kinetic operator is squared, the way to write the product is to place $\overleftrightarrow{m}^{-1}$ between the two operators. The GL-equation for the anisotropic, continuous case is called the LD-equation and now reads

$$\alpha\psi + \beta |\psi|^2 \, \psi - \frac{\hbar^2}{2} \left(\nabla - i\frac{2e}{\hbar} A \right) \cdot \overleftrightarrow{m}^{-1} \cdot \left(\nabla - i\frac{2e}{\hbar} A \right) \psi = 0 \tag{7.47}$$

The inverse mass tensor has principal values: $\frac{1}{m_{ab}}, \frac{1}{m_{ab}}, \frac{1}{m_c}$.

The anisotropy encountered here is extremely important in the high-T_c cuprates, and its consequences must be dealt with realistically in any serious approach. We immediately come upon this fundamental issue in the description of the coherence length, ξ. Clearly, we now have to write the relationship

---

[1]The Josephson coupling energy expresses the energy stored in the junction, and can be determined by integrating up the electrical work which the current source does in changing the phase φ, i.e. by $\int I_s V \, dt = \int I_s (\hbar/2e) d(\Delta\varphi)$.

between ξ and the parameter α in a new form, with corresponding labels i on ξ and m:

$$\xi_i^2 = \frac{\hbar^2}{2m_i\alpha(T)} \tag{7.48}$$

And since ξ_i is anisotropic, so is λ_i as expressed by

$$\xi_{ab}\lambda_{ab} = \xi_c\lambda_c = \xi_i\lambda_i = \frac{\Phi_0}{2\sqrt{2\pi}\,B_c} \tag{7.49}$$

Thermodynamic quantities like B_c, of course, should have no anisotropy label. The inverse relationship between ξ_i and λ_i is noteworthy. Remembering that the index i on ξ_i represents the direction of the corresponding axis and the direction of movement of the mass m_i, the index i on λ_i takes a different meaning: It refers to supercurrent shielding by currents flowing in the plane normal to the i-th axis. The anisotropy was introduced above through the masses m_i, and these are different by a factor of the order of 10 in high-T_c compounds. The lattice anisotropy has a direct influence on masses m_i, and via the mass it propagates to all other directionally distinguishable quantities.

One other notable consequence is in the critical fields, where we now have

$$B_{c2}^{\|c} = \frac{\Phi_0}{2\pi\xi_{ab}^2} \quad ; \quad B_{c2}^{\|ab} = \frac{\Phi_0}{2\pi\xi_{ab}\xi_c} \tag{7.50}$$

The reason why we have to divide by the product $\xi_{ab}\xi_c$ in the second case is that the currents defining the vortices and flux lines flow around the axis lying in the ab-plane and therefore flow both in the ab-plane and in the c-direction. Since $\xi_{ab} \gg \xi_c$ we get $B_{c2}^{\|c} < B_{c2}^{\|ab}$. To summarize, the anisotropy factor γ can be written as

$$\gamma \equiv \left(\frac{m_c}{m_{ab}}\right)^{\frac{1}{2}} = \frac{\xi_{ab}}{\xi_c} = \frac{\lambda_c}{\lambda_{ab}} = \frac{B_{c2}^{\|ab}}{B_{c2}^{\|c}} \tag{7.51}$$

An illustration of the definitions of the length parameters λ and ξ in anisotropic superconductors of the type discussed above is given in Figure 7.7. In Figure 7.8 we have illustrated how the vortices created in a layered superconductor are broken up into a stack of 'pancakes'. When the field is tilted, the pancakes are connected by horizontal Josephson vortices, shown in the figure only as horizontal lines. An important distinction is the fact that pancakes have a normal core, while the Josephson vortices have no such core.

To study H_{c2}, or B_{c2} in the strongly anisotropic case we return to the discretised version Eq. 7.45, but with appropriate terms included to allow a magnetic field through a vector potential A. We choose to let $A_z = Bx$, where B is along the y-axis, in the ab-plane.

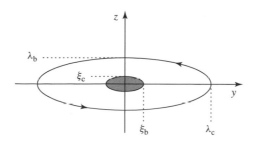

Figure 7.7   Sketch of a vortex of elliptical cross section aligned parallel to the *a* axis in an anisotropic superconductor described by the anisotropic Ginzburg–Landau theory.

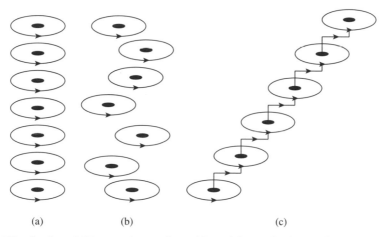

(a)   (b)   (c)

Figure 7.8   Stacks of 2D pancake vortices: (a) straight stack, the configuration of lowest energy at zero temperature, and (b) disordered stack, which occurs at higher temperatures. (c) Sketch of a tilted stack of pancake vortices in successive superconducting layers, connected by interlayer Josephson strings.

We will now express the contribution to the LD-equation in the n'th super-conducting plane by the continuous model as

$$\alpha \psi_n + \beta \,|\psi_n|^2\, \psi_n - \frac{\hbar^2}{2m_{ab}} \left(\nabla - i\frac{2e}{\hbar} A \right)^2 \psi_n \qquad (7.52)$$

A contribution must be added here to include the effect of coupling to planes $n-1$ and $n+1$. Since ψ_n has neighbours ψ_{n-1} and ψ_{n+1} the phases relative to these with respect to the phase of ψ_n must be expressed explicitly. The three terms in the summation over three planes ψ_{n+1}, ψ_n, ψ_{n-1} must be modified to take account of the phase contribution from the vector potential A. The phase difference to look for is indicated by the second term in the operator $(\nabla - i\frac{2e}{\hbar} A)$. Here $i\frac{2e}{\hbar} A$ is equivalent to the derivative of the phase difference, so it must be multiplied by the distances between planes to give the phase difference.

And finally, the difference between wave functions on the corresponding two neighbouring planes has to be divided by s^2 to give the discretized second derivative. Upon adding these terms to those written down before, recalling also the direction of the field we have chosen, we find

$$\alpha \psi_n + \beta |\psi_n|^2 \psi_n - \frac{\hbar^2}{2m_{ab}} \nabla^2_{x,y} \psi_n$$

$$- \frac{\hbar^2}{2m_c s^2} \left(\psi_{n+1} e^{-\frac{2ieA_z s}{\hbar}} - 2\psi_n - \psi_{n-1} e^{\frac{2ieA_z s}{\hbar}} \right) = 0 \qquad (7.53)$$

This equation can be simplified enormously by linearization, i.e. by removing $\beta |\psi_n|^2_z \psi_n$, and by having $A = (0, 0, A_z)$, and finally by writing the last parenthesis in the form of a coupling energy. Also, here again we make the perfectly reasonable assumption that all superconducting planes are physically equivalent. We can now omit the index n on the various ψ-functions. We find

$$\alpha \psi - \frac{\hbar^2}{2m_{ab}} \frac{\partial^2 \psi}{\partial x^2} - \frac{\hbar^2}{2m_c s^2} \left(1 - \cos \left(\frac{2e A_z s}{\hbar} \right) \right) \psi = 0 \qquad (7.54)$$

Inserting the expression $\alpha = \frac{\hbar^2}{m_{ab} \xi^2_{ab}}$ and $A_z = Bx$, and expanding the cos-function for small argument, corresponding to small s we find the equation

$$\frac{d^2 \psi}{dx^2} + \left[\frac{m_{ab}}{m_c} \left(\frac{2\pi Bx}{\Phi_0} \right)^2 - \frac{1}{\xi^2_{ab}(T)} \right] \psi = 0 \qquad (7.55)$$

Using a trial solution of the form $\psi = e^{ik_y y} e^{ik_z z} f(x)$ one finds

$$B_{c2} = \frac{\Phi_0}{2\pi \xi_{ab} \xi_c} \qquad (7.56)$$

7.6 Ginzburg–Landau–Abrikosov (GLA) predictions for B_{c2}/B_{c1}

The most important parameters to be determined by measurements in type II superconductors are the lower and upper critical fields, B_{c1} and B_{c2}, and the ratio $\kappa = \lambda/\xi$.

Difficulties were always connected with the determination of the lower critical field B_{c1}, because of the presence of pinning, which can mask completely the predicted sharp drop in the magnetization at B_{c1}. Therefore, experimental data on B_{c1} are often quite uncertain. The test of consistency with GLA theory is usually to compare data according to the relation $(B_{c1} B_{c2})^{1/2}/B_c = \ln \kappa$. Due to the $\ln \kappa$ factor, this is a quite insensitive test of the consistency of data on critical fields and κ. A much more sensitive test can be made by taking the ratio of critical fields B_{c1} and B_{c2} rather than their product, as we will now show.

By deriving the well-known formula for B_{c1}

$$B_{c1} = \frac{\Phi_0}{4\pi\lambda^2}(\ln\kappa + \varepsilon) \tag{7.57}$$

one has in reality derived the κ-dependence of the ratio B_{c2}/B_{c1}, or the inverse thereof. This is easily shown by writing B_{c1} in the form

$$B_{c1} = \frac{\Phi_0}{2\pi\xi^2 2\lambda^2/\xi^2}(\ln(\lambda/\xi) + \varepsilon) = \frac{\Phi_0}{2\pi\xi^2}\frac{(\ln\kappa + \varepsilon)}{2\kappa^2} = B_{c2}\frac{(\ln\kappa + \varepsilon)}{2\kappa^2} \tag{7.58}$$

We define the function $K(\kappa)$ by the relation

$$\frac{B_{c2}}{B_{c1}} = \frac{2\kappa^2}{(\ln\kappa + \varepsilon)} \equiv K(\kappa) \tag{7.59}$$

Taking the first and second derivatives of $K(\kappa)$ one finds this function to have a minimum, occurring at a κ-value $\kappa e^{1/2-\varepsilon}$.

In order to carry out a test procedure, we need to distinguish between isotropic systems, where Eqs 7.57, 7.58 and 7.59 are the predictions, and layered structures as in the high-T_c cuprate materials. The results are remarkably simple [12]: The $K(\kappa)$-functions have precisely the same form as in Eqs 7.58 and 7.59. With the field along the c-axis one obtains

$$B_{c2}^c/B_{c1}^c = \frac{2\kappa_c^2}{(\ln\kappa_c + \varepsilon)} \equiv K(\kappa_c); \quad \kappa_c = \lambda_{ab}/\xi_{ab} \tag{7.60}$$

With the field in the ab-plane we get

$$B_{c2}^{ab}/B_{c1}^{ab} = \frac{2\kappa_{ab}^2}{(\ln\kappa_{ab} + \varepsilon)} \equiv K(\kappa_{ab}); \quad \kappa_{ab} = \left[\frac{\lambda_{ab}\lambda_c}{\xi_{ab}\xi_c}\right]^{\frac{1}{2}} \tag{7.61}$$

We then have

$$K(\kappa) = K(\kappa_c) = K(\kappa_{ab}) = 2\kappa^2/(\ln\kappa + \varepsilon) \tag{7.62}$$

Therefore, in all these cases the same maximum of B_{c2}/B_{c1} is predicted to exist, given by the criterion $\kappa = e^{1/2-\varepsilon}$.

Data in the literature on independent measurements of *all* parameters B_{c1}, B_{c2}, λ, and ξ turn out to be rather scarce. It is common practice to use the GLA-relations to calculate for instance the coherence length on the basis of a measurement of B_{c2}, using the relation we have found in Section 7.5.1. Figure 7.9 shows a plot of the function $K(\kappa)$ given in Eq. 7.62.

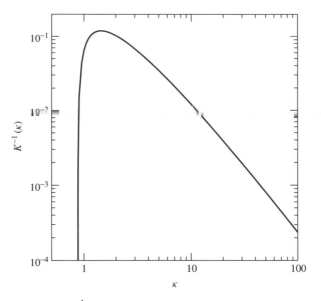

Figure 7.9 The function $K^{-1}(\kappa) = B_{c1}/B_{c2}$ derived in Eq. 7.62 from GLA theory, with $\epsilon = 0.13$.

GLA theory has been an enormously powerful tool by which to work out the physics of the vortex state of superconductors, even well beyond the limits of expected applicability.

7.7   Surface superconductivity and B_{c3}

So far our treatment of the Ginzburg-Landau equations at the mean-field level has not taken into account the surface of the sample. It has been implicitly assumed that we are working in the bulk. At the surface of the superconductor, however, some additional boundary conditions need to be imposed on the solutions, since one can quite reasonably expect the presence of an interface between the superconductor an a non-superconducting material such as a normal metal or an insulating material, to affect the nucleation of superconductivity in the material.

For the purposes of studying the nucleation (i.e. onset) of superconductivity, it suffices to consider the linearized version of the first Ginzburg-Landau equation, namely

$$\frac{1}{2m}(-i\hbar\nabla - 2e\mathbf{A})^2\psi = -\alpha\psi \tag{7.63}$$

or equivalently

$$\left(-i\nabla - \frac{2e}{\hbar}\mathbf{A}\right)^2\psi = -\frac{2m\alpha}{\hbar^2}\psi \equiv \frac{1}{\xi^2}\psi \tag{7.64}$$

where we have introduced the characteristic length

$$\xi^2 \equiv -\frac{\hbar^2}{2m\alpha} \qquad (7.65)$$

We now make a couple of simplifying assumptions. a) We consider ξ to be small compared to the bulk dimension of the sample. b) We assume that the surface is such that the local radii of curvature of the boundaries are large compared to ξ. This allows us to consider the problem of a plane boundary. We first take the external magnetic field to be parallel to the surface, which is located in the (y, z)-plane, with magnetic field $\mathbf{B} = B\hat{z}$. The superconducting sample is located in the half-space $x > 0$, while we take the non-superconducting material to be located in the half-space $x < 0$. The latter material is here chosen to be either vacuum or an insulating material, we will only briefly consider the case of a metallic coating of the superconductor, where the boundary conditions to be imposed are different. Then, the boundary condition to be imposed is that the component of the gauge-invariant supercurrent normal to the surface is zero, i.e.

$$\left(-i\hbar\frac{\partial}{\partial x} - 2eA_x\right)\psi|_{x=0} = 0 \qquad (7.66)$$

We choose to work in the gauge $\mathbf{A} = Bx\hat{y}$ and look for a solution to equation of the form

$$\psi = f(x)e^{iky} \qquad (7.67)$$

Inserting this Ansatz into the GL-equation and the boundary condition we find

$$-\frac{\hbar^2}{2m}\frac{d^2 f}{dx^2} + \frac{1}{2m}(\hbar k - 2eBx)^2 f = -\alpha f \qquad (7.68)$$

with the constraints that $df/dx = 0$ for $x = 0$ and $x = \infty$. The above equation is the Schrödinger equation for a harmonic oscillator with frequency

$$\omega = \frac{2eB}{m} \qquad (7.69)$$

and with a minimum of the quadratic potential located at

$$X_0 = \frac{\hbar k}{2eB} \qquad (7.70)$$

The complication arises because the boundary condition states that the solution must be flat at a position $x = 0$ while the minimum of the potential is located at $x = X_0$. If $X_0 = \infty$ or $X_0 = 0$, this is simple, because when the minimum of the potential is located very far from the surface ($X_0 = \infty$) we can ignore the boundary condition, and when $X_0 = 0$ the boundary condition is satisfied by the standard solution to the Schrödinger equation in terms of a Gaussian multiplied by Hermite polynomials with Eigenvalues

$$\omega = \hbar\omega\left(n + \frac{1}{2}\right) = \frac{2e\hbar B}{m}\left(n + \frac{1}{2}\right); n = 0, 1, 2, \ldots \tag{7.71}$$

The lowest Eigenvalue for these two cases ($X_0 = (0, \infty)$) quite clearly corresponds to the case where *nucleation* of superconductivity is energetically most favourable (for $X_0 = (0, \infty)$), and corresponds to

$$\frac{2e\hbar B}{m}\frac{1}{2} = -\alpha = \frac{\hbar^2}{2m\xi^2}$$

$$B = \frac{\hbar}{2e\xi^2} = \frac{(h/2e)}{2\pi\xi^2} = \frac{\Phi_0}{2\pi\xi^2} = B_{c2} \tag{7.72}$$

Hence, we rederive the well-known result that when the surface of the boundary does not constitute a complicating factor in the GL equations (which happens deep in the specimen or right at the surface) then superconductivity is nucleated at the upper critical magnetic field B_{c2}. (But note that this type of sharp onset of superconductivity at a sharply defined field is an artifact of mean-field theory! In reality, B_{c2} is a crossover line).

Hence, we reach the conclusion that the surface has consequences for the solution to the GL equations only at *intermediate* values of X_0. Let us first see precisely what we mean by this statement. We start by considering the lowest energy solution for X_0 very large, which is given by the well-known harmonic oscillator function for $n = 0$, namely

$$f(x) = A\exp\left[-\frac{m\omega}{2\hbar}(x - X_0)^2\right]$$

$$= A\exp\left[-\left(\frac{x - X_0}{\xi}\right)^2\right] \tag{7.73}$$

Thus, we see that the solution is located within a distance ξ of the position X_0. Hence, we conclude that *within a sheeth of thickness ξ of the surface of the sample, nucleation of superconductivity is altered by the presence of a surface.*

We must now think of the Schrödinger equation

$$-\frac{\hbar^2}{2m}\frac{d^2 f}{dx^2} + \frac{1}{2m}(\hbar k - 2eBx)^2 f = -\alpha(X_0)f \tag{7.74}$$

as en Eigenvalue problem where the Eigenvalue is X_0-dependent, and our task is to minimize this with respect to X_0 subject to the boundary condition on f at the surface. This may in principle be done by writing down the *general* solution to this equation in terms of a linear combination of two degenerate hypergeometric functions, finding the expansion coefficients from the boundary conditions at the surface and at infinity, thus determining the lowest eigenvalue. This requires numerical work. We proceed by a different analytical method which yields the correct answer to within 2%. We start by rewriting the equation as

$$-\frac{d^2 f}{dx^2} + \left(\frac{m\omega}{\hbar}\right)^2 (x - X_0)^2 f = -\frac{2m\alpha(X_0)}{\hbar^2} f \tag{7.75}$$

Introducing $z = \sqrt{m\omega/\hbar}x$, $z_0 = \sqrt{m\omega/\hbar}X_0$, and $\beta = -2\alpha/\hbar\omega$, we find

$$-\frac{d^2 f}{dz^2} + (z - z_0)^2 f = \beta f \tag{7.76}$$

where the task now is to find the lowest possible value of β subject to the boundary conditions $df/dz = 0$ at $z = (0, \infty)$. This problem may be phrased as the following variational problem of minimizing the functional

$$\beta = \frac{\displaystyle\int_0^\infty dx \left[\left(\frac{df}{dz}\right)^2 + (z - z_0)^2 f^2\right]}{\displaystyle\int_0^\infty dx f^2} \tag{7.77}$$

with respect to variations in f. The Euler-Lagrange equation for this variational problem is precisely the above scaled differential equation. To do the minimization, we use a trial wavefunction of a form similar to the lowest eigenvalue state of the harmonic oscillator, but now shifted with respect to the minimum of the oscillator potential

$$f = \exp\left[-\frac{1}{2}bz^2\right] \tag{7.78}$$

which we note satisfies the correct boundary conditions. It is a simple matter to perform the necessary integrals, and we obtain

$$\beta = \frac{b}{2} + \frac{1}{2b} - \frac{2z_0}{\sqrt{\pi b}} + z_0^2 \tag{7.79}$$

This is now to be minimized with respect to b, which yields b in terms of z_0. This particular value of b is then to be substituted back in the expression for β, which then is minimized with respect to z_0. One may just as well do the minimization in reverse order, first minimize with respect to z_0 and then with respect to b. We choose the latter.

$$\frac{\partial\beta}{\partial z_0} = -\frac{2}{\sqrt{\pi b}} + 2z_0 = 0 \tag{7.80}$$

yielding $z_0 = 1/\sqrt{\pi b}$. Substituting this back in β and minimizing with respect to b, we find

$$\frac{\partial \beta}{\partial b} = \frac{1}{2} - \frac{1}{2b^2} + \frac{1}{\pi b^2} = 0 \tag{7.81}$$

which yields $b = \sqrt{1 - 2/\pi}$. Substituting this back into β yields

$$\beta_{\min} = \sqrt{1 - \frac{2}{\pi}} \tag{7.82}$$

We now go back to the definition of $\beta = -2\alpha/(\hbar\omega)$, and reintroduce the characteristic length $1/\xi^2 = -2m\alpha/\hbar^2$ as well as the frequency $\omega = 2eB/m$ to find

$$\beta_{\min} = \sqrt{1 - \frac{2}{\pi}} = \frac{\hbar}{2eB\xi^2} = \frac{h/2e}{2\pi\xi^2 B} = \frac{B_{c2}}{B} \tag{7.83}$$

Hence, we find that in the presence of a surface, i.e. when nucleation of superconductivity occurs within a length ξ from the surface plane, the upper critical field at which it starts to nucleate as the field is lowered from above, is given by

$$B = B_{c3} = \frac{1}{\sqrt{1 - 2/\pi}} B_{c2} = 1.66 B_{c2} \tag{7.84}$$

A more careful numerical analysis based on hypergeometric functions yields the slightly larger value for the surface nucleation field

$$B_{c3} = 1.69 B_{c2} \tag{7.85}$$

We see that our analytical estimate is in excellent agreement with this, differing by less than 2%. It is straightforward, but a bit tedious, to obtain essentially perfect agreement with the numerical result if we use a slightly refined variational function $f(x) = (1 + cz^2)\exp(-bz^2/2)$.

Physically, the boundary condition $df/dx = 0; x = 0$ means that the order parameter ψ is *enhanced* at the surface, since the solution to the problem ignoring boundary conditions has a negative derivative. Thus, we may view this as pinning of Cooper-pairs to the surface of the sample. We should, however, be cautious about regarding B_{c3} as some sort of critical field, it is after all a surface phenomena we are talking about, and superconductivity with long-range order in the superconducting order parameter does not exist at any finite temperature in two dimensions, as we shall see in Chapter 9. Moreover, as the above results show, B_{c3} is tied to the notion of the bulk upper critical field B_{c2}, which itself is not a critical field either, but rather defines a crossover line. B_{c3} is therefore best thought of as a surface enhanced crossover line where one sees the first nucleations of locally superconducting islands in the system. The main point about the above results, is that when superconductivity occurs in a system, *it tends to nucleate at the surface of an ideal (defect-free) sample, not in the interior of it*. Alternatively, if we have material defects in the interior of the system, superconductivity may nucleate in the vicinity of such defects as well.

Let us also consider the case of an inclined magnetic field with respect to the sample surface. We imagine tilting the field in the (x, z) plane and angle θ with respect to the sample surface located in the (y, z)-plane. hence we have

$$\mathbf{B} = B \cos(\theta)\hat{z} + B \sin(\theta)\hat{x} \tag{7.86}$$

We now work in a gauge such that the vector potential is given by

$$\mathbf{A} = B[x \cos(\theta) - z \sin(\theta)]\hat{y} \tag{7.87}$$

The first Ginzburg-Landau equation now becomes

$$-\frac{\hbar^2}{2m}\frac{\partial^2 \psi}{\partial x^2} - \frac{\hbar^2}{2m}\frac{\partial^2 \psi}{\partial z^2} + \frac{\hbar^2}{2m}\left[i\frac{\partial}{\partial y} + \frac{2eB}{\hbar}[x \cos(\theta) - y \sin(\theta)]\right]^2 \psi = -\alpha\psi \tag{7.88}$$

Again, introducing $1/\xi^2 = -2m\alpha/\hbar^2$ and new coordinates $\zeta = \sqrt{m\omega/\hbar}x$, $\eta = \sqrt{m\omega/\hbar}y$, $\rho = \sqrt{m\omega/\hbar}z$ with $\omega = 2eB/m$, and defining $\beta = -2\alpha/(\hbar\omega)$ the equation may be written on the form

$$-\frac{\partial^2 \psi}{\partial \zeta^2} - \frac{\partial^2 \psi}{\partial \rho^2} + \left[i\frac{\partial}{\partial \eta} + [\zeta \cos(\theta) - \rho \sin(\theta)]\right]^2 \psi = \beta\psi \tag{7.89}$$

We now try a solution of the form

$$\psi(\zeta, \eta, \rho) = g(\zeta, \rho) \exp[i\zeta_0\eta] \tag{7.90}$$

and obtain the following differential equation

$$-\frac{\partial^2 g}{\partial \zeta^2} - \frac{\partial^2 g}{\partial \rho^2} + [\zeta \cos(\theta) - \zeta_0 - \rho \sin(\theta)]^2 g = \beta g \tag{7.91}$$

This equation is the Euler-Lagrange equation for the variational problem of minimizing the following functional

$$\beta = \frac{\int_0^\infty d\zeta \int_{-\infty}^\infty d\rho \left[\left(\frac{\partial g}{\partial \zeta}\right)^2 + \left(\frac{\partial g}{\partial \rho}\right)^2 + (\zeta \cos(\theta) - \zeta_0 - \rho \sin(\theta))^2 g^2\right]}{\int_0^\infty d\zeta \int_{-\infty}^\infty d\rho g^2} \tag{7.92}$$

Here, we follow the strategy of the previous treatment and try a variational Ansatz of the form

$$g(\zeta, \rho) = \exp\left[-\frac{1}{2}a\zeta^2 - \frac{1}{2}b\rho^2\right]$$ (7.93)

Performing the necessary Gaussian integrals, we find

$$\beta = \frac{1}{2}\left(b + \frac{1}{b}\cos^2(\theta)\right) + \frac{1}{2}\left(a + \frac{1}{a}\sin^2(\theta)\right) + \cos^2(\theta)\left(\zeta_0^2 - \frac{2\zeta_0}{\sqrt{\pi b}}\right)$$ (7.94)

Minimizing this with respect to a, b, and ζ, we find

$$\beta = \cos(\theta)\sqrt{1 - \frac{2}{\pi}} + \sin(\theta)$$

$$b = \cos(\theta)\sqrt{1 - \frac{2}{\pi}}$$ (7.95)

with $a = \sin(\theta)$ and $\zeta_0 = 1/\sqrt{\pi b}$. Now, in analogy with the $\theta = 0$ case, where we had

$$\beta_{\min} = \sqrt{1 - \frac{2}{\pi}} = \frac{B_{c2}}{B_{c3}(0)}$$ (7.96)

we may define

$$\beta = \frac{B_{c2}}{B_{c3}(\theta)}$$ (7.97)

which leads to

$$\frac{1}{B_{c3}(\theta)} = \frac{1}{B_{c3}(0)}\cos(\theta) + \frac{1}{B_{c2}}\sin(\theta)$$ (7.98)

or equivalently

$$B_{c3}(\theta) = \frac{B_{c3}(0)}{\cos(\theta) + \frac{\sin(\theta)}{\sqrt{1 - 2/\pi}}}$$ (7.99)

Hence, we see that when the field is perpendicular to the surface, $B_{c3} = B_{c2}$, and the variation in B_{c3} is smooth in between the limiting cases $\theta = 0$ and $\theta = \pi/2$. In particular

$$-\left(\frac{1}{B_{c3}}\frac{dB_{c3}}{d\theta}\right)\bigg|_{\theta=0} = \frac{B_{c3}(0)}{B_{c2}} = \frac{1}{\sqrt{1 - \frac{2}{\pi}}} = 1.66$$ (7.100)

A more careful numerical analysis of the problem of tilted magnetic fields yields for the above slope the result 1.35, which is not too far away from our analytical estimate. A more accurate variational result may be obtained by refining the variational function, to a form like $g(\zeta, \rho) = (1 + c\zeta^2 + d\rho^2) \exp(-(a\zeta^2 + b\rho^2)/2)$.

In the case of a metallic coating of the superconductor, a current can in principle go across the interface at $x = 0$, such that the boundary conditions are different. Consider again the case of a magnetic field parallel to the surface, with the vector potential given by $\mathbf{A} = Bx\hat{y}$. The boundary condition to be used is then given by, crudely

$$\frac{d\psi}{dx}\Big|_{x=0} = -\frac{\psi}{l} \tag{7.101}$$

where l is some characteristic length that provides a proximity effect of superconductivity into the metal. This reduces the flatness of ψ at the surface, i.e. the pinning of Cooper pairs to the surface is reduced. Hence one can expect Cooper pairs at the surface to diffuse into the metal and be destroyed. A metallic overlayer thus acts as a pair-breaker and this will impede surface nucleation of superconductivity and suppress surface superconductivity above B_{c2}.

Finally, a cautionary remark is in order concerning hard superconductors, i.e. those with large bulk values of κ. The treatment in this section has been exclusively at the mean field level. As the surface of the superconductor is approached from within one must expect a continuous depletion of superfluid density. As we shall demonstrate in Chapters 9 and 10, the superfluid density is connected to the phase-stiffness of the order parameter. This will also reduce the correlation length ξ, since this is basically a measure of the length over which the local phases of the order parameter are correlated. A reduction in ξ would lead to an increased estimate for the upper critical field compared to the bulk value.

8

More on the Flux-line System

8.1 Elementary pinning forces and simple models

8.1.1 The concept of a pinning force

We have previously derived the force per unit length from one vortex, labelled as number 1, on another parallel vortex labelled as number 2, in Chapter 6. This situation is illustrated in Figure 8.1. We found in Chapter 6:

$$f_{2,x} = j_y \Phi_0 \tag{8.1}$$

which we can rewrite as

$$f_2 = J_1 \times \Phi_0 \tag{8.2}$$

where J_1 is the supercurrent density from vortex number 1 at the position of vortex number 2. If there is a current from other vortices, or an additional external current, giving a total current density J_s, this will determine the total force per length, f on the reference vortex, i.e.

$$f = J_s \times \Phi_0. \tag{8.3}$$

f becomes a volume force, F, if we divide both sides by unit area, and we have

$$F = J_s \times B \tag{8.4}$$

This force will, if acting alone, cause a continuous displacement of flux lines as long as the force F is acting upon the flux line lattice. This is a basic problem encountered when a type II superconductor is to be used for current and energy transport. The forced motion of the vortices against the friction of the electron gas, normal and superfluid, causes dissipation of energy, and the superconductor

Superconductivity: Physics and Applications   Kristian Fossheim and Asle Sudbø
© 2004 John Wiley & Sons, Ltd   ISBN 0-470-84452-3

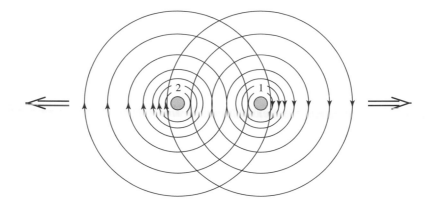

Figure 8.1 Two interacting flux lines, repulsive forces shown.

behaves essentially like a normally conducting material. To stop or prevent this motion, a pinning force F_p of magnitude at least equal to F is necessary. When $|F| < F_p$ the flux-line lattice will be at rest; whereas the lattice will move if $|F| > F_p$. The precise value of F corresponding to the volume pinning force F_p is specially significant. When the applied force density is equal to F_p the corresponding current density is defined as the critical current density, J_c. We now have

$$F_p = J_c B$$

or

$$J_c \equiv F_p / B \tag{8.5}$$

This is the conceptual definition of J_c. How it is to be determined is not obvious. F_p is not usually measured directly. Rather, by judicious choice of a *criterion* for determination of J_c, one can *calculate* the pinning force based on that definition.

8.1.2 Pinning force and flux gradient

Equation 8.4 could equally well be written by means of Maxwell's equation $\nabla \times H = J$, as

$$F = (\nabla \times H) \times B \tag{8.6}$$

This points to the significant fact that a current corresponds to the existence of a gradient of H. We have here the curl of H, $\nabla \times H$, but in well defined cases this is just a negative gradient. Take for example $\boldsymbol{H} = \hat{k}H$ along the z-axis and the current along the y-axis. We get, using $\hat{i}$, $\hat{j}$, $\hat{k}$ as unit vectors

$$\boldsymbol{J} = \nabla \times \boldsymbol{H} = -\hat{j}\frac{\partial}{\partial x}H = J_y\hat{j} \tag{8.7}$$

The volume force acting on the flux lattice is now, on introducing the B-field

$$|\boldsymbol{F}| = \frac{B}{\mu_0}\frac{\partial B}{\partial x} \tag{8.8}$$

This relation provides a basis for a definition of critical current density J_c in the Bean model.

8.2   Critical state and the Bean model

The critical current density J_c will in general depend on both $\boldsymbol{r}$ and $\boldsymbol{B}$. This should be regarded not only as a possibility, but as a reality under usual conditions. This fact tends to complicate the analysis of the critical current condition substantially. Even so, simplified models are quite often useful. Bean [87] suggested to model J_c as a constant value corresponding to regarding $J_c = \mu_0^{-1}(\mathrm{d}B/\mathrm{d}x)$ as constant. Figure 8.2 shows sketches of the correspondence between J_c and the flux gradient $\partial B/\partial x$ in several situations, including the reversal of flux gradient on reducing the applied field B_a after it has been raised to a maximum value B_m, larger than B^*, which is the field where the flux first penetrates to the center. In all these sketches $|J_c|$ is taken to be a constant value in space and time. This demonstrates the properties of the simplest form of Bean model. The situation where the flux front reaches the center, and the current density is J_c everywhere, is the *Bean critical state*. It allows a simple and often quite useful analysis of what is going on in the sample under varying external applied field.

The Bean model, as shown in Figure 8.2, takes J_c to be independent of B. This is a condition which is never quite correct. In cases where the pinning force is independent of B, J_c will be proportional to B^{-1}. We have also to keep in mind that higher temperatures necessarily will influence J_c negatively, and that a spatial dependence of J_c is also present unless the material is perfectly homogeneous with regard to pinning force. Experience tells us that relatively simple modifications of the Bean model may be sufficient to account for more complicated situations. One modification, often referred to as the Kim–Anderson model [88], is described by the following expression for J_c

$$J_c(T, B) = \frac{J_0(T)}{1 + B/B_0} \tag{8.9}$$

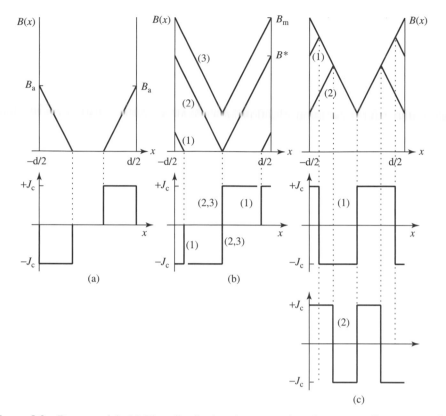

Figure 8.2   Bean model. (a) Flux distribution (upper part) and corresponding current distribution (lower part) after application of an external field B_a which is subsequently kept constant. (b) Similar to situation (a), labelled as (1); then (2) after raising B_a to B^* where the flux front reaches the centre of the sample, the situation referred to as the critical state, and (3) after increasing B_a to B_m above B^*. (c) Shows the effect of lowering the applied field, case (1) and case (2), by different amounts.

which has been shown to agree well with experimental results in some cases. Here $J_0(T)$ is the critical current density at zero field and B is the local flux density. B_0 is a model parameter to be determined by experiment.

Next we derive a simple relationship between J_c and the magnetization, implied to exist in Figure 8.3. Imagine that we have measured a complete magnetization curve $M(H)$, looking like the one sketched in Figure 8.3a. In Figure 8.3b we indicate how the area A represents the difference in magnetisation of the sample at point (1) and point (2) in Figure 8.3a.

The total magnetization difference at H_1 is $\Delta M = M_\uparrow - M_\downarrow = \Delta M(H_1)$, where we have once arrived at H_1 from lower field (1), and once from higher field (2), as illustrated in Figure 8.3b. The two approaches to H_1 result in different magnetisations $M_\uparrow$ and $M_\downarrow$, and with a difference $M_\uparrow - M_\downarrow$. This construction can also be made via graphical display of the Bean model. Now,

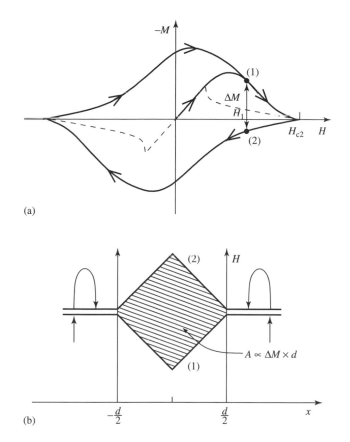

Figure 8.3   (a) is a sketch of magnetization versus field H in a type II superconductor with pinning. Magnetization starts out from zero. Points (1) and (2) refer to the same magnetic field with different magnetization. The dotted line is the magnetization without pinning. In (b) we sketch how the magnetisation difference ΔM is obtained.

since $J_c = \frac{M(x)}{x}$, we have $M(x) = J_c x$ in the Bean model. We calculate

$$M_\downarrow = -\frac{1}{d/2} \int_0^{d/2} J_c x \, dx = -\frac{1}{4} J_c d \qquad (8.10)$$

Correspondingly we get $M_\uparrow = \frac{1}{4} J_c d$. The difference is then $\Delta M = M_\uparrow - M_\downarrow = \frac{1}{2} J_c d$, or

$$J_c = 2\frac{\Delta M}{d} \qquad (8.11)$$

This allows an estimate of J_c from a half cycle measurements of $M(H)$, ('Magnetization J_c'). In other sample geometries, a numerical, geometry dependent factor of the order of unity will modify the right-hand side of Eq. 8.11.

The simple, and very useful rule we glance from this, is that the wider the magnetization loop is, the higher is J_c. This kind of data analysis is widely used to obtain J_c. Before computing the J_c-value, one should also attempt to remove the equilibrium part of the magnetization curve, i.e. the $M(H)$ curve which is obtained without pinning. This magnetisation method for measurement of J_c is often easier to use than the transport method.

8.3 Flux-line dynamics, thermal effects, depinning, creep and flow

8.3.1 TAFF, flow and creep: Definitions

Current transport measurement are always central in the study of superconductivity. In Figure 8.4 we have sketched measured voltage V versus current I, (alternatively electric field E versus current density J) in the presence of an external magnetic field which has created a certain vortex density in the superconductor. The definitions which apply to the 3 different parts of the I–V curve are: 'thermally assisted flux flow' (TAFF) in the low current end, 'flux flow' in the high current region, and 'flux creep' in the transition region between the two. While the latter two terms were commonly used in low-T_c research, the new term, TAFF, was coined for this new aspect of high-T_c, I–V characteristics.

The particularly important feature of this plot is the highly nonlinear region, where the critical current I_c is marked in the region crossing over between TAFF and flux flow. What has been agreed on as a practical measure is to define I_c at the point where $V = 1\,\mu V$, or similarly, $E = 1\,\mu V/cm$. The point is that

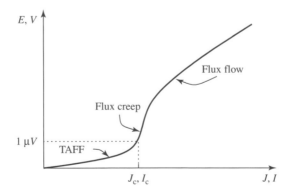

Figure 8.4 A sketch for the definition of I_c or J_c by transport measurement. This curve has a linear V versus I at low I, followed by a highly nonlinear region at intermediate values of I, and finally a linear relation again. The effective resistance is quite different in the two regions where ohmic behaviour is found. The plot can be read either as an $E-J$ characteristic or as an $I-V$ characteristic.

there is no way to define sharply what I_c is. One has to agree on a criterion related directly to measurements. However, it should be noted that flux flow is a phenomenon which arises when the current is high enough to essentially tear loose the whole flux lattice, and set it in sliding motion. It is then quite reasonable and conceptually sound to define I_c and the corresponding J_c as the point where this flux motion is just about to begin.

All of the above implies the presence of various kinds of flux line dynamics present during the measurement.

8.3.2 Flux flow

If the superconductor is pin-free, or if the Lorentz force is so strong that it exceeds the pinning force, the vortex velocity v is limited by the viscous drag exerted by the surrounding medium, through the force balance $\Phi_0 J = \eta v$ giving

$$v = \frac{1}{\eta}\Phi_0 J \tag{8.12}$$

Multiplying the above equation with B gives the E-field

$$E = Bv = \frac{B}{\eta}\Phi_0 J = \rho_{ff} J \tag{8.13}$$

with a flux flow resistivity

$$\rho_{ff} = \frac{1}{\eta}B\Phi_0 \tag{8.14}$$

The most widely accepted theory of flux-flow was worked out by Bardeen and Stephen [89]. They considered a vortex core of radius ξ, which can be treated as if it were in the normal state. When a vortex is set into linear motion the moving flux pattern will generate an electric field in its surroundings, the vortex core included. The induced resistive currents in the vortex core creates a loss $W = \rho_{ff} J^2$, and thereby a flux-flow resistivity. The result is

$$\rho_{ff} = \rho_n \frac{B}{B_{c2}} \tag{8.15}$$

where ρ_n is the normal state resistivity. It is not difficult to guess such a relation: when B reaches B_{c2}, ρ_{ff} becomes ρ_n as it should. When B goes to zero, ρ_{ff} goes to zero as it should, and in between the factor B/B_{c2} is simply the fraction of the material occupied by the vortex cores. ρ_{ff} is therefore equal to ρ_n times this fraction. The result is the same as if a uniformly distributed current were passed through the material and losses would appear from the fraction of the current which passed through the core regions.

8.3.3   Thermally activated flux creep: Anderson model

Random thermal forces may induce escape of a vortex from its potential minimum at the pinned position. Depinned vortices will move, or 'creep' in order to
relax the critical state field gradient. We outline the theory of flux creep which
was put forward by Anderson and Kim [88]. The escape rate R from pin sites
is determined by the Boltzmann factor,

$$R = \nu_0 e^{-U/kT} \tag{8.16}$$

where U is the height of the activation barrier and ν_0 is some microscopic attempt
frequency, typically of order $10^{-8} - 10^{-10}\,\mathrm{s}^{-1}$. As indicated in Figure 8.5, the
effective height of the barrier will be reduced with increasing current. We can
model this by writing $U = U_0(1 - J/J_c)$. Here J_c is the critical current density, and is proportional to the maximum energy gradient dU/dx. The decay of a
finite (non-equilibrium) magnetization as well as its associated current density J,
proportional to R, for forward jumps is

$$\frac{dJ}{dt} \propto e^{-\frac{U_0}{kT}(1-J/J_c)} \tag{8.17}$$

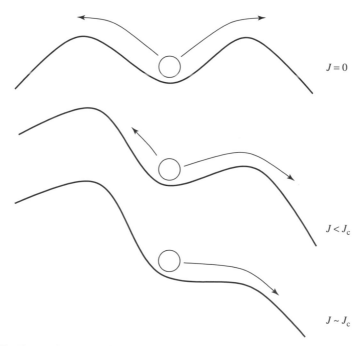

Figure 8.5   Energy landscape for a flux line in a pinning environment and a uniform external
force.

with the long-time solution

$$J(t) \propto \text{const.} - \ln(t).$$ (8.18)

Hence, at long times the magnetization current and therefore also the magnetic moment decay proportional to $\ln t$. In a transport measurement the $E-J$-characteristics will be exponential, since the force equation $eE = ev \times B$ can be written as

$$E = Blv_0 e^{-\frac{U_0}{kT}(1-J/J_c)}$$ (8.19)

where l is the effective jump distance. This equation is of course the equivalent of an $I-V$ characteristic. It is highly non-linear as required by experimental observation sketched in Figure 8.4, and the model contains simple creep behaviour.

The flux-creep theory described here is a single particle model, i.e. depinning events occur by escape of individual vortices, independently at individual pins. The simplest way of accounting for collective dynamics is to assume that bundles of correlated flux-lines move simultaneously. Anderson suggested that all flux-lines within a distance λ would constitute a correlated flux bundle. However, the bundle concept does not introduce truly new dynamics, as its effect can be shown to alter only phenomenological parameters like U, J_c, l etc. A more advanced approach to collective dynamics is the theory of collective creep, a subject we return to below.

8.4 Single particle TAFF

The creep rate as given above has a finite value even for $J = 0$. This unphysical result occurs because we did not take into account the asymmetry of forward and backward flux jumps. If one subtracts jumps in the backwards direction from forward jumps in the rate equation, one obtains the E-field,

$$E = Blv_0 \left(e^{-\frac{U_0}{kT}(1-J/J_c)} - e^{-\frac{U_0}{kT}(1+J/J_c)} \right)$$

$$= 2Blv_0 e^{-U_0/kT} \sinh\left(\frac{U_0 J}{kT J_c}\right)$$ (8.20)

which gives the same result as earlier at large currents, $J \gg J_c kT/U_0$, but at low current it gives a linear resistance, the so-called TAFF resistance (thermally assisted flux flow)

$$E = 2Blv_0 \frac{U_0}{J_c kT} J = \rho_{TAFF} J$$ (8.21)

According to this analysis, magnetization current and magnetic moment will decay exponentially with time in a way analogous to skin effect in normal superconductors. Linear resistance has been found at low current densities in a large range of temperatures below the transition. The relevance of the TAFF diffusive dynamics for this regime was discussed by Kes and coworkers [90].

8.5 FLL elasticity and pinning

8.5.1 Collective effects

Among the more important recent developments is a deeper insight into the wavevector dispersion of the elastic constants, which is important for melting as well as for pinning of the flux line lattice. In this Section we give an elementary discussion of pinning effects based on the elasticity theory worked out before in Chapter 6. An elastic energy is always written as $(1/2)c\varepsilon^2$, where c is an elastic constant, and ε is a corresponding elastic strain. This is the case whether the strain is applied to the crystal lattice as well as to the flux line lattice. What concerns us in the following is the flux line lattice.

To visualize the meaning of typical strains and elastic constants in the FLL, we refer to Figure 8.6 which illustrates the three important strains, with corresponding elastic constants c_{11} for a strain ε_1, c_{44} for a strain ε_4, and c_{66} for a strain ε_6.

Calculations in Section 6.7 show that an important k-vector dispersion occurs at high k due to non-local response. By high k we mean k-vectors that are comparable to the inverse screening length λ^{-1}. We have three lengths to compare: (i) The lattice constant of the FLL, $a = [(2/\sqrt{3})\Phi_0/B]^{\frac{1}{2}}$, (ii) the screening length λ and (iii) the wavelength λ_ε of the strain. Due to the high density of vortices giving a short length a at a typical magnetization, we usually have $\lambda \gg a$ except very close to B_{c1}. What is important for the elastic response under such conditions is whether the applied strain has a k-vector which is of the order of λ^{-1} or greater. Under such conditions the flux line interactions are substantially weakened, since the denominator in the expression for c_{ii} typically goes like $1 + k^2\lambda^2$, with an exception for c_{66}. We refer to Chapter 6 regarding the elastic properties of the FLL. In the isotropic case the results are:

$$c_{11}(k) \approx \frac{B^2}{\mu_0(1 + k^2\lambda^2)}; \quad c_{66} \approx \frac{B}{8\mu_0} \cdot \frac{\Phi_0}{2\pi\lambda^2} \approx \frac{B B_{c1}}{8\mu_0};$$

$$c_{44}(k) \approx c_{11}(k) + 2c_{66} \ln \frac{\kappa^2}{1 + k_z^2\lambda^2} \tag{8.22}$$

We notice that c_{66} is non-dispersive and very small compared to the two others. In high-T_c superconductors where λ is very long (≈ 150 nm) the softening of the

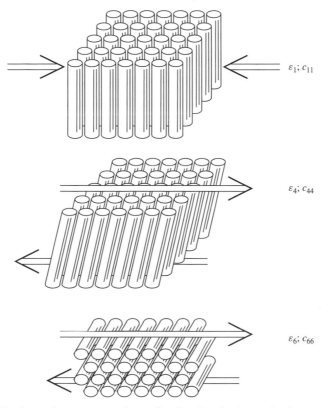

Figure 8.6 Elastic strains ε_1, ε_4, and ε_6 of a flux line lattice under homogeneous strain, i.e. $k = 0$. Corresponding elastic moduli are indicated in each case.

FLL becomes important at k-vectors with wavelength even larger than 150 nm, and all shorter wavelengths. The lattice constant a at 0.2 T is about 100 nm. From there on up in field, screening currents from surrounding flux lines overlap more and more, with the gradual breakdown of net screening, and accompanied by softer and softer FLL.

Let us estimate the effects on thermal phonons at the Brillouin zone boundary of the crystal lattice. These have a k-vector in the ab-plane of approximately $\pi(\mathrm{nm})^{-1}$. The product $k\lambda$ is then about 300. This means that the elastic constants c_{44} and c_{11} for tilt and compression are two orders of magnitude softer as seen by Brillouin zone phonons than by zone centre, $k = 0$ modes. In other words, acoustic waves with low wavevector will propagate faster in the presence of a magnetic field, while at the zone boundary it makes almost no difference. (Recall here that the elastic constant of the FLL is to be added to that of the crystal lattice.) We note that at $k = 0$, c_{66} typically is weaker than c_{11} and c_{44} by a factor $\frac{1}{8}\frac{B_{c1}}{B}$ as seen from the approximation $c_{66} \approx \frac{BB_{c1}}{8\mu_0}$. The above discussion underscores the susceptibility of the FLL of high-T_c materials to thermal fluctuations.

We refer here for completeness the results for c_{44} and c_{66} in the anisotropic case with the field along the c-axis, from which c_{11} can also be found.

$$c_{44}(\boldsymbol{k}) \approx \frac{B^2}{\mu_0(1 + (k_x^2 + k_y^2)\lambda_c^2 + k_z^2\lambda_{ab}^2)} + \frac{2c_{66}}{\Gamma^2} \ln \frac{\kappa^2\Gamma^2}{1 + k_z^2\lambda_{ab}^2} \tag{8.23}$$

$$c_{66} \approx \frac{B}{8\mu_0} \times \frac{\Phi_0}{2\pi\lambda_{ab}^2} \tag{0.24}$$

Here $\kappa = \lambda_{ab}/\xi_c$ and $\Gamma = \lambda_c/\lambda_{ab} = \xi_{ab}/\xi_c$.

Except for differences arising in symmetry properties, we recognize the results from the isotropic case: c_{66} remains low and non-dispersive, and its magnitude relative to c_{44} is still of the order of $B_{c1}/8B$; more precisely $(\Phi_0/2\pi\lambda_{ab}^2)/8B$.

The key idea in collective weak pinning is to describe the distortion of the vortex lattice in terms of a correlation volume, $V_c = R_c^2 L_c$ where R_c and L_c are the transverse and longitudinal correlation lengths, respectively. A correlated volume may be thought of as a volume $R_c^2 L_c$, which appears approximately like a piece of regular flux-line lattice, but with an average lattice constant which may be slightly different from that of the unpinned lattice. The whole flux-line system may be thought of as a collection of such correlated volumes, and R_c and L_c are average lengths defining the typical correlation volume V_c. The underlying energy principle which determines the positions of flux lines, and hence the distortions in each correlation-volume, is the balance between the elastic energy cost of deforming the flux-line lattice locally within V_c versus. the energy gained by letting the flux lines pass through several point pins within V_c. The collective pinning theory was originally developed by Larkin and Ovchinnikov [91].

For a brief analysis of these effects on J_c, following Tinkham [53] we disregard the strain ε_1. To calculate the increased elastic energy connected with strains ε_4 and ε_6 we need to estimate

$$\delta F_e = \tfrac{1}{2}c_{44}\varepsilon_4^2 + \tfrac{1}{2}c_{66}\varepsilon_6^2 \tag{8.25}$$

Here we can divide the range ξ of the pinning force by the corresponding correlation lengths to obtain the strains.

$$\varepsilon_4 = \frac{\xi}{L_c}; \quad \varepsilon_6 = \frac{\xi}{R_c} \tag{8.26}$$

This gives

$$\delta F_e = \frac{1}{2}c_{44}\left(\frac{\xi}{L_c}\right)^2 + \frac{1}{2}c_{66}\left(\frac{\xi}{R_c}\right)^2 \tag{8.27}$$

Next we consider the pinning energy δF_p. When N elementary pinning forces f are added randomly in a volume V_c, the total force is $N^{\frac{1}{2}}f$. Here $N = nV_c$, where n is the number of forces f per volume, and V_c is the correlation volume. This force acts, on the average, over a distance ξ, after which it becomes random. The potential energy is force times distance:

$$\delta F_p = \xi N^{\frac{1}{2}} f / V_c = \xi f n^{\frac{1}{2}} / V_c^{\frac{1}{2}} = \xi f n^{\frac{1}{2}} / L_c^{\frac{1}{2}} R_c \tag{8.28}$$

The total energy $\delta F = \delta F_e + \delta F_p$ is

$$\delta F = \frac{1}{2} c_{44} \left(\frac{\xi}{L_c} \right)^2 + \frac{1}{2} c_{66} \left(\frac{\xi}{R_c} \right)^2 - \frac{\xi f n^{\frac{1}{2}}}{L_c^{\frac{1}{2}} R_c} \tag{8.29}$$

By the minimisation of δF one finds approximate expressions for L_c, R_c, V_c, as functions of $c_{44}, c_{66}, \xi, n, f$.

The minimum of δF is found [53] to be

$$\delta F_{\min} = -\frac{n^2 f^4}{8 c_{44} c_{66}^2 \xi^2} \tag{8.30}$$

On equating the pinning force per unit volume $f(n/V_c)^{\frac{1}{2}}$ with the maximum Lorenz force density Tinkham found for the critical current density:

$$J_c = \frac{n^2 f^4}{2 c_{44} c_{66}^2 \xi^3 B} \tag{8.31}$$

This expression has implications borne out by experiment: J_c increases with softening of the lattice, i.e. with smaller elastic moduli, because the flux line system can more easily adjust the positions of individual flux lines to fit the random position of pins. This is often observed near the melting temperature of the flux-line lattice.

It is interesting to imagine the opposite extreme; that of a completely rigid FLL in a random arrangement of (weak) pins of high density. In this case the pinning force will be zero: The total pinning energy V_p is in this case necessarily independent of the position of the FLL. The force is given by the negative of the gradient of U, and since U does not vary with position, the force must be zero. This sends a message that the density of pins should be high enough to contain a wide k-spectrum in the Fourier expansion $n(k)$ of the pin density $n(r)$, to take advantage of the wide range of k-values where the elastic constants c_{11} and c_{44} are soft. A Fourier density peaking at the FLL Brillouin zone boundary wavevector might seem ideal. This corresponds to a pinpoint density with an average distance like the FLL lattice constant. However, this would be ideal for that particular field only. If pinning is to be effective over a wide range of fields, there has to be a broad variation in pin distance, even for the use of the Brillouin zone boundary matching criterion to be good. Hence we end up, again, with a broad spectrum of Fourier density k-values as a good guideline. Only when optimizing pinning for a particular value of the magnetic field, i.e. a particular FL density, can a pin distance corresponding to the Brillouin zone wavelength or its vicinity, be the optimal solution.

8.5.2 Collective creep: inverse power law $U(J)$

From the previous discussion of collective pinning it becomes evident that a full description of creep phenomena needs to take into account collective effects. This has been attempted [91, 92], and has led to interesting new predictions for collective creep, i.e. for the behaviour of an elastic flux line lattice in the presence of random, weak point-pin sites. In this theory the collective creep regime is characterized by diverging time and length scales in the limit of zero driving force which is also the signature of so-called vortex glass behaviour.

We review here only briefly some aspects of collective creep theory without full justification, and refer to the original literature for details. The correlated volume $V_c = R_c^2 L_c$ was defined in the previous subsection as the characteristic volume which maintains the quality of a regular flux-line lattice even in the presence of distortions due to the interaction with a high density of point pins. Theory predicts that the volume V_c grows with the displacement ω of flux lines, raised to the power D, the dimensionality of the correlated volume, divided by the so-called wandering exponents ξ. In other words $V_c \propto \omega^{D/\xi}$. Here $D = 0$ for single pancake vortices, considered as point vortices, $D = 1$ for a single vortex line, $D = 2$ for a bundle of point vortices (pancakes), and $D = 3$ for a bundle of vortex lines. The wandering exponent ξ is determined by equating the energy of the elastic vortex lattice deformation to the fluctuation in pinning energy. The next requirement is that both these energies should be of the order of the energy gain due to the driving force $jBV_c\omega$, the Lorentz force density multiplied by volume and displacement.

Through this requirement the correlated volume becomes current dependent, and results in a current-dependent flux creep activation barrier

$$U = U_c \left(\frac{J_c}{J}\right)^{\mu} \tag{8.32}$$

with $\mu = (D + 2\xi - 2)/(2 - \xi) > 0$. It has the surprising and important consequence that as $J \to 0$, the activation energy $U \to \infty$. This leads to vanishing of both flux motion and linear resistance in that limit.

In the single vortex regime in three spatial dimensions $\mu = 1/7$, and the divergence of U is weak (near logarithmic). As the current density decreases one crosses over from single vortex regime to a regime of small bundles with $\mu = 5/2$, then to larger bundles with $\mu = 1$, and finally to large bundles with

$\mu = 7/9$. Other values of the exponent μ are obtained when a layered structure becomes dominant like in some high-T_c cuprates.

The collective creep theory applies at current densities much lower than the critical one, i.e. $J \ll J_c$. However, U is expected to go to zero at $J = J_c$ (in the Anderson–Kim model as $U_0(1 - J/J_c)$). To incorporate this into the collective creep model one often rewrites Eq. 8.32 as $U(J) = U_c[(J_c/J)^\mu - 1]$, where $U_c = U_0/\mu$ [92]. This provides an interpolation between the low current ('glassy') phase and the high current (linear A-K) phase.

8.5.3 Logarithmic $U(J)$

The various models used for $U(J)$ have often turned out to be difficult to distinguish on the basis of experimental data. This situation is compounded by the discovery that in many cases a logarithmic $U(J)$ is found to give the most adequate interpretation of data, i.e.

$$U(J) = U_c \ln \frac{J_c}{J} \tag{8.33}$$

Due to the short coherence length and the strong anisotropy of high-temperature superconductors, pinning energies are small, and thermally activated flux motion is a significant source of dissipation in the current carrying state, even when the Lorentz force on the vortices is less than the depinning force. The resulting decay of a screening current corresponds to a decay of the induced magnetic moment. For this reason, measurement of the magnetic relaxation serves as a powerful tool to investigate the dynamics of vortices in superconductors. In the thermally activated regime of flux creep the decay rate of the magnetisation current density is given by the rate equation,

$$\frac{dJ}{dt} = -A \times \exp\left(-\frac{U(T, B, J)}{T}\right) \tag{8.34}$$

Where A is a constant. In Figures 8.7, 8.8 and 8.9 [93], [94] we illustrate the procedures involved in analysing magnetization current decay data for possible logarithmic activation energy. The substance studied was $HgBa_2CaCu_2O_{6+x}$ (Hg-1212) with a $T_c = 127$ K. In Figure 8.7 the time dependence of magnetisation was measured over two decades in time in 1 T magnetic field parallel to the c-axis using a SQUID detector. The magnetic moment due to screening currents caused by the flux gradient set up by application of the external field is shown here. Notice the great differences observed at different temperatures. Data have been corrected for the equilibrium magnetisation which would be present without pinning.

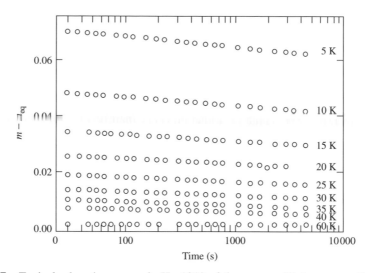

Figure 8.7   Typical relaxation curves in Hg-1212 of the non-equilibrium magnetic moment in a field of 1 T, at different temperatures as indicated by the numbers. The magnetic field is along the c axis. After Gjölmesli *et al.* [94].

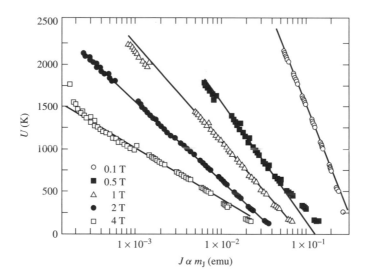

Figure 8.8   The barrier energy of Hg-1212 as calculated from Eq. 8.35. Each curve consists of data sets obtained at the temperatures indicated in Figure 8.7. The numbers indicate the values of the external magnetic fields, in tesla. B parallel to the c-axis. After Gjölmesli *et al.* [94].

In Figure 8.8 data were analysed by the Maley method [95]. This consists first in inverting Eq. 8.34, to write it as

$$U(T, B, m_J) = T \left(\ln A - \ln \left| \frac{dm_J}{dt} \right| \right) \tag{8.35}$$

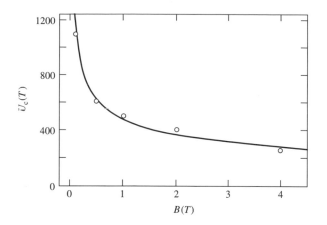

Figure 8.9 Resulting U_c versus T from the analysis of data in Figure 8.8.

after replacing current J in Eq. 8.34 with the associated magnetic moment $m_j = m - m_{eq}$. Here, the quantity m_{eq} is determined from the experimental data by the asymmetry in the magnetization during increasing versus decreasing field. Next, all measurement taken at different temperatures in the same magnetic field are regarded as giving access to different segments of $U(J)$ since J associated with any magnetisation is proportional to that magnetisation. The data in Figure 8.8 immediately attests to a current-dependent activation energy close to logarithmic in its dependence on J, according to Eq. 8.33. The experiment also gives further information about the B-field dependence of the activation energy, while it was not necessary to assume T-dependence of U.

A logarithmic $U(J)$ could possibly be expected from a model [96] in which a logarithmic spatial variation of pinning energy is involved. Alternatively, it can be a special case of collective creep for 2D pinning in the extreme low-current limit [97]. Alternatively again, it may be viewed as a special case of the inverse power law Eq. 8.32 with vanishing μ. Furthermore, in the critical region near a vortex glass transition (see later), power law E–J curves are expected for intermediate currents, irrespective of the value of μ, often referred to as the vortex glass exponent. Such power-law E–J would suggest logarithmic $U(J)$ for intermediate currents, and could fill the gap between the low current limit and high current limit in the interpolation formula for the collective creep.

We note finally, that the experiment just described gave indication of a different type of creep at very low temperatures, a temperature independent limit creep due to quantum mechanical tunnelling between pin sites.

8.5.4 The vortex solid–liquid transition

A much debated subject in the scientific literature on the properties of the vortex system in superconductors, and in high-T_c superconductors in particular, is the question regarding the nature of the transition between a vortex liquid and a vortex solid, following the works of Gammel and Bishop [98], Nelson [99], and Fisher *et al.* [100]. Two views have been advocated, that the solid phase may be in the nature of a vortex glass, or that it may be a more regular vortex lattice. In the first case a continuous transition is expected, in the latter a first order discontinuous one. Over the years the conclusion has been reached that both are possible. Depending on the amount of pinning, at low pin density freezing of a regular lattice may occur. We refer here to the Topical contribution of Kadowaki (Section 13.8) showing in a convincing manner that a first order transition to a regular vortex solid can indeed take place. At high density of pins the "vortex matter" is sufficiently distorted from a regular lattice that it freezes into what has become known as a vortex glass state. The analysis of this state was made earlier than the first order transition, largely due to the difficulty of obtaining superconducting material with low pin density. The fact that the frozen vortex system may appear as a regular lattice has been amply demonstrated by various decoration and imaging techniques. We refer to Figure 8.10 for an example.

The other alternative, the vortex glass, was concluded to exist from very detailed I–V (or E–J) characteristic measurements [102] in YBCO thin film (see Figure 8.11 [103]).

The glass line (T_g-line) in the phase diagram may be determined by scaling of E–J data. Just at T_g the current–voltage characteristic is expected to show

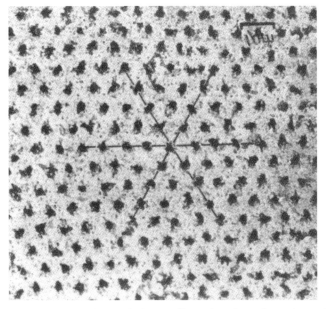

Figure 8.10   Magnetic decoration techniques allows imaging of nearly perfect triangular vortex lattice. After Träuble and Essmann [101].

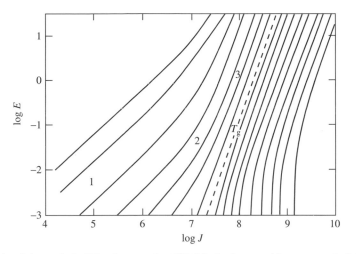

Figure 8.11 Schematic I–V isotherms of an YBCO single crystal in a magnetic field applied parallel to the *c*-axis. The dashed line marks the glass transition temperature below which the curves decrease exponentially. The temperature decreases in equal steps from top to bottom or from left to right. Region 1 marks the linear I–V behaviour, region 2 indicates non-linear behavior above T_g and region 3 is when $d(\log E)/d(\log I) = 3$ both above and below T_g. [103]. Simplified from Blatter *et al.* [97].

power law behaviour,

$$E \propto J^\alpha \tag{8.36}$$

with $\alpha = (z+1)/(D-1)$, where z is the dynamic exponent with typical value 4–5, and D is the relevant spatial dimension. Above T_g the characteristic should change from ohmic

$$\rho(T) \propto (T - T_g)^{\nu(z+2-D)} \tag{8.37}$$

at very low J where ν is the temperature exponent for the coherence length, to power law at high J.

Below T_g a glass-type response at small current densities is expected as

$$E \propto \exp\left[-\frac{U_0}{kT} \left(\frac{J_c}{J} \right)^\mu \right] \tag{8.38}$$

Again a power law critical behaviour is expected at high current densities. The crossover between the two regimes of power law at small scales and large current, and liquid/glassy state at small scales (probed by low current density) scale as

$$J_x^\pm \propto |T - T_g|^{\nu(D-1)} \tag{8.39}$$

8.5.5   Lindemann criterion and melting of a clean flux-line system

In this subsection, we will present a calculation that determines the rough position of the line in the (B, T) phase-diagram where the flux-line lattice may be expected to melt into a liquid state, i.e. the transition line from an Abrikosov flux-line lattice to a vortex state with zero shear stiffness. Crucial in our considerations will be the observation that conventional melting of a three dimensional solid is expected to be a first order phase transition. This means that long-wavelength fluctuations do not occur at the point of the transition, unlike in a second order phase transition. Under such circumstances, short-wavelength fluctuations are the ones responsible for inducing the transition from the solid to the liquid state. For our purposes this may be rephrased as follows. The Fourier modes of the fluctuations that dominate at the transition point are those that live at the boundary of the first Brillouin zone of the vortex lattice, i.e. those Fourier modes of fluctuations that have the largest wave vectors. Thus, it is reasonable to conjecture that when we want to estimate the position of the melting line using a description of the vortex lattice in terms of elasticity theory, we need to consider wave-vector dependent elastic moduli, i.e. non-local elastic moduli. In fact, as we shall see, this is absolutely crucial in order to obtain a reasonable estimate of the location of the melting line, since it will turn out that the elastic moduli of the flux-line lattice soften significantly as the wave vectors increase, and this dramatically enhances the susceptibility of the flux-line system to thermal fluctuations. Moreover, the variation of the position of the melting line with degree of layeredness of the material, can only be brought out through a non-local formulation.

Before we go into the details, let us outline the structure of the the calculation that determines the position of the melting line, $T_m(B)$. We consider a situation where we have an external magnetic field along the $\hat{z}$-direction, and we consider also the case where we have uniaxial anisotropy. This will make our results directly applicable to the study of the melting transition in the high-temperature cuprates, where the issue of flux-line lattice melting was first seriously considered as an experimental reality. (Many years earlier, Eilenberger had considered the melting of the flux-line lattice theoretically in moderate type II superconductors, and concluded that a melting line would be practically indistinguishable from the upper critical field line $H_{c2}(T)$) [104]. The ground state of the system is the hexagonal Abrikosov vortex lattice, and we now consider the Hamiltonian that governs the elastic fluctuations around this ground state. Hence, we write the positions of the flux lines as

$$\mathbf{r}_\nu = \mathbf{R}_\nu + \mathbf{s}_\nu(\mathbf{z})$$

Here, $\mathbf{R}_\nu$ is the position of flux line number ν, while $\mathbf{s}_\nu(\mathbf{z}) = (s_{x\nu}(z), s_{y\nu}(z), 0)$ is the displacement of the νth flux line away from its position in the perfect

Abrikosov flux-line lattice. Note how this displacement vector has two components, which vary with the distance along each flux line. In order to consider the fluctuations of the flux-line lattice in the continuum limit, we define an effective elastic Hamiltonian governing these fluctuations. For an ideal triangular lattice the Hamiltonian in the harmonic approximation takes the simple form

$$H = \frac{1}{2}\sum_{\mathbf{k}} s_i(-\mathbf{k}) \left[c_L(\mathbf{k})k_ik_j + \delta_{ij}\left(c_{66}(\mathbf{k})k_\perp^2 + c_{44}(\mathbf{k})k_z^2\right)\right] s_j(\mathbf{k}) \quad (8.40)$$

Here, c_L, c_{66} and c_{44} are bulk-, shear-, and tilting elastic moduli, respectively, and $s_i(\mathbf{k})$ are the Fourier modes of $s_\nu(\mathbf{z})$. Given the Hamiltonian Eq. 8.40, the elastic propagator can be written in the form

$$G_{ij}(\mathbf{k}) = k_BT \left[\frac{P_T}{c_{66}(\mathbf{k})k_\perp^2 + c_{44}(\mathbf{k})k_z^2} + \frac{P_L}{c_{11}(\mathbf{k})k_\perp^2 + c_{44}(\mathbf{k})k_z^2} \right] \quad (8.41)$$

where $c_{11}(\mathbf{k}) \equiv c_L(\mathbf{k}) + c_{66}(\mathbf{k})$. Moreover, $P_T = \delta_{ij} - k_ik_j/k_\perp^2$ is the transverse projection operator, while $P_L = k_ik_j/k_\perp^2$ is the longitudinal projection operator.

To examine the stability of the flux-line lattice against thermal fluctuations we will determine the mean square displacement of the flux lines from their equilibrium positions

$$d^2(T) = \left\langle \sum_\nu s_\nu^2 \right\rangle = \sum_{\mathbf{k}} \mathrm{Tr}G_{ij}(\mathbf{k}) \quad (8.42)$$

Here $\langle .. \rangle$ indicates a thermal average with respect to the Hamiltonian Eq. 8.40. Using Eq. 8.41 in Eq. 8.42

$$d^2(T) = k_BT \int_0^\infty \frac{dk_z}{2\pi} \int_0^{\Lambda^2} \frac{dk_\perp^2}{2\pi}$$

$$\times \left[\frac{1}{c_{66}(\mathbf{k})k_\perp^2 + c_{44}(\mathbf{k})k_z^2} + \frac{1}{c_{11}(\mathbf{k})k_\perp^2 + c_{44}(\mathbf{k})k_z^2} \right] \quad (8.43)$$

The Lindemann criterion now states that the flux-line lattice will melt when thermal fluctuations are such that the thermally induced root mean square displacements of each flux line is some fraction of the intervortex distance in the ground state of the Abrikosov vortex lattice, i.e. the melting criterion is given by $d(T) = cl$, where l is the lattice parameter of the flux-line lattice, and c is some fraction typically of order $c = 1/10$. In the triangular Abrikosov vortex lattice, we have $\sqrt{3}l^2/2 = \Phi_0/B$, i.e. l is determined by a flux quantization condition. In the above expression, we have approximated the k-space integral over the first Brillouin-zone by assuming a circular Brillouin-zone of radius Λ, where $\Lambda^2 = 2b/\xi_{ab}^2$, and $b = \langle H \rangle_{sp}/H_{c2}$. Here $\langle H \rangle_{sp}$ is the spatial average of the field, averaged over the sample, and H_{c2} is the upper critical field. Moreover, $\xi_\perp$ is the ab-plane coherence length.

The k-dependant elastic moduli can only be obtained from the full Ginzburg–Landau theory after a very arduous and lengthy calculation, there really appears to be no easy way of obtaining them. However, a detailed derivation may be found in the paper by Houghton and colleagues , [105], and the interested reader is referred to this paper, see also the work of Sardella [106]. The expressions for the elastic moduli, despite their involved derivations, are relatively simple and given by

$$c_{44}(\mathbf{k}) = \frac{B^2}{\mu_0} \frac{m_{ab}}{m_c} \langle \omega_0 \rangle \left[\frac{1}{k_\perp^2 + (m_{ab}/m_c)(k_z^2 + \langle \omega_0 \rangle)} + \frac{1}{2b\kappa^2} \right] \quad (8.44a)$$

$$c_{11}(\mathbf{k}) = \frac{B^2}{\mu_0} \langle \omega_0 \rangle \left[\frac{k^2 + (m_{ab}/m_c)\langle \omega_0 \rangle}{(k^2 + \langle \omega_0 \rangle)(k_\perp^2 + (m_{ab}/m_c)(k_z^2 + \langle \omega_0 \rangle))} \right.$$

$$\left. - \frac{1}{k_\perp^2 + (m_{ab}/m_c)k_z^2 + k_\psi^2} \right] \quad (8.44b)$$

where $k_\psi^2 = 2(1 - b)/\xi_{ab}^2$. Compare these expression with those given in Eq. 6.129 in Chapter 6, for the case where the magnetic field is parallel to the $\hat{c}$-axis. These expressions are a bit more precise than those given there and in Subsection 8.5.1, in that they explicitly also incorporate the core-contribution to the elastic moduli. Generalization of the above Eqs 8.44a and 8.44b for parallel flux lines ($k_z = 0$) for the case where the magnetic field is tilted away from the $\hat{c}$-axis was given by Sardella [106]. The corresponding generalization of Eqs 8.44a and 8.44b for the case $k_z \neq 0$ has not yet been achieved.

In the above, $\langle \omega_0 \rangle$ is the spatial average of the square of the Ginzburg–Landau function in the ground state of the Abrikosov flux-line lattice. Since we will not need an explicit expression for this quantity here, we do not specify it further. Moreover, the uniaxial anisotropy is manifest through the parameter m_c/m_{ab}, where m_c is the normal-state quasiparticle mass in the z-direction, while m_{ab} is the quasiparticle mass in the ab-plane. Another way of writing this is $m_c/m_{ab} = \lambda_c^2/\lambda_{ab}^2$.

The shear modulus of the Abrikosov flux-line lattice is for all practical purposes k-independent. In the present geometry, where the external magnetic field is directed along the axis of uniaxial anisotropy, the shear modulus will be that of an isotropic superconductor, since only shearing in the ab-plane makes any physical sense in this case. Thus, the shear modulus is given by

$$c_{66} = \frac{B_{c2}}{\mu_0} \frac{b(1 - b)^2}{8\kappa^2} \quad (8.45)$$

Several things are worth noting about these expressions. First of all, when $k \to 0$, all effects of uniaxial anisotropy, manifest through the parameter m_c/m_{ab}, vanishes. Note also that when $k = 0$, the maximum value of c_{44} and c_{11} occurs at the upper critical field, as a function of magnetic field. Using the dimensionless wavevector $q = k/\Lambda$, the above expressions for c_{44} and c_{11} may be written

$$c_{44}(\mathbf{q}) = \frac{B^2}{\mu_0} \frac{(1-b)}{2b\kappa^2} \frac{m_{ab}}{m_c} \left[\frac{1}{q_\perp^2 + (m_{ab}/m_c)(q_z^2 + m_\lambda^2)} + \frac{1}{2b\kappa^2} \right] \quad (8.46)$$

$$c_{11}(\mathbf{q}) = \frac{B^2}{\mu_0} \frac{(1-b)}{2b\kappa^2} \left[\frac{q^2 + (m_{ab}/m_c)m_\lambda^2}{(q^2 + m_\lambda^2)(q_\perp^2 + (m_{ab}/m_c)(q_z^2 + m_\lambda^2))} \right.$$

$$\left. - \frac{1}{q_\perp^2 + (m_{ab}/m_c)q_z^2 + m_\xi^2} \right] \quad (8.47)$$

where we have introduced the 'mass'-parameters $m_\xi^2 = (1-b)/b$ and $m_\lambda^2 = (1-b)/2b\kappa^2$. Now, another interesting feature of the wave-vector dependant moduli becomes evident. For finite $k(q)$, they *vanish* at the upper critical field, in contrast to the $k = 0$ moduli, which acquire their maximum at that value of the magnetic field. This means that those Fourier modes that are most relevant for melting the flux-line lattice, become particularly soft as the magnetic field approaches the upper critical field limit. This is true for the case of uniaxial anisotropy, as well as for the isotropic case. It substantially contributes to melting the flux-line lattice well below the upper critical field line, and is a key feature which leads to a suppression of the melting line below the upper critical field even for moderately small Lindemann parameters c.

To determine $d^2(T)$ from Eq. 8.42, we have to integrate over the entire circularized Brillouin-zone $0 < q_\perp < 1$. Therefore, it is evident that a fully non-local theory, i.e. an elastic description with wave-vector dependant moduli, is essential. For arbitrary field and temperature, the necessary integrals in Eq. 8.42 can only be done numerically, however with some limitations (that turn out to be not very restrictive in strong type II superconductors), analytic results can be obtained that provide key insights into the main physical parameters controlling the position of the melting line. We proceed by writing

$$d^2(T) = d_1^2(T) + d_2^2(T) \quad (8.48)$$

where $d_1^2(T)$ is the contribution to Eq. 8.42 containing the shear modulus, while $d_2^2(T)$ is the contribution to Eq. 8.42 containing the bulk modulus. For $b > 1/2\kappa^2$ we find

$$d_1^2(T) = \left[\frac{k_B T}{4\pi} \sqrt{\frac{\Lambda^2}{c_{44}^0 c_{66}^0}} \right] \sqrt{\frac{m_c}{m_{ab}}} \sqrt{\frac{2b\kappa^2}{1-b}} (\sqrt{2} - 1) \quad (8.49)$$

while under the condition $b > 1/2\kappa^2$ and $(1 - b)/b \gg 1$ we find

$$d_2^2(T) = \left[\frac{k_B T}{4\pi} \sqrt{\frac{\Lambda^2}{c_{44}^0 c_{11}^0}} \right] \sqrt{\frac{m_c}{m_{ab}}} \frac{b\kappa^2}{1 - b} \qquad (8.50)$$

Here, c_{44}^0 and c_{11}^0 are the elastic moduli given above taken at $k = 0$. The factors in square brackets in each term is what we would have obtained had we used k-independent elastic moduli in the calculation. The other factors enter into the expressions only because the k-dependence of the elastic moduli is taken into account. Note that if we had ignored the k-dependence, the enhancement factors both coming from the large value of κ in extreme type II superconductors, as well as the enhancement factor coming from the mass anisotropy, would have been missed. Below, we shall identify a dimensionless parameter that explicitly shows how dramatic the effect of including non-local elastic moduli really is.

To see this, we introduce explicit expressions for c_{44}^0 and c_{11}^0, thus finding

$$d^2(T) = \frac{1}{2\pi} \sqrt{\frac{\epsilon m_c}{m_{ab}}} \frac{t}{\sqrt{1 - t}} \frac{\sqrt{b}}{1 - b} \left(\frac{4(\sqrt{2} - 1)}{\sqrt{1 - b}} + 1 \right) l^2 \qquad (8.51)$$

Hence, the Lindemann melting criterion is given by

$$\frac{t}{\sqrt{1 - t}} \frac{\sqrt{b}}{1 - b} \left(\frac{4(\sqrt{2} - 1)}{\sqrt{1 - b}} + 1 \right) \geq 2\pi \sqrt{\frac{m_{ab}}{\epsilon m_c}} c^2 \equiv \alpha \qquad (8.52)$$

Here, we have introduced the Ginzburg criterion parameter

$$\epsilon = \left(\frac{k_B T_c \mu_0 x^2}{4\pi B_{c2}^{02} \xi_{ab}} \right)^2 \qquad (8.53)$$

and $t = T/T_c$. The dimensionless parameter α is given by

$$\alpha = \left[2 \times 10^7 c^2 \sqrt{\frac{B_{c2}^0}{T_c^2}} \right] \sqrt{\frac{m_{ab}}{m_c}} \frac{1}{\kappa^2} \qquad (8.54)$$

This parameter essentially controls the position of the melting line in the (B, T) phase diagram. In Eq. 8.54 we have again singled out in square brackets what we would have obtained for α had we only used local elastic moduli, the other two factors involving the mass ratio and the Ginzburg–Landau parameter κ appear exclusively due to non-locality in the elastic description of the flux-line lattice. If α is very large, it is clear from Eq. 8.52 that t and b need to be very close to unity for the melting criterion to be satisfied. A reduction in α facilitates a suppression of the melting line below the upper critical field line. In conventional superconductors, typically we have $\kappa \approx 1$, $B_{c2}^0 = 1\,\text{T}$, $T_c = 10\,\text{K}$, and $m_c/m_{ab} \approx 1$. If we set $c = 0.1$, we have $\alpha \approx 10^4$. In high-temperature superconductors, on the other hand, we have $\kappa \approx 100$, $B_{c2}^0 = 100\,\text{T}$, $T_c = 100\,\text{K}$, and $m_c/m_{ab} \approx 100$. This gives $\alpha \approx 10^{-1}$, *which is a reduction of nothing less than five orders of magnitude*. Notice that the factor in square brackets in the expression for α is about the same in conventional moderate type II superconductors and extreme type II high-temperature superconductors, namely of order $10^7 c^2$. The reduction in α originates exclusively in mass anisotropy and large value of κ. *It is therefore an intriguing fact that the enhancement of T_c in high-temperature superconductors compared to conventional low-temperature superconductors in itself does not contribute substantially to suppressing the melting line below the B_{c2} line, since the enhancement of T_c is cancelled by the enhancement of B_{c2}^0.*

However, once non-locality is included and mass anisotropy and large values of κ explicitly come into play, then the melting line is suppressed well below the upper critical field line, and the liquid phase of the flux-line system should be easily experimentally accessible. The one dominant feature which makes this possible, is not first and foremost the mass anisotropy, but rather the large value of the Ginzburg–Landau parameter κ, notice how it is the square root of the mass anisotropy that enters in the melting criterion while it is the square of the Ginzburg–Landau parameter which is relevant. *Once more, it is the use of a non-local elastic theory of the flux-line lattice which brings out the dependence of both the mass anisotropy and the Ginzburg–Landau parameter.* It is in particular the factor $1/\kappa^2$ which brings the melting criterion above in reasonable agreement with experiments [98] for a sensible Lindemann parameter $c \approx 0.4$, and the factor $\sqrt{m_{ab}/m_c}$ which distinguishes between flux-line lattice melting in YBCO and BSCCO in a manner consistent with experiments [98].

In a deep sense, the sensitivity of the melting criterion to κ reflects the softness of *phase-fluctuations* of the superconducting order parameter in strong type II superconductors. This is because large values of κ are due to small values of the superfluid density in these compounds. On the other hand, the superfluid density is nothing but a measure of the stiffness of phase fluctuations. We shall have much more to say about such fluctuations in Chapters 9 and 10.

In closing this subsection, we emphasize that the above is not a theory of melting of the flux-line lattice, the Lindemann criterion is after all just that, a criterion. It does however provide useful insight into what parameters that determine the position of the melting line, provided the assumption of a first order melting transition is correct. In fact, numerous Monte Carlo simulations and experiments now strongly indicate that this is indeed so. See for instance the work of Hetzel and co-workers [107] for the first demonstration through large-scale Monte Carlo simulations that the Abrikosov vortex lattice melts in a first order melting transition, and the contemporary experimental work of Safar and colleagues [108], showing pronounced hysteresis in resistivity data at what was interpreted as a melting transition of the flux-line lattice.

8.5.6 Modelling non-linear vortex diffusion

The voltage–current characteristics in many high-T_c superconductors are well described by a power-law over many orders of magnitude in the electric field. This has been found directly from transport measurements, and indirectly from the magnetic relaxation. It seems to be the general behaviour for YBCO- and BISCCO-based compounds, as well as for the Hg-based cuprate superconductors. As pointed out before, the power-law E–J-characteristics is most commonly interpreted as a result of thermally activated flux creep with a pinning barrier diverging slowly as the current vanishes, i.e. $U(J) \propto \log(J_c/J)$ or $U(J) \propto (J_c/J)^\mu$, with μ small. Although there is evidence for some curvature in $\log(E)$ versus $\log(J)$, the power-law may still serve as a first order approximation and as a model description of the voltage–current characteristic of high temperature superconductors. An analysis of the magnetic relaxation due to a power-law, with static boundary conditions, has been given by Vinokur and Feigel'man [109] and by Gilchrist and van der Beek [110]. Analysis of the AC-losses due to a power-law has been given by Rhyner [111] who considered an AC magnetic field at the surface of a superconducting semi-space. Such a configuration would apply to a real sample when the characteristic penetration depth of the field is small compared to the sample thickness or radius. In the following we consider the situation from zero to full penetration of a cylindrical sample, infinite in the axial direction. The boundary condition is a sinusoidally time-varying magnetic field applied at the surface, and the AC-susceptibility is calculated for different n-values. The numerical results are compared to experimental susceptibility data on a YBCO single crystal.

The power-law can be written in dimensionless units as

$$\frac{E}{E_0} = \left(\frac{J}{J_0}\right)^n \tag{8.55}$$

By letting n vary from $n = 1$ to $n = \infty$, one covers a full class of cases ranging from ohmic with $\rho = E_0/J_0$, to the ideal critical state superconductor with a well-defined critical current density $J_c = J_0$, as illustrated in Figure 8.12. In terms of vortices and vortex dynamics, the limit $n = \infty$ represents the static critical state in which the vortex density gradient is balanced by the pinning force $F_p = J_0 B$. When n has some finite value, this is equivalent to a relaxation of the critical state, e.g. by thermally activated creep. Now consider an infinitely long cylindrical sample of radius a in an axial magnetic field $B = B_{ac}\sin(\omega t)$. Maxwell's equations for the field in the sample in cylindrical coordinates read

$$\frac{1}{r}\frac{\partial}{\partial r}(rE) = -\frac{\partial B}{\partial t} \tag{8.56}$$

$$\frac{\partial}{\partial r}B = -\mu_0 J \tag{8.57}$$

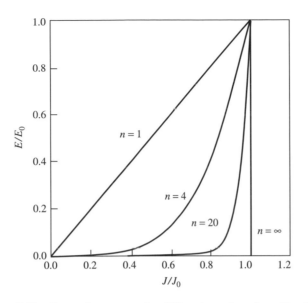

Figure 8.12 Power-law curves for different n-values from Eq. 8.55.

where E and J are azimuthal field and currents, while the B-field is in the axial direction. Next we introduce the dimensionless variables $b = B/B_{ac}$, $\zeta = r/a$ and $\tau = \omega t/2$, and aualyze the above equations with respect to the reduced magnetic field b, obtaining

$$\frac{1}{\zeta}\frac{\partial}{\partial \zeta}\left[\zeta\left(\frac{\partial b}{\partial \zeta}\right)^n\right] = \left(\frac{a}{\delta}\right)^{n+1}\frac{\partial b}{\partial \tau} \times \mathrm{sgn}\left(\frac{\partial b}{\partial \zeta}\right) \qquad (8.58)$$

Here sgn denotes the sign function, and the characteristic length-scale δ represents a generalized penetration depth which can be written as

$$\delta = \delta_{cl}^{\frac{2}{(n+1)}} \times \delta_{\mathrm{Bean}}^{\frac{(n-1)}{(n+1)}} \qquad (8.59)$$

where $\delta_{cl} = 2E_0/\mu_0\omega J_0$ is the classical skin depth in the linear limit ($n = 1$) and $\delta_{\mathrm{Bean}} = B_{ac}/\mu_0 J_0$ is the Bean penetration depth in the critical state (static) limit ($n = \infty$). The combined depth δ reduces to its appropriate limits $\delta = \delta_{cl}$ and $\delta = \delta_{\mathrm{Bean}}$ when $n = 1$ and $n = \infty$ respectively.

The total magnetic flux in a cylindrical sample of radius a is

$$\Phi(t) = 2\pi \int_0^a B(r)r\,dr, \qquad (8.60)$$

and the complex susceptibility, $\chi = \chi' + i\chi''$, is a measure of the first harmonic Fourier components of the spatially averaged flux density,

$$\chi' + 1 = \frac{1}{\pi a^2 B_{ac}} \int_{\omega t=0}^{2\pi} \Phi(t) \sin \omega t \, d(\omega t)$$

$$\chi'' = \frac{1}{\pi a^2 B_{ac}} \int_{\omega t=0}^{2\pi} \Phi(t) \cos \omega t \, d(\omega t). \tag{8.61}$$

The susceptibility can be calculated exactly in both the linear [112] and the critical state [113] limits. In the linear limit one finds

$$1 + \chi = 1 + \chi' + i\chi'' = \frac{2J_1(ka)}{ka J_0(ka)} \tag{8.62}$$

with $k = (1+i)\delta_{cl}$ and $J_0(ka)$ and $J_1(ka)$ denote Bessel functions of zero'th and first order respectively. (For a London superconductor $k = 1/\lambda$, which results in loss-free screening currents and $\chi'' = 0$.) In the limit $n = \infty$ Eq. 8.58 reduces to the critical state equation, $\partial B/\partial r = \pm\mu_0 J_0$ [111]. The susceptibility can then be found by calculating the critical state profiles for a cycle of the surface field and integrating over space and time. This has been done by Clem [113], and the solution reads

$$\chi' = g_1(x)$$

$$\chi'' = g_2(x) \tag{8.63}$$

where $x = B_{ac}/B^* = B_{ac}/\mu_0 J_0 a$, and $g_1(x)$ and $g_2(x)$ are known polynomials of x and $\sin^{-1}(x^{-1/2})$. In the linear case the susceptibility will depend on the frequency of the AC field, but not on the amplitude, while in the critical state limit the situation is the opposite, a behaviour which is also reflected in δ. Moreover, in the linear limit the response is purely harmonic while for $n > 1$ the response is anharmonic, resulting in an infinite set of higher order susceptibility terms.

The AC susceptibility can be measured over a wide range of penetration, often by varying either the temperature or the static background magnetic field, or both. A maximum in the loss component χ'' is observed when the penetration depth is of the order of the sample radius, i.e. when

$$\delta_{cl}^{\frac{2}{(n+1)}} \delta_{\text{Bean}}^{\frac{(n-1)}{(n+1)}} \sim a \tag{8.64}$$

or

$$\left(\frac{2E_0}{\mu_0 J_0 \omega}\right)^{\frac{1}{(n+1)}} \left(\frac{B_{ac}}{\mu_0 J_0}\right)^{\frac{(n-1)}{(n+1)}} \sim a \tag{8.65}$$

In the exact limits the conditions for a peak maximum in χ'' is $\delta_{\text{Bean}} = a$ and $\delta_{cl} \simeq a/1.77$, respectively. By taking the differential of Eq. 8.65 we find the relation $d\ln(\omega)/d\ln(B_{ac}) = n - 1$. These differentials can be found experimentally by measuring the shift in the loss maxima (e.g. with temperature) as a function of both amplitude and frequency of the AC field. Such experiments may therefore be used to obtain the value of n directly.

Numerical values of χ' and χ'' were calculated [93] for different values of n by letting the ratio δ/a vary from zero to some large number $\gg 1$. $B(r, t)$ was found by solving Eq. 8.58 by means of finite differences, and χ' and χ'' were found by integrating in space and time according to Eqs 8.60 and 8.61.

A convenient way of plotting susceptibility results is the Cole–Cole plot, in which χ'' is plotted as a function of χ'. We now make a comparison of the numerical results with experimental susceptibility data obtained on a single crystal YBCO. The data used here are only a small subset from a wider experimental study [114]. The crystal sample was cubic with sides 1 mm, and due to such a non-ideal sample shape, a comparison is only of a semi-quantitative nature. The susceptibility was measured as a function of temperature in different DC fields ranging from 1 to 8 T and with an AC field of amplitude 10^{-4} T, and frequency 121 Hz. The measured $\chi'(T)$ and $\chi''(T)$ are shown in Figure 8.13 and in a Cole–Cole plot in Figure 8.14.

The data collapse in Figure 8.14 shows that the same process is governing the susceptibility behaviour for all curves, occurring at different temperatures

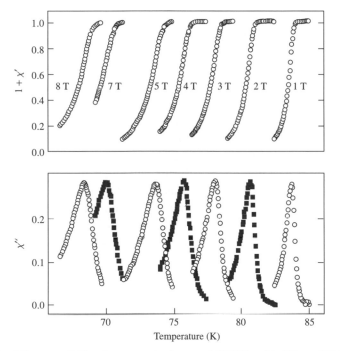

Figure 8.13 AC susceptibility data on a single crystal YBCO, with $B\|c$ at 121 Hz. Applied fields have corresponding values in the upper and lower panel. After Gjölmesli [93] and Karkut *et al.* [103].

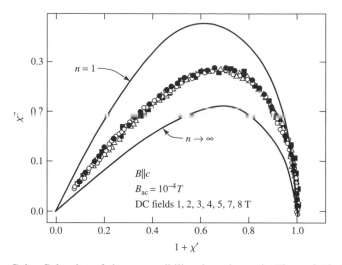

Figure 8.14 Cole–Cole plot of the susceptibility data shown in Figure 8.13. The plotted data follow the $n = 1$ curve at the highest values of $1 + \chi'$, corresponding to the highest temperatures. But at some point they switch over to a non-linear domain with $n \gg 1$. This is the freezing point of the lattice. Data Analysis: Ref [93] from Karkut *et al.* [103].

determined by the DC field. In other words, the susceptibility can be written as a function of a single variable which changes monotonically with temperature and field. At large penetration, $\chi' + 1 \simeq 1$ the data follow closely the linear curve $n = 1$. Then at some particular value $1 + \chi' \simeq 0.95$ and $\chi'' \simeq 0.2$, the curve makes a small jump into the non-linear region with $n > 1$. Moreover, the linear-nonlinear jump is rather sharp, occurring over less than 0.2 K, indicative of a phase transition of the vortex state, possibly from a vortex liquid to a vortex glass [93].

We conclude that the non-linear diffusion due to a power-law E–J-characteristic gives a generalized penetration depth δ which reduces to the classic skin-depth and to the Bean penetration depth in the appropriate limits. The peak shift in χ'' with respect to amplitude and frequency can be used to determine the value of the exponent n. The non-linear diffusion equation was solved numerically and the complex magnetic susceptibility was calculated, in order to be compared to experimental data. By plotting χ'' versus χ' the experimental data collapse occurred, with indications of a sharp transition from a vortex liquid state with linear dynamics into a state of non-linear dynamics, possibly a vortex glass.

8.6 Flux-line entry at B_{c1}: thermodynamic and geometric restrictions

8.6.1 The critical field B_{c1}

The mixed state of a superconductor is one where the superfluid phase coexists with the flux-line lattice, a vortex glass, or with the flux-line liquid. When an applied magnetic field is increased from zero, the superconductor is first in the Meissner phase, with complete shielding. At a certain value of the applied field $H_{c1} = \frac{1}{\mu_0} B_{c1}$ it becomes energetically favourable to have flux-lines created near the surface, and for these to penetrate the body of the super-conductor.

A thermodynamic analysis to determine H_{c1} starts by writing down the Gibbs free energy G when n vortices per area are inside and far apart so that their mutual interaction can be neglected. We now have:

$$G = nF - BH \tag{8.66}$$

where F is the vortex line energy previously determined in Chapter 6. This is an excellent approximation very near H_{c1} where in principle even a single flux-line can penetrate alone. Since the average field is $B = n\Phi_0$, we make this substitution, and find

$$G = B \left(\frac{F}{\Phi_0} - H \right) \tag{8.67}$$

The lowest value for G as long as the expression in the parenthesis is positive, is $G = 0$, corresponding to $B = 0$, as expected for a system in the Meissner phase. A simultaneously positive value of the expression in parenthesis signifies stability against vortex entrance, consistent with Meissner screening. However, when the expression in the parenthesis is negative, i.e. when $H > F/\Phi_0$, G can be lowered below the Meissner-state value $G = 0$ by $B \neq 0$. A finite B means that flux has entered, and that the energy G is lower than in the Meissner state, which is consistent with instability towards flux penetration. The critical field at which the instability occurs clearly is

$$B_{c1} = \frac{F}{\Phi_0} = \frac{\Phi_0}{4\pi\lambda^2} \left(\ln\left(\frac{\lambda}{\xi}\right) + \varepsilon \right) \tag{8.68}$$

as derived before.

8.6.2 The Bean–Livingston barrier

When a vortex is created near the surface of a superconductor in the Meissner state two opposing forces act on it. The screening currents create a Lorentz force normal to a flat surface, attempting to move the vortex into the material according to

$$f_{x,L} = \Phi_0 j_y \tag{8.69}$$

as shown in Eq. 6.110. We can write this as

$$f_{x,L} = \Phi_0 \frac{H}{\lambda} e^{-\frac{x}{\lambda}} \tag{8.70}$$

reflecting the decay of the screening current into the superconductor.

The second force is the image force attracting the vortex to the surface. Using the result in Eq. 6.106 where the distance between the vortex and its image is $2x$, the interaction energy which is now negative, becomes

$$U'_{12} = -\frac{\Phi_0^2}{2\pi\mu_0\lambda^2} K_0\left(\frac{2x}{\lambda}\right) \tag{8.71}$$

Taking the derivative, and making use of the relation $\frac{d}{dx}K_0(x) = -K_1(x)$ for these Bessel functions the resulting image force is

$$f_{x,im} = -\frac{\Phi_0^2}{2\pi\mu_0\lambda^3} K_1\left(\frac{2x}{\lambda}\right) \tag{8.72}$$

We now evaluate the force at $x = \xi$, since this is as close as we can approach the surface using the vortex model. The sum of opposing forces is

$$f_x = f_{x,L} + f_{x,im} = \Phi_0 \frac{H}{\lambda} e^{-\xi/\lambda} - \frac{\Phi_0^2}{2\pi\lambda^3} K_1\left(\frac{2\xi}{\lambda}\right) \tag{8.73}$$

The total force is zero at field

$$H_{sb} = \frac{\Phi_0}{2\pi\mu_0\lambda^2} K_1\left(\frac{2\xi}{\lambda}\right) e^{\frac{\xi}{\lambda}} \tag{8.74}$$

which is called the surface barrier field for obvious reasons.

Furthermore, using the property of the Bessel function

$$\lim_{x\to 0} K_1(x) = \frac{1}{x} \tag{8.75}$$

for $\xi \ll \lambda$ one finds

$$H_{sb}^{lim} = \frac{\Phi_0}{4\pi\mu_0\xi\lambda} = \frac{1}{\sqrt{2}}H_c \tag{8.76}$$

This is the surface barrier field at which the net force against flux entry is zero. This barrier field has important implications. As an example, in microwave applications where flux entry would represent a loss factor it is important that H_c in type II superconductors like Nb is much greater than H_{c1}. To take full advantage of this situation the surface should be smooth, at least on the scale of λ,

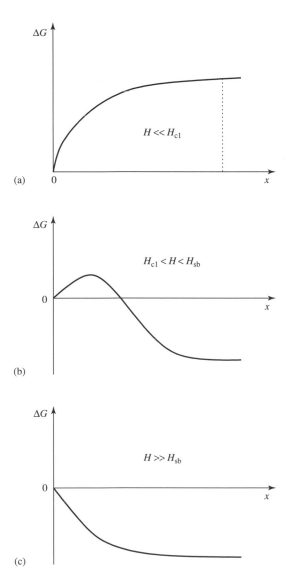

Figure 8.15 Sketch of the Gibbs energy from which the two opposing forces active in the Bean–Livingston barrier are derived. At $H = H_{sb}$ the slope of the curve at the origin will be zero, providing free entry of flux.

since the force found above is calculated under the assumption of a flat surface. Figure 8.15 shows how the Bean–Livingston barrier [115] can be viewed and understood to operate.

In most real situations sample shapes and surface conditions will allow flux entry of static fields near H_{c1}. Normally H_{sb} will be the entry field only when special precautions are taken.

8.6.3   Geometric barriers

The concept of a geometric barrier was introduced by Zeldov and coworkers [116] and discussed also extensively by Brandt [117]. This is a shape dependent barrier which can be explained qualitatively by reference to the sketch in Figure 8.16.

A sample S of thickness d, with sharp corners is placed in a magnetic field. Near the sharp corners the local field is enhanced considerably compared to the applied field H, and therefore penetrates first at these points. This undermines the opposing forces due to the Bean–Livingston barrier. However, for the flux line to enter, the length of flux line L_f as seen in the figure, has to be considerably longer than just $L_s = d$, which would be the length corresponding to penetration along the flat outer surface. If we approximate the length L_f by a half circle at the point when it just leaves the surface we estimate that $L_f \sim \pi \frac{d}{2}$. The flux line therefore was stretched by an amount

$$\Delta L = L_f - L_s \approx (\pi - 2)\frac{d}{2} \sim \frac{d}{2} \tag{8.77}$$

compared to the length it would have in the centre of the specimen. Notice also that Figure 8.16 indicates that a vortex line has jumped directly to the centre of the sample once it entered. This is because here again the length inside the sample is only equal to d. Alternatively we may say that the flux line was driven to the centre by the Meissner currents discussed before. Hence it is energetically advantageous to travel all the way to the centre in case of no pinning. The energy cost of creating the flux line near the surface is therefore raised by $\sim 50\%$ due to sample geometry in this simple estimate. Consequently we may expect the flux entry field to be raised by an amount which is of the same order of magnitude as H_{c1} itself. Obviously, the effect is strongly shape dependent, hence its name. In case of elliptical crossection it may be shown that there is no geometrical barrier.

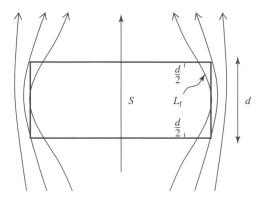

Figure 8.16   Sketch illustrating flux entry in the presence of a geometrical barrier.

This effect also has an associated hysteresis, since there is no corresponding barrier against flux exit. The geometric barrier therefore is recognized experimentally by hysteresis in magnetisation experiments even in the absence of pinning. The additional effect of pinning was discussed by Brandt [117].

8.7   Critical current issues

8.7.1   Critical current in the Meissner state

According to the London equation there is a relation between the supercurrent density and the magnetic field at any point in the superconductor. In the Meissner phase, due to screening, the current and field can only penetrate to a depth measured by λ. Due to the correspondence between J and H it does not matter whether J is the sum of a transport current and a screening current. Experiments show that there is a critical current density J_c above which screening and superconductivity break down. Due to the equivalence of transport current and screening current it is to be expected that superconductivity breaks down when the total current density—the sum of transport and screening current—exceeds the critical current density J_c. The famous Silsbee hypothesis states the converse: A superconductor looses its state of zero resistance when the total magnetic field strength due to transport current and the applied field exceeds the critical field strength H_c.

Mainly because of large variations in charge density, λ may vary by an order of magnitude, from ≈ 15 nm in good polyvalent metals to 150 nm in YBCO.

What we have then, in a surface layer of thickness 15–150 nm, is a decaying screening current and an associated magnetic field which are locally related in the London approximation. With a field H created by a current J along a cylindrical rod, the H-field is circular in the plane normal to the rod with radius a. We get $\frac{\partial}{\partial r} H_\theta(r) = J$ from Maxwell's equations $\nabla \times H = J$, leading to

$$J = \frac{H_\theta(a)}{\lambda} \tag{8.78}$$

where $H_\theta(a)$ is the value of H_θ at the surface. Now, if we allow J to reach J_c, this corresponds to $H_\theta(a) = H_c$. In other words:

$$J_c = \frac{H_c}{\lambda} \tag{8.79}$$

This is the relation which holds in the Meissner state, and therefore is valid in type I as well as type II superconductors. It is worth noting that this rule is very similar to the one we derived for J_c in the Bean critical state. In that case J_c was also given by a field divided by a length, namely $\Delta M/d$. Since H and

M are in the same units, what these expressions have in common, is that the gradient of the field gives the critical current.

Does the relation $J_c = H_c/\lambda$ reconcile with a calculation based on Ampere's law?

If the superconducting rod is of radius a, and the total current is I, the field at the surface, $H_\theta(a)$ is given by

$$I = 2\pi a H_\theta(a) \tag{8.80}$$

or, if the current is strong enough to reach H_c:

$$I_c = 2\pi a H_c \tag{8.81}$$

Equation 8.81 looks different from Eq. 8.78. But for real comparison we have to introduce the current density,

$$J = I/A_{\text{eff}} = I/2\pi a\lambda; \quad I = 2\pi a\lambda J \tag{8.82}$$

Here, the effective area was written as $A_{\text{eff}} = 2\pi a\lambda$, since $a \gg \lambda$. We now find $2\pi a\lambda J_c = 2\pi a H_c$, which leads to

$$J_c = \frac{H_c}{\lambda} \tag{8.83}$$

and we have recovered the answer we had before, in Eq. 8.79.

8.7.2   Depairing critical current

Calculations of critical current density can be carried out using Ginzburg–Landau as well as BCS theory. Except for a numerical factor they come out with essentially the same result:

$$J_c \approx \frac{H_c}{\lambda} \tag{8.84}$$

In Ginzburg–Landau-theory the exact result is

$$J_c = \frac{2\sqrt{2}}{3\sqrt{3}} \frac{H_c}{\lambda} \simeq 0.54 \frac{H_c}{\lambda} \tag{8.85}$$

In BCS theory the kinetic energy added to the system by turning on a current J, has to be compared to the energy gained by forming the Cooper-pairs. When the kinetic energy equals the pair condensation energy we have the critical current density. The condensation energy can be expressed on the basis of thermodynamics as $(\frac{1}{2}\mu_0)H_c^2$, and from BCS theory as $\frac{1}{2}N(0)\Delta^2$, where $N(0)$ is the normal density of states at the Fermi level, and Δ is the BCS energy gap. This would give

$$H_c^2 = \mu_0 N(0)\Delta^2 \tag{8.86}$$

Using $J_c = \frac{H_c}{\lambda}$ we find $J_c = \frac{\sqrt{\mu_0 N(0)\Delta^2}}{\lambda}$ as a mixed expression between phenomenology and BCS.

Example

Measurements in a 1-mm diameter wire have given critical current at $0\,K$: $I_c = 75\,A$. With $a = 0.5\,mm$, $\lambda = 5.1 \times 10^{-6}\,cm$; $J_c(0) = \frac{I_c}{2\pi a \lambda} = 4.7 \times 10^7\,A/cm^2$.

A simple estimate of J_c based on the basic formula $J_c = H_c/\lambda$ gives typically, with $\mu_0 H_c \approx 0.1\,T$ and $\lambda \approx 100\,nm$:

$$J_c = \frac{B_c}{\mu_0 \lambda} = 10^{12}\,A/m^2$$

Often such values are quoted in A/cm^2, in this case: $J_c \approx 10^8\,A/cm^2$.

This is the maximum ideal current density the superconductor can sustain under any circumstance. There are many reasons why one does not reach this value in practical cases. The most important obstacles are typically flux motion and grain boundary problems, particularly in high-T_c cuprates. Yet, often one finds that real materials can come within a factor of $2-3$ of this pair-breaking current.

8.7.3   Reduction of J_c at grain boundaries

So far we have discussed the idealized situation as far as critical current density J_c is concerned. These considerations give a good idea of what maximum of critical current density we can expect in a superconductor. However, there are many mechanisms by which J_c may be reduced, even substantially. Of course, if in a granular material the grains do not join together well, for instance due to a transition layer of different composition between the grains, this has a strongly derogative effect. High-T_c materials are known for their short coherence length ξ. In a transition layer the wavefunction ψ may be depressed over a distance ξ without bridging properly across the boundary, in which case little current can pass across the interface, except by tunnelling which may be very weak.

In addition, due to the $d_{x^2-y^2}$ character of the wavefunction in high-T_c cuprates several misorientation effects at the microscopic level are possible, each representing an obstacle to current transport, and hence to reduction of J_c [118]. These mechanisms are illustrated in Figure 8.17. Referring to this figure

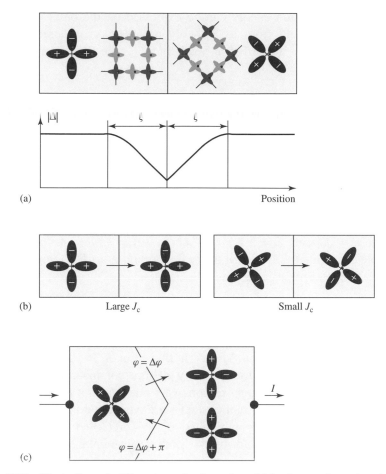

Figure 8.17   Illustration of different mechanisms by which J_c may be reduced at grain boundaries in high-T_c $d_{x^2-y^2}$ type superconductors. After Hilgenkamp and Mannhart [118].

the mechanisms are from top down: In Figure 8.17a two adjoining grains are differently oriented, causing a depression of the wavefunction over a distance ξ due to frustration caused by misorientation of CuO_2-planes. In Figure 8.17b one sees a case of alignment of the lobes of the $d_{x^2-y^2}$ function allowing a high J_c, and a misalignment causing a small J_c. Finally in Figure 8.17c a faceted grain boundary facilitates the current in one channel, and opposes it in the other. The result is lower J_c.

8.7.4   Relaxation of magnetic moment and the irreversibility line

Soon after the discovery of high-T_c superconductors by Bednorz and Müller, the IBM team also found an important difference between low-T_c and high-T_c

superconductors when exposed to a magnetic field [119]. Essentially the discovery was that in a broad temperature region below $T_c(H)$ the superconducting properties were reversible, corresponding to critical current $J_c = 0$. Only below a much lower temperature $T_{irr}(H)$ did the behaviour shift to irreversible, corresponding to $J_c \neq 0$. This created a dividing line in the (H, T)-plane, later called the *irreversibility line*. Furthermore, this line was *not* a fixed line, determined once and for all for a given material. Rather, in a given sample the experimentally determined location of the irreversibility line $T_{irr}(H)$ depended on the characteristic measuring time of the experiment, and in samples of the same nominal composition it would depend also on sample history. These effects are present to some extent in all superconductors, but to a far less degree in low-T_c than in the cuprate high-T_c materials.

What is going on can briefly be described as follows: After a certain arrangement of vortices has been established, typically after application of a magnetic field, and subsequent removal of the field to observe the vortex dynamics, vortices will leak out of the sample, leaving the most shallow pinning sites first. As time passes, the depinning process goes slower since the vortices have to climb higher and higher barriers in order to leave their pinning sites. The process goes on indefinitely, all the time at a slower rate. This effect is most clearly demonstrated by observing the decay of the magnetic moment $m(t)$ originally present as m_i at a time t_i when the field is switched off, and then monitoring the relaxation process over decades of time. The observed time-dependent magnetization is typically such that the magnetic moment, when plotted versus time on a double logarithmic scale, decays linearly, i.e. the functional dependence is a power law. Due to the relatively strong decay observed in high-T_c materials the behavior in these was termed 'giant flux creep' [120]. In Figure 8.18 we give an example of magnetization decay measurements at several temperatures in a wide range, observed in a thin film of Bi2223. The decay was monitored by a SQUID, located in a superconducting chamber, shielded against outside signals except for the signal picked up by a gardiometer coil surrounding the sample. Measurements started immediately after switching off a 10 mT field from a small solenoid surrounding the sample, starting the flux leakage from the sample [121]. Below 50 K the decay is seen to be quite linear in the double logarithmic plot over 4 decades in time, except for a small deviation in the long time limit ($t \geq 100\,\text{s}$) at the highest temperatures. The dominating decay behaviour may be expressed simply as

$$m = m_i \left(\frac{t}{t_i}\right)^{-S} \tag{8.87}$$

where the exponent S is $S = \partial(\ln m)/\partial(\ln t)$. For temperatures $T \geq 50\,\text{K}$ a crossover to faster dynamics is apparent, as would be expected in a crossover to vortex liquid. In this range the data are relatively noisy, due to the low magnetic moment.

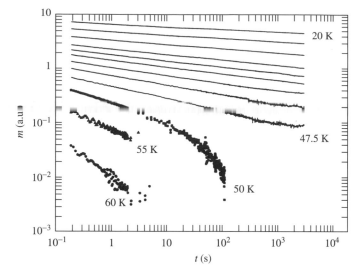

Figure 8.18 Relaxing irreversible magnetic moment in a Bi2223 film after a field step of $-10\,\mathrm{mT}$ is applied at different temperatures. The flux is leaving the sample, and shielding currents decay proportionally to the decrease of the magnetic moment. The temperatures are from top to bottom: 20, 25, 30, 35, 37.5, 40, 42.5, 45, 47.5, 50, 55, and 60 K. The c-axis oriented thin film with $T_c = 97\,\mathrm{K}$ was grown by metalorganic chemical vapor deposition (MOCVD) by Endo and coworkers at the Electrotechnical Laboratory in Japan. The measurements are by Tuset *et al.* [122].

The above experiment makes very clear why the so-called irreversibility line depends on the timescale of the experiment: The longer timescale is allowed in a given experiment, the more of the magnetization decay is integrated into the observation. To illustrate the consequences of this, we refer to Figure 8.19 where the position of the peak in the loss component of the dynamic susceptibility was measured as a function of temperature at different frequencies and field directions. The excitation field was parallel to the static magnetic field.

Figure 8.19 shows the irreversibility line as measured by AC susceptibility in YBCO crystal at a range of frequencies. Here the line is determined as the locus of points where the loss peak in the imaginary part of the susceptibility χ'' appears in the (H, T)-plane. We clearly see the effect of changing the time-scale of the experiment: With increasing frequency the $T_{\mathrm{irr}}(H)$ line moves up in temperature. The question is now how this line can be described quantitatively, or at least semi-quantitatively. A crude argument which gives some insight, but without taking into account the collective effects we discussed before, goes as follows:

Assuming that pinning occurs at pin points of volume V_c, the energy gained by having a flux line pass through this volume is equal to the condensation energy

$$U = \tfrac{1}{2}\mu_0 H_c^2(T)V_c \tag{8.88}$$

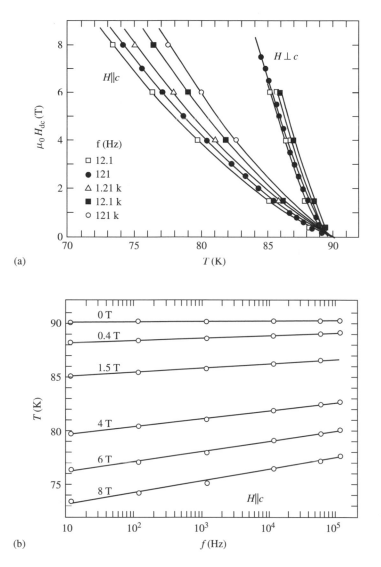

Figure 8.19 Irreversibility line measurements (a) at various frequencies in YBCO with $H\|c$-axis and $H\perp c$-axis. The bottom, (b), figure shows how the irreversibility line with $H\|c$ is shifted to higher temperature with increasing frequency. After Karkut *et al.* [103].

Assume further that V_c may be written as $a_0^2 \xi(T)$ where a_0^2 is the correlation area in the transverse direction given by the flux-line lattice constant. The critical field has almost linear temperature dependence in the field range we are discussing in high-T_c materials, i.e. $H_c \propto (1 - T/T_c)^1$ and $\xi(T) \propto (1 - T/T_c)^{-\frac{1}{2}}$ in the mean field approximation, as in Ginzburg–Landau theory. The result is, when inserting the expression for $a_0 = 1.075(\Phi_0/B)^{1/2}$ that

$$U = \frac{U_0 \left(1 - \frac{T}{T_c}\right)^n}{B}; \quad n = 1.5 \tag{8.89}$$

In order to test this line of thinking, Sagdahl [123] analysed a relaxation model for the permeability, writing the two components on the usual Debye form

$$\mu' = \frac{1}{1 + (\omega\tau)^2} \tag{8.90}$$

$$\mu'' = \frac{\omega\tau}{1 + (\omega\tau)^2} \tag{8.91}$$

The relaxation time is expected to be $\tau = a/Rl$ where l is the typical jump distance, R is the flux jump rate and a is a macroscopic distance describing the total depth of penetration of flux. We rewrite this as

$$\tau = (a/f_0 l)\exp(U/k_B T) \tag{8.92}$$

Here f_0 is an attempt frequency, and U is the pinning potential already introduced.

The maximum in μ'' occurs at $\omega\tau = 1$, setting the criterion for B and T to be on the irreversibility line as B^* and T^* at a selected frequency and AC amplitude. Using this criterion we arrive at the following condition for the irreversibility line:

$$\frac{U}{k_B T^*} = -\ln(\omega\tau_0) \tag{8.93}$$

where $\tau_0 = (a/f_0 l)$.

Next, inserting the expression for U arrived at previously we find that Eq. 8.93 converts to

$$U_0(1 - T^*/T_c)^n / k_B T^* B^* = -\ln(\omega\tau_0) \tag{8.94}$$

where A_j is a parameter which we leave undetermined. Finally, we rewrite Eq. 8.94 as

$$B^* = C(\omega)(1 - T^*/T_c)^n / T^* \tag{8.95}$$

where $C(\omega) = -(U_0/k_B)\ln(\omega\tau_0)$. This constitutes an expression for the irreversibility line in the present analysis. In [123] this expression was tested, and found to agree with experiments, with $n = 1.5$ at the fields above 1.5 T, while $n = 1$ below. Furthermore, in fitting the expression the parameter T_c had to be taken with a value well below the zero-field value $T_c(0)$. This shows that the model does not apply in the vicinity of $T_c(0)$. The irreversibility line was found to be almost flat over several degrees below $T_c(0)$. In other words very weak irreversibility was found in that temperature interval. Clearly pinning very close to $T_c(0)$ at very low fields is quite different from that at high fields. This can at least partly be argued to be caused by twin boundaries taking a more prominent role as pinning centres when the flux line density is low, in which case the flux line dynamics may be altered substantially.

It turns out that Eq. 8.89 agrees surprisingly well with experiments, considering the simplicity of the model, and the complexity of the real problem. Figure 8.20 gives an example of such overall agreement. Part of the reason should be ascribed to the fact that point pinning is important in YBCO due to the presence of oxygen vacancies as pinning sites. In the vicinity of the melting line the softening of the FLL strengthens the role of individual pinning.

The irreversibility lines of high-T_c compounds are located well below the H_{c2} line, and the difference is larger the more 2D-character the material possesses. This has represented a great challenge in the search for material processing methods suitable for attainment of high-J_c at reasonably elevated temperatures and in high magnetic fields. It is natural to ask how the irreversibility line is related to the melting line. The answer is that these are of different origin and largely independent, but not unrelated. The irreversibility line is determined for the decay of magnetisation current, in other words governed by flux line dynamics, while the melting line is a thermodynamic line and as such an equilibrium phase boundary, fixed in the (H, T)-plane for a given piece of material. The position of the melting line should be expected to depend somewhat on pinning

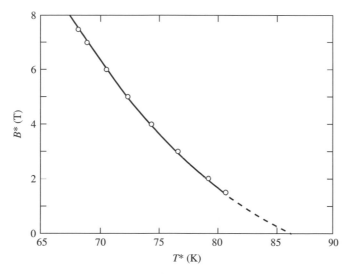

Figure 8.20   The irreversibility line $B^*(T^*)$ in the field range 0 to 7.5 T ($H \| c$) determined from AC permeability measurements on an YBCO single crystal with $f = 121$ Hz and $B_{ac} = 3 \times 10^{-4}$. The solid and dashed lines represent theoretical fits using Eq. 8.95, with exponents $n = 1.5$ for fields higher than 1.5 T and $n = 1$ for fields lower than 1.5 T. After Sagdahl [123].

properties of the material. However, melting can only be observed in materials with relatively low density of pins. In pin-free materials each superconductor should possess a unique melting line.

The irreversibility line is highly anisotropy dependent, as is the melting line as was shown in Figure 8.19 in two directions in YBCO. Clearly, both flux line dynamics and melting properties should be expected to be different in the two cases: When flux lines are along the c-axis they are free to displace sideways in two directions, while in the ab-plane they are packed parallel between superconducting CuO_2 sheets and can only move sideways parallel to the ab-plane. This means that a thermal energy $k_B T$ has much less influence on the configuration of flux lines when the field is in the ab-plane than when it is along the c-axis. Therefore it is much harder to melt the FLL with the field in the ab-plane than along the c-axis. The melting line for field in the ab-plane is therefore much higher than with field along the c-axis.

The statement made above that the irreversibility lines are independent but not unrelated can now be given more content: We see from Figure 8.19 that the irreversibility line is much steeper and higher up in temperature with field in the ab-plane than when it is in the c-direction. In this sense they are clearly related. But the irreversibility lines move as a function of frequency of the AC field, in this sense they are independent.

The relationship between the two types of lines is due to the fact that time constants near phase boundaries depend stronger on temperature relative to $T(H)$ and tend to diverge at the phase boundary. A wide range of time constants is therefore available in the vicinity of the T_m-line. As it turns out the experimentally accessible time constants are in that same range. Hence the irreversibility line will appear not too far away from the T_m-line. A sketch of this is given in Figure 8.21.

8.7.5   How can J_c be increased?

The irreversibility line problem in high-T_c materials has been a serious obstacle against the development of wires and cables for many kinds of superconductor

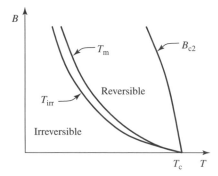

Figure 8.21   Sketch of the irreversibility line T_{irr} and the melting line T_m.

applications. Huge research efforts have been made to overcome it. The improvements which have been achieved are very impressive. Compared to the critical current densities obtained in the early stages of high-T_c research these systematic efforts have led to increased critical current density by many orders of magnitude. For a good example we refer to the topical contribution by Muralidhar and Murakami (Section 13.4) on a modification of YBCO-type material, where extremely strong magnetic fields may be stored in bulk material. The highest fields reached are so strong as to challenge the physical strength of the material itself due to the strong forces that the pinned vortex lattice causes on the structure.

In the following we give some some more examples of results where the *improvements* are substantial.

The first example is from the work of Krusin-Elbaum and coworkers [124]. In this case the mercury-based superconductor $HgBa_2Ca_2Cu_2O_{6+\delta}$, also referred to as Hg-1212, was subjected to 0.8 GeV proton bombardment which induced fission processes in Hg, leaving tracks in the material. These tracks created what is often referred to as 'columnar defects' in a 'pinning landscape' which reduced the vortex motion (creep) substantially. Figure 8.22 shows the resulting improvements in critical current density, as well as a big shift of the irreversibility line,

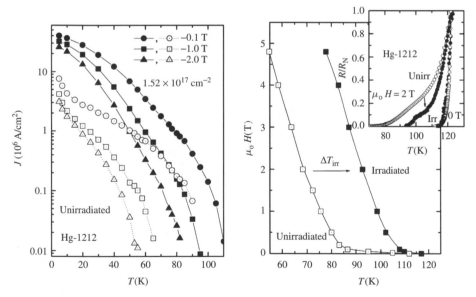

Figure 8.22   The effect of proton bombardment at 0.8 GeV, with resulting fission fragments creating columnar defects in superconducting $HgBa_2CaCu_2O_{6+\delta}$. On the left the improvement of critical current is shown at various fields, with open and closed symbols showing the results before and after irradiation, respectively. The proton fluence of the irradiation was 1.52×10^{17} cm^{-2}. The figure on the right shows the corresponding shift of the irreversibility line, with the insert giving simultaneous resistivity changes [124].

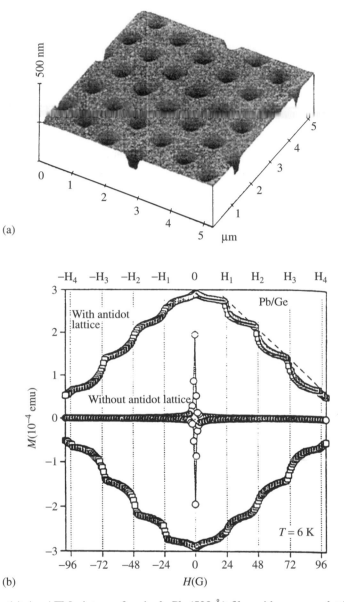

(a)

(b)

Figure 8.23 (a) An AFM picture of a single Pb (500 Å) film with a square lattice of sub-micron antidots (periods $d = 1\,\mu$m, radius $r = 0.2\,\mu$m. (b) Magnetization loop $M(H)$ at $T = 6$ K of a (Pb (100 Å))/(Ge (50 Å))$_2$ multilayers with and without the triangular anti-dot lattice ($d = 1\,\mu$m). The solid line is a fit following the logarithmic dependence of the magnetization between the matching fields H_m and H_{m+1}. The dashed line is demonstrating the validity of the linear behaviour of $M(H_m)$. The loops were measured for $M > 0$ and symmetrized for clarity for $M < 0$ [125].

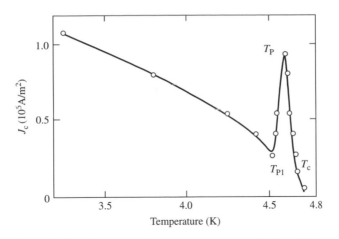

Figure 8.24   The 'peak effect' demonstrated in NbSe$_2$ [126].

moving it up by about 20 K at the highest field employed [124]. Since the irreversibility line is a borderline between regions where $J_c \neq 0$ and $J_c = 0$ this means that a much larger area of the (H, T)-plane is potentially useful after the irradiation.

A different strategy was followed by Fossheim and coworkers [127] who were able to embed carbon nanotubes into Bi$_2$Sr$_2$CaCu$_2$O$_{8+x}$. Due to the small diameter of the nanotubes, in the 10 nm range, they are suitable as columnar defects and can provide strong pinning.

The next example, Figure 8.23 is from a thin film superconductor by V. Moshchalkov and coworkers [125]. In this case a thin film superconductor was modified by making a large number of holes, "antidots", in the films, creating a regular lattice of holes. Each hole is capable of pinning one or more vortices. The measurements shown are magnetization *vs.* applied field in both the positive and the negative directions. The narrow inner curve shows, for comparison, the magnetization before making the holes in the film. The magnetization is very low, corresponding to a very low critical current density. In the upper curve, corresponding to measurements in the film with pinning arrays, the situation has changed dramatically: As the magnetic field is turned on, one finds a magnetization which is orders of magnitude higher than in the intrinsic case. In addition, some structure appears. The first peak corresponds to a matching of the number of holes with the number of vortices available for pinning, at the so-called 'matching field B_ϕ'. On further increase of the applied field, new matching conditions are found, representing two, and three vortices per hole, etc. A number of different geometries and variations in pinning strategies have been investigated.

An interesting effect which arises in many superconductors when the temperature approaches T_c from below, is the appearance of a sharp maximum in the critical current density just before it goes to zero at T_c. This is the so-called 'peak effect', and is ascribed to collective pinning. It arises because the flux line system softens on approaching T_c, as can be seen from the analysis we have done of the elastic properties of the FLL previously in this chapter. Since a softer vortex lattice is able to adjust more easily to the random positions of pinning centra, the vortices are now better pinned. However, as T_c is approached further, the Cooper pair density and the wavefunction diminish even more, and pinning vanishes altogether. J_c then must go to zero. The example given in Figure 8.24 is from Kes and coworkers on $NbSe_2$ [126], with $T_c = 7.1\,K$. Three characteristic temperatures are indicated in the figure, representing the minimum of critical current density reached below the peak temperature, the position of the maximum, and finally T_c where J_c goes to zero.

PART II

Advanced Topics

9

Two-dimensional Superconductivity. Vortex-pair Unbinding

9.1   Introduction

In this chapter, we will go through the case of how superfluidity is destroyed in thin film superconductors or thin films of superfluids with a scalar order parameter. The starting point is the Ginzburg–Landau theory, now applied to two-dimensional systems. In this case, the distinction between superconductors (charged condensates) and superfluids (neutral condensates) vanishes. The reason is that in the case of superconductors, the *effective* London penetration depth is given by the bulk penetration depth λ and the thickness of the film d as follows

$$\lambda_{\text{eff}} = \lambda \, \frac{\lambda}{d}$$

$$\lambda = \sqrt{\frac{m_q}{\mu_0 n_s q^2}} \tag{9.1}$$

where q is the charge of the carriers in the superconducting state, m_q their mass, and n_s their number density. We may thus write the effective penetration depth in the form

$$\lambda_{\text{eff}} = \sqrt{\frac{m_q}{\mu_0 n_s q_{\text{eff}}^2}}. \tag{9.2}$$

Thus, when $d \to 0$, this is equivalent to letting $q_{\text{eff}} \to 0$, and we thus effectively have a neutral condensate in this case. The physics of this is simply that in the case of an infinitely thin film $\lambda_{\text{eff}} \to \infty$, thus precluding gauge-field fluctuations in the interior of the film. We therefore can ignore the vector potential $\mathbf{A}$ as an independent fluctuating variable in the action. In the presence of an external

Superconductivity: Physics and Applications   Kristian Fossheim and Asle Sudbø
© 2004 John Wiley & Sons, Ltd   ISBN 0-470-84452-3

field inducing vortices in the thin film, this would simply enter as appropriately twisted, but non-fluctuating, boundary conditions on the sample. Here, we shall for the moment ignore an externally imposed magnetic field.

9.2   Ginzburg–Landau description

The Ginzburg–Landau theory in the two-dimensional case, regardless of the charge of the condensate, takes the form

$$F_s(\boldsymbol{r}, T) = F_n(\boldsymbol{r}, T) + \alpha \, |\psi|^2 + \frac{\beta}{2} \, |\psi|^4 + \frac{\hbar^2}{2m} \, |\boldsymbol{\nabla} \psi|^2 \qquad (9.3)$$

where $\psi(\mathbf{r}) = |\psi(\mathbf{r})| e^{i\theta(\mathbf{r})}$ is the complex scalar order parameter of the condensate. For the subsequent discussion, we will find it convenient to *regularize* this continuum theory by placing it on a lattice, i.e. we introduce a short distance cutoff in the problem, which in physical terms is taken to be of the order of the coherence length ξ. Moreover, we ignore the normal state free energy, measuring free energies in the superconducting state relative to the normal state. Such a lattice regularization is tantamount to introducing finite differences instead of gradients everywhere. Also, the order parameter will now be defined on lattice sites with labels given by the set $\{i\}$, which runs through $1, \ldots, N = L^2$, where L is the length of the sample in each direction. Ultimately, the $L \to \infty$ limit needs to be taken.

Note that in the end, it is really the continuum version of the Ginzburg–Landau theory we will be interested in, more precisely the continuum version at the *critical point of the theory*. When a phase transition happens to be a critical phenomenon, this means that there is a diverging length scale (ξ) in the problem. We shall identify this length in the present chapter. Under such circumstances, the physics we are interested in are critical fluctuations on a length scale set by the diverging length. Hence, the lattice constant in our regularized theory is an irrelevant variable and the physics at the critical point of the lattice superconductor is expected to capture the physics of the continuum theory. Lattice regularizations, however, do require a bit of caution whenever they are applied, since it is not in any way obvious that such discretizations of space are not singular perturbations. Moreover, even if the lattice constant 'disappears' from the problem in this case, it does not necessarily leave the scene quitely, but may endow the system with an exponent η.

The lattice regularized Ginzburg–Landau theory for a two-dimensional lattice superconductor takes the form

$$F_s = \sum_i \left[\alpha \, |\psi_i|^2 + \frac{\beta}{2} \, |\psi_i|^4 + \sum_{\mu=x,y} \frac{\hbar^2}{2m} \, |D_\mu \psi|^2 \right]$$

$$D_\mu \psi = \psi_{i+\mu} - \psi_i \qquad (9.4)$$

where $\psi_{i+\mu}$ is the order parameter evaluated at a lattice site which is shifted *one* lattice constant (set equal to unity) in direction μ. Expressing the above in terms of the amplitude and phase of the order parameter we find

$$F_s = J_1 \sum_i \left[\frac{2\alpha(T)}{\alpha(0)} |\psi_i'|^2 + |\psi_i'|^4 \right.$$

$$\left. + \sum_{\mu=x,y} \frac{\xi_\mu^2}{a_\mu^2} |\psi_{i+\mu}'||\psi_i'|(1 - \cos(\Delta_\mu\theta)) \right] \qquad (9.5)$$

where

$$\Delta_\mu\theta = \theta_{i+\mu} - \theta_i$$

$$J_1 = 2\alpha(0)\beta$$

$$\psi_i' = \frac{\psi_i}{\sqrt{\frac{\alpha(0)}{\beta}}} \qquad (9.6)$$

9.3 Critical fluctuations in two-dimensional superfluids

In principle, this formulation allows for both amplitude and phase fluctuations. However, we will ignore the amplitude fluctuations in the following. We simply set the amplitudes $|\psi_i'| = 1$. This is very much like what we do in spin models where the length of the spin is taken to be frozen, and only the direction of the spin is taken to be important in investigating whether the magnet orders or not. Although we will not go into the details here, it is by now very well established that phase-only approximation to the fluctuations in ψ_i is on solid ground. This means that we can write the energy for latticized two-dimensional superconductor/superfluid on the form

$$F_s = J_0 \sum_{i,\mu} \left[1 - \cos(\Delta_\mu\theta) \right]$$

$$J_0 = J_1 \frac{\xi_\mu^2}{a_\mu^2} \qquad (9.7)$$

when all constant terms are ignored. This is the well-known two-dimensional *XY*-model. It may be recast slightly into the following form (in addition, we introduce H as the Hamiltonian instead of F_s)

$$H = J_0 \sum_{\langle i,j \rangle} \left[1 - \cos(\theta_i - \theta_j) \right] \qquad (9.8)$$

Here, $\langle i, j \rangle$ denotes nearest neighbour coupling, which emerges naturally by putting gradient terms in the Ginzburg–Landau theory on the lattice, and i runs over all lattice sites in the problem. The entire Ginzburg–Landau theory is now expressed solely in terms of the *phase* of the superconducting order parameter, and the resulting statistical mechanics is entirely governed by phase fluctuations.

Phase differences between lattice points are expressed as lattice gradients of the phase. Let us for simplicity denote them by $\nabla\theta$, which is a vector. A vector may always be split into a transverse part and a longitudinal part. The transverse part is divergence free, the longitudinal part is curl-free, hence we have

$$\nabla \cdot (\nabla\theta)_{\mathbf{T}} = 0,$$

$$\nabla \times (\nabla\theta)_{\mathbf{L}} = 0 \tag{9.9}$$

Consider first this model at low temperatures. We then expect that thermally generated fluctuations in θ are, in some sense, small. The situation we have in mind in this case is depicted in Figure (9.1).

This means that we can expand the cosine-term to leading order in the phase-gradients, arriving at

$$H = \frac{J_0}{2} \sum_{\langle i,j \rangle} (\theta_i - \theta_j)^2 \tag{9.10}$$

where J_0 now can be identified as a bare spin-wave stiffness parameter. The correlation function that probes the phase-coherence in the system is

$$\Gamma(\mathbf{r}) = \langle e^{i\theta_i} e^{-i\theta_j} \rangle \tag{9.11}$$

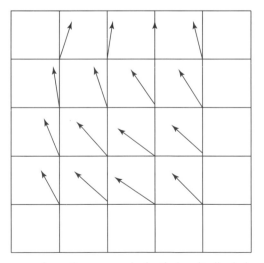

Figure 9.1 Spin-wave configurations, or equivalently longitudinal phase-fluctuations, in the two-dimensional XY-model on a square lattice. The arrows indicate values of θ_i, analogous to an XY spin direction, on the lattice points where the order parameter is defined.

where $\mathbf{r} = \mathbf{i} - \mathbf{j}$. At low temperatures, the statistical average can safely be computed using the Hamiltonian Eq. 9.10. Since this is quadratic in the phase fluctuations, the average may be evaluated exactly. First of all, this leads to the simplification that

$$\Gamma(\mathbf{r}) = e^{-\langle (\theta(\mathbf{r}-\theta(0))^2/2\rangle} = e^{-\langle (\theta(\mathbf{r})-\theta(0))\,\theta(0)\rangle} \qquad (9.12)$$

Here, $\mathbf{r}$ is the lattice vector between points i and j. In the last equality above, we have used the fact that the system is assumed to be translationally invariant. This means that we are neglecting effects of disorder in what follows. Thus, we are faced with the task of computing the correlation function

$$G(\mathbf{r}) \equiv \langle [\theta(\mathbf{r}) - \theta(0)]\theta(0)\rangle \qquad (9.13)$$

The computation of this quantity is facilitated by the equipartition theorem applied to Eq. 9.10, expressed in terms of the Fourier-modes $\theta_{\mathbf{q}}$ of the phase-fluctuations as follows

$$H = \frac{J_0}{2} \sum_{\mathbf{q}} q^2\, \theta_{\mathbf{q}}\, \theta_{-\mathbf{q}} \qquad (9.14)$$

This is quadratic in each of the Fourier modes and thus we have, by the equipartition theorem

$$\frac{J_0}{2} q^2 \langle \theta_{\mathbf{q}} \theta_{-\mathbf{q}}\rangle = \frac{k_B T}{2} \qquad (9.15)$$

which implies that the Fourier transform of the correlation function $\langle \phi(\mathbf{r})\phi(0)\rangle$ is given by

$$\langle \theta_{\mathbf{q}} \theta_{-\mathbf{q}}\rangle = \frac{k_B T}{J_0} \frac{1}{q^2} \qquad (9.16)$$

Thus, we get

$$G(\mathbf{r}) = \frac{k_B T}{J_0} \int \frac{d^2 q}{(2\pi)^2} \frac{e^{i\mathbf{q}\cdot\mathbf{r}} - 1}{q^2} \qquad (9.17)$$

Note how the phase-fluctuations vanish when $T \to 0$. This means that phase-correlations become truly long-ranged when $T = 0$, so at least in this limit we have established long-range order. What happens as we switch on T? The above Fourier-integral may be evaluated straightforwardly, to obtain

$$G(\mathbf{r}) = -\frac{k_B T}{2\pi J_0} \ln \frac{r}{a} \qquad (9.18)$$

where a is some short-distance cutoff $a = \mathrm{e}^{-\gamma}/2\sqrt{2}$ and $\gamma = 0.577\ldots$ is the Euler–Mascheroni constant. Inserting this back into Eq. 9.12, we get

$$\Gamma(\mathbf{r}) = \left(\frac{r}{a}\right)^{-\eta} \tag{9.19}$$

where $\eta = k_{\mathrm{B}}T/2\pi J_0$. Note how this correlation function fits in with a general Ansatz for the two-point correlator $\Gamma_2(\mathbf{r})$ close to a critical point, which on general grounds may be written as

$$\Gamma_2(\mathbf{r}) = \frac{1}{r^{d-2+\eta}}\, \mathcal{G}(r/\xi) \tag{9.20}$$

where ξ is some correlation length of the problem, d is dimensionality, $\mathcal{G}$ is some scaling function, and η is a so-called anomalous scaling dimension of the superfluid matter field ψ. Eq. 9.20 is of the form of Eq. 9.19, provided we set $d = 2$ and $\xi = \infty$. The latter is the hallmark of a system which is *critical*. Thus, what we learn from this is that a two-dimensional superfluid/superconductor has an entire low-temperature phase, which is critical. This is also reflected in the anomalous scaling dimension $\eta = k_{\mathrm{B}}T/2\pi J_0$, which is *non-universal*. Critical exponents associated with critical points are universal. Non-universality of exponents means that we are dealing here with a critical line, not a critical point.

An important feature of $\Gamma(\mathbf{r})$ is that $\lim_{\mathbf{r}\to\infty}\Gamma(\mathbf{r}) \to 0$ provided $\eta \neq 0$, i.e. as soon as $T \neq 0$. Hence, there is never true long-range order at finite temperature in a two-dimensional superfluid/superconductor. In other words, the $U(1)$ symmetry that is spontaneously broken in the three-dimensional case is never broken at finite temperature in two dimensions. This is a specific example of a somewhat more general statement, namely the Hohenberg–Mermin–Wagner theorem that states that at finite T, no continuous symmetry can be spontaneously broken in dimensions $d \leq 2$ [128, 129][1]. What we at most can have is power-law decay (which is much slower than exponential decay characteristic of short-range order). Power laws mean that there is no length scale in the problem any longer, $\xi \to \infty$ in Eq. 9.20. The system is thus scale-invariant, which is exactly the hallmark of criticality. This is often referred to as *quasi-long-range order*.

The above result is based on the presumption that we can define local order of the phase-fields, and expand in small fluctuations around local order. A minute or two of reflection reveals that this cannot possibly be correct at high temperatures. It is obvious that in the high-temperature limit, even short range order must break

---

[1]Often just referred to as the Mermin–Wagner theorem. Both the papers of Hohenberg and of Mermin and Wagner demonstrate the *absence* of long-range order in low dimensions for systems with continuous symmetries. Hohenberg considers neutral superfluids, Mermin and Wagner consider ferromagnets. Ironically, the title of the paper by Hohenberg is nonetheless *Existence of long range order in one and two dimensions*! A humorous remark by Halperin at Hohenberg's 60th birthday suggested that perhaps the title is the reason for the much more frequent reference to Mermin–Wagner, rather than Hohenberg–Mermin–Wagner! Let it be said at once that both papers arrive at correct conclusions.

down, and the above analysis is invalid. What have we missed? The process of elimination reveals the answer. Our starting point only has phase-fluctuations to begin with. The smooth phase-fluctuations we have included so far correspond to longitudinal phase-fluctuations. What we have left out, are transverse phase-fluctuations. Their inclusion in the description, and their physical consequences, will be the topic of the next sections.

We conclude this section by emphasizing that what we have demonstrated here, is that smooth phase-fluctuations (analogues of spin-waves in magnetic systems) suffice to destroy long range order in two dimensions and render the system critical, at any finite temperature. Spin-waves are therefore critical fluctuations in two dimensions (spin waves are not critical in three dimensions). Nonetheless, this does not mean that it does not cost energy to 'twist' the phase of the superconducting order parameter at low temperatures. Even if the system does not exhibit true long-range phase order, it still has a phase stiffness associated with it. On the other hand, at very high temperatures, phases are randomly oriented relative to each other even on the shortest length scales, and a local twist is expected to come at no cost in the free energy. Hence, somewhere in between low and high temperatures some phenomenon must occur such that the phase stiffness vanishes. If this happens at a finite temperature, then this phenomenon must in fact be a phase-transition, not a crossover. This is because a finite phase stiffness cannot be brought to precisely zero on a finite temperature interval in an analytic manner. It must therefore vanish in a phase transition. That this putative phase transition must be a subtle one, is clear from the simple observation that it cannot be a phase transition from an ordered low-temperature state to a disordered high-temperature state. It must be a phase transition from a quasi-ordered state to a disordered state. Under such circumstances we do not have the luxury, as we have in three dimensions, of probing the phase transition by monitoring the expectation value of a local ordering field (a local order parameter) and asking when it vanishes. In fact, when we are dealing with systems with a continuous symmetry in two dimensions we are forced to invoke a much more general (but often unnecessary) method of detecting a phase transition: We must identify a generalized stiffness of the system, which is a global quantity, and investigate when this stiffness vanishes through a proliferation of topological defects facilitated by entropy production in the free energy. In fact, this is a method we can always use, also in three dimensions, to detect phase transitions. We shall rephrase the three-dimensional superconductor to normal metal transition in this language in the next chapter.

9.4 Vortex–antivortex pairs

Transverse phase fluctuations are more violent than the smooth longitudinal ones considered in the previous section. Consider the quantity $\nabla\theta$ and imagine integrating this around some arbitrary closed contour in the plane. The *vorticity* q of the phase-distribution is measured by the quantity

$$\oint \mathbf{dl} \times \nabla\theta = 2\pi q \tag{9.21}$$

Here, $q \in \mathbb{Z}$. It is impossible to continually deform a phase distribution with $q = 0$ where only longitudinal, smooth phase fluctuations are excited, into a distribution where $q \neq 0$. Those phase distributions where $q = 0$ and those where $q \neq 0$ are *topologically* distinct. A vortex with a vorticity q is a topological defect superimposed on the phase distribution where no vortices are present. It is obvious that for high enough temperatures, transverse phase fluctuations will be excited thermally, not only smooth longitudinal ones. Vortices will therefore be generated spontaneously at high enough temperatures since this will increase the configurational entropy of the system, thereby lowering the free energy. Vortices are topological defects of the superconducting ordering field $e^{i\theta(\mathbf{r})}$. When they proliferate, the phase stiffness of the system vanishes, and this defines the phase transition. In Figure 9.2, we illustrate a vortex configuration in the two-dimensional XY-model on a square lattice, while in Figure 9.3 we illustrate a vortex–antivortex pair.

Note that for a charged superfluid, $\nabla\theta$ gives rise to an electric current. The curl of this current is a magnetic field, and by Eq. 9.21 we thus see that q

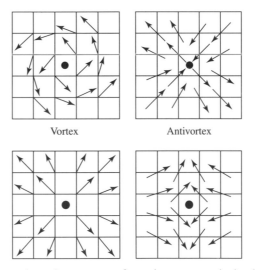

Vortex   Antivortex

Figure 9.2 Four examples of vortex configurations, or equivalently transverse phase-fluctuations, in the two-dimensional XY-model on a square lattice. Notice how each vortex corresponds to a collective phase-mode. The black dot in the centre is meant to illustrate the core of the vortex in this particular configuration. Thermally excited transverse phase fluctuations always give a vortex and an antivortex. The arrows indicate values of θ_i, analogous to an XY spin direction, on the lattice points where the order parameter is defined.

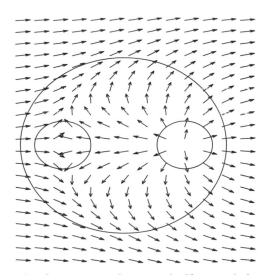

Figure 9.3 An example of a vortex–antivortex pair. If we encircle one of these vortices along one of the small paths, we pick up a phase-difference $\pm 2\pi$. The phase-difference along the large path is zero. The fluctuations outside of this path correspond to spin waves.

may be interpreted essentially as the quantum of a magnetic field penetrating through an area enclosed by a contour C which the integral in Eq. 9.21 is taken around. It is intuitively perhaps obvious that no net magnetic field can be generated throughout the system by thermal fluctuations. Hence, vortices must always be generated in pairs of opposite vorticity, a vortex–antivortex pair. At low temperatures where vortices are expected to be unimportant, the above statement of the existence of vortex–antivortex pairs translates into a statement that at low temperatures only very tightly bound pairs exist. The proliferation of free topological defects at high temperatures, responsible for destroying the phase stiffness of the system, is equivalent to saying that the vortex–antivortex pair has dissociated into two free vortices of opposite vorticity. As we shall see, the above may be precisely rephrased into the language of a metal-insulator transition in a two-dimensional Coulomb gas, where the low-temperature phase corresponds to an insulating dielectric phase and the high-temperature phase corresponds to a metallic plasma phase.

9.5 Mapping to the 2D Coulomb gas

When vortex configurations are thermally excited in the two-dimensional super-fluid, this is equivalent to generating a set of fluctuating 'charges' $\{q_i\}$ with overall charge-neutrality imposed, $\sum_i q_i = 0$. The loss of phase-stiffness in the system, which we have argued above to be the phase-transition, must be related somehow to the statistical mechanics of this ensemble of charges. This follows from the fact that these charges are manifestations of transverse phase fluctuations in the superconducting ordering field, and that the longitudinal phase fluctuations are incapable of driving the system into the high-temperature phase. What we need to do, is to include explicitly the feature that the phases of the order parameter are *angular* variables defined on the interval $\theta_i \in [0, 2\pi)$.

For the purposes of studying the phase transition in the two-dimensional superfluid on a lattice, i.e. where we have introduced a short-distance cutoff in the problem, we will perform a number of standard steps essentially amounting to what is called a *duality transformation*. Such a transformation is basically the following. When we focus on studying the critical properties of a system exhibiting a critical phenomenon, we try to identify precisely the thermal excitations that are responsible for destroying order in this system. These excitations may be viewed in general as topological defects in some ordering field, in our case the superfluid/superconducting condensate wave function ψ_i. Next, we attempt to express the partition function, more precisely the part of the partition function giving a singular contribution to the free energy, in terms of these topological defects. When this has been done, we have effectively executed a duality transformation. Strictly speaking the final step in the duality transformation is to write down a field theory for the action appearing in the partition function when it is written in terms of the topological objects of the original system. This field theory is the dual to the original field theory we started from.

In this context, let us make an important remark on duality before we start doing computations in earnest. Suppose we have arrived at some dual field theory starting from an original theory such as the Ginburg–Landau theory of the two-dimensional superfluid. This dual theory itself has topological defects in its (dual) ordering field, which in principle we may identify. The partition function for the dual model may thus also be written in terms of the topological defects of the dual ordering field, and a field theory can be written for the corresponding action. This field theory is the dual of the dual model, and must in fact correspond to the original starting model. Duality squared must be equal to one. This is a statement which holds for all systems with only one phase transition. It is much weaker than the statement of self-duality, which means that the dual model has an identical form as the original model (but with inverted coupling constants). For instance, the dual model of the two-dimensional XY model does not have the same form as the two-dimensional XY model. The dual of the dual of the two-dimensional XY model is the XY model itself. However, the two-dimensional Ising model is selfdual, while the dual of the three-dimensional Ising model is a three-dimensional Ising gauge-theory, i.e. a lattice model with a local Ising theory. This is probably the simplest example where two models with the same symmetry, but where one has a global symmetry and the other has a local one, are related by a duality transformation. As it turns out, a similar situation is true for three-dimensional systems with local and global $U(1)$ symmetries, which we will consider in the next chapter.

The model with a global $U(1)$ symmetry is a description of a three-dimensional charge-neutral superfluid, while the model with a local symmetry is a model of a three-dimensional superconductor. Hence, three-dimensional superfluids and superconductors are duals of each other.

The partition function for the system is *a priori* given by

$$Z_{XY} = \int D\theta e^{-\beta H_{XY}}$$

$$H_{XY} = J \sum_{\langle i,j \rangle} \left[1 - \cos(\theta_i - \theta_j) \right] \tag{9.22}$$

We next ignore the constant term in the Hamiltonian, and introduce the Fourier expansion of $e^{\alpha \cos(x)}$ as follows

$$e^{\alpha \cos(x)} = \sum_{n=-\infty}^{\infty} I_n(\alpha) e^{inx} \tag{9.23}$$

where $I_n(\alpha)$ is a Bessel function of order n. Inserting this in Eq. 9.22, we get

$$Z_{XY} = \sum_{n=-\infty}^{\infty} I_{n_{\mu j}}(\beta) \int D\theta \exp\left(\sum (in_{\mu j}\Delta_\mu \theta_j) \right) \tag{9.24}$$

Carrying out the angular integrations yields a constraint on the integer-valued field **n** given by

$$\Delta_\mu n_{\mu j} = 0 \tag{9.25}$$

This is solved by introducing the two-dimensional curl

$$n_{\mu j} = \varepsilon_{\mu\nu} \Delta_\nu \phi_j \tag{9.26}$$

where $\varepsilon_{\mu\nu}$ is the two-dimensional Levi–Civita tensor, and ϕ_j is some scalar integer-valued field, which constitutes the actual degrees of freedom of the system. Up to now, everything is exact. The partition may now be written in terms of our new degrees of freedom, $\{\phi_j\}$, as follows

$$Z = \sum_{\{\phi_j\}} I_{\phi_j}(\beta) \tag{9.27}$$

This may be approximated, using standard properties of the Bessel functions, as follows

$$Z = \sum_{\{\phi_j\}} \exp\left(\sum_{j,\mu} \left(-\frac{1}{2\beta} [\Delta_\mu \phi_j]^2 \right) \right) \tag{9.28}$$

which we now want to write in a form that explicitly exhibits the topological excitations of the theory. Strictly speaking, the approximation we have made in passing from Eq. 9.27 to Eq. 9.28 is quantitatively useful at large values of β, i.e. it is naively expected to be a good low-temperature approximation. However, it is also qualitatively useful outside of the low-temperature domain. First, the symmetries of the exact partition function Eq. 9.27 and the approximate partition Eq. 9.28 are the same. This means that that the topological excitations of these two objects are the same. Hence, we expect that for the purpose of studying the critical properties of the system, these two objects essentially produce results which are in the same universality class. Less important details, such as the exact value of the critical temperature, will of course differ. To proceed further, we first introduce the Poisson-summation formula

$$\sum_{n=-\infty}^{\infty} f(n) = \sum_{k=-\infty}^{\infty} \int_{-\infty}^{\infty} dz\, f(z) e^{i2\pi kz} \qquad (9.29)$$

Thus, the partition function is written in the following form, when we introduce a scalar auxiliary field p_i

$$Z = \sum_{\{p_i\}} \int D\{\phi\} \exp\left(\sum \left(-\frac{1}{2\beta}[\Delta_\mu \phi_j]^2 + i2\pi p_j \phi_j\right)\right) \qquad (9.30)$$

The next step is now to perform the ϕ_j integrations, which gives

$$Z = Z_0 \sum_{\{p_i\}} \exp\left[-4\pi^2 \beta \sum_{i,j} p_i D(|i - j|) p_j\right]$$

$$-\Delta_\mu^2 D(|i - j|) = \delta_{ij} \qquad (9.31)$$

where Z_0 is the partition function for spin waves. Note how the partition function factorizes (in the approximation of Eq. 9.28) in the product of a spin-wave part and a part which is *isomorphic to the partition function of a two-dimensional Coulomb gas*. We will see below that the quantities p_j may be thought of as vortices in the original phase fields θ_i. The above factorization shows that spin waves and vortices do not interact within the approximation made in Eq. 9.28. Explicitly, we have

$$D(|i - j|) = \int_{-\pi}^{\pi} \frac{dq_x}{2\pi} \int_{-\pi}^{\pi} \frac{dq_y}{2\pi} \frac{e^{iq\cdot(i-j)}}{2\sum_\mu [1 - \cos(q_\mu)]} \qquad (9.32)$$

which is just the Coulomb-potential on the two-dimensional lattice. It has the asymptotic value $D(|i - j|) \propto \ln(|i - j|)$ as $|i - j| \to \infty$, moreover the only allowed vortex configurations are those where $\sum_i p_i = 0$. This 'charge-neutrality' condition follows from the fact that the above describes a Coulomb system with no screening, and hence it is incompressible. This means that any finite charging of the system will cost an infinite amount of energy in the thermodynamic limit such that configurations with $\sum_i p_i \neq 0$ are suppressed.

Let us now see how the fields p_i correspond to vortex configurations of the original phase fields. In order to interpret the p_i-excitation in as easy a fashion as possible, it is convenient to retrace our above steps, using the following approximation for the exponential of the cosine-potential in the XY action

$$e^{\beta \cos(x)} \approx e^{\beta} \sum_{n=-\infty}^{\infty} \exp\left[-\frac{\beta}{2}(x + 2\pi n)^2\right] \qquad (9.33)$$

where we allow x to vary on the interval $\langle-\infty, \infty\rangle$ on the right-hand side. This approximation is usually referred to as the Villain-approximation, and is extremely useful in the current context (as well as in higher dimensions, as we shall see in Chapter 10). Essentially, the partition function for the XY-model is approximated by the partition function for a 2π periodic Gaussian model as follows

$$Z = \sum_{\{a_\mu\}} \int D\theta \exp\left(-\frac{\beta}{2} \sum_{\mu,j} (\Delta_\mu \theta_j - 2\pi a_{\mu,j})^2\right) \qquad (9.34)$$

where $a_{\mu,j}$ is an integer-valued field, and we are instructed to sum over all such integers. The next step is to introduce the Fourier transform for each of the exponentials appearing in the partition function using

$$\exp\left[-\frac{\beta}{2}(x + 2\pi n)^2\right] = \frac{1}{\sqrt{2\pi\beta}} \int_{-\infty}^{\infty} dy \exp\left[-\frac{1}{2\beta}y^2 + iy(x + 2\pi n)\right] (9.35)$$

Thus we get, up to multiplicative factors which do not contribute to singular parts of the free energy,

$$Z = \sum_{\{a_\mu\}} \int D\theta \int Dv_\mu \exp\left[\sum_{\mu,j} -\frac{1}{2\beta}v_{\mu j}^2 + iv_{\mu j}(\Delta_\mu \theta_j + 2\pi a_{\mu,j})\right] (9.36)$$

The next step is to carry out the integrations over the phase-field θ_j, thus producing the constraint

$$\Delta_\mu v_{\mu,j} = 0 \qquad (9.37)$$

which is satisfied provided

$$v_{\mu,j} = \varepsilon_{\mu\nu} \Delta_\nu \phi_j \qquad (9.38)$$

Let us insert this representation into the partition function Eq. 9.36. Upon a partial integration in the exponent of Eq. 9.36, it may be written on form

$$
Z = \sum_{\{a_\mu\}} \int D\phi \exp\left[\sum_{\mu, i} -\frac{1}{2\beta} v_{\mu j}^2 - 2\pi i \varepsilon_{\mu\nu} \Delta_\mu a_{\nu, j} \phi_j \right] \qquad (9.39)
$$

Now, comparing this with Eq. 9.30, we see that it has an identical form to Eq. 9.39, but we can identify the p-fields, namely $p_j = \varepsilon_{\mu\nu} \Delta_\mu a_{\nu, j}$. If we now think of $a_{\mu j}$ as the integer part of the phase-difference $\Delta_\mu \theta_j$, which we may think of as a current, then we see that p_j represents the vorticity of the original fields.

It is also useful to rederive all of the above in a continuum version using a particular representation of the gradients of the phase which also takes into account the singularities arising from the fact that the phase is an angular variable, i.e. is defined with compact support, $\theta_i \in [0, 2\pi)$. We write the gradient of the phase (a vector) as a longitudinal and transverse part, cf. Eq. 9.9. The longitudinal part, corresponding to spin waves, is non-singular, while the transverse part gives rise to vortices. Moreover, since they are orthogonal components of the phase gradient, we have

$$
(\partial\theta)^2 = (\partial\theta_L)^2 + (\partial\theta_T)^2 \qquad (9.40)
$$

Recall also the equation relating $\partial\theta_T$ to vorticity, namely Eq. 9.21, which we may also write, using Stoke's theorem, as

$$
\partial \times (\partial\theta_T) = 2\pi q
$$

or equivalently

$$
\varepsilon_{\mu\nu} \partial_\mu \partial_\nu \theta_T = 2\pi q(r) \qquad (9.41)
$$

where $q(r)$ is the local vortex density. This equation is satisfied if the following representation is used for $(\partial_\nu \theta)_T$

$$
(\partial_\nu \theta)_T = -2\pi \varepsilon_{\mu\lambda} \partial_\lambda \int d^2 r' G(r - r') q(r') \qquad (9.42)
$$

provided that the function $G(r - r')$ satisfies the Laplace equation

$$
-\partial^2 G(r) = \delta(r)
$$

From this, we obtain

$$(\partial_\nu \theta_T)^2 = (2\pi)^2 \varepsilon_{\nu\lambda} \varepsilon_{\nu\rho} \int d^2r_1 \int d^2r_2 \partial_\lambda G(r - r_1) \partial_\rho G(r - r_2) q(r_1) q(r_2)$$

$$= (2\pi)^2 \delta_{\lambda\rho} \int d^2r_1 \int d^2r_2 \partial_\lambda G(r - r_1) \partial_\rho G(r - r_2) q(r_1) q(r_2)$$

$$= -(2\pi)^2 \int d^2r_1 \int d^2r_2 G(r - r_1)[\partial^2 G(r - r_2)] q(r_1) q(r_2)$$

$$= +(2\pi)^2 \int d^2r_1 q(r_1) G(r - r_1) q(r) \tag{9.43}$$

Here, we have used a property of the two-dimensional Levi–Civita tensor, namely that $\varepsilon_{\nu\lambda} \varepsilon_{\nu\rho} = \delta_{\lambda\rho}$. If we now write the XY model in the continuum limit as follows

$$H = \frac{K}{2} \int d^2r (\partial\theta)^2$$

$$= \frac{K}{2} \int d^2r [(\partial\theta_L)^2 + (\partial\theta_T)^2] \tag{9.44}$$

and then utilize Eq. 9.43, we obtain

$$H = \frac{K}{2} \int d^2r (\partial\theta_L)^2 + 2\pi^2 K \int d^2r \int d^2r' q(r) G(r - r') q(r') \tag{9.45}$$

which again describes a spin-wave system and a Coulomb gas in two dimensions, since G is the Green's function for the electrostatic potential between point charges in two dimensions. Again, note how spin-waves and vortices decouple in Eq. 9.45, which in this derivation is a consequence of expanding the cosine of the gradient up to quadratic order only. Coupling between spin waves and vortices is seen to be an effect due to higher order gradient terms, and is therefore irrelevant in renormalization group sense.

To summarize this section, we have shown that the physics of the two-dimensional XY model, i.e. that of a two-dimensional superfluid or superconductor, corresponds to that of a spin wave system coupled to vortex configurations of the XY model. In one particular approximation the spin waves and vortices decouple. Since the spin waves render the system critical at any finite temperature, but do no suffice to induce short-range phase correlations, it follows that the phase-transition from quasi-long range order to short-range order must be encoded in the Coulomb-piece of the partition function. Precisely what sort of phase transition such a system is capable of sustaining, will be the topic of the next section.

9.6   Vortex-pair unbinding and Kosterlitz–Thouless transition

In this section, we study the phase transition in the two-dimensional Coulomb gas. We emphasize that we will be using nomenclature appropriate for a system of charges, but want to stress that the connection between the two-dimensional Coulomb gas and the two-dimensional XY model presented in the previous section (i.e. phase-only approximation to the Ginzburg–Landau theory of a two-dimensional superfluid) must always be kept in mind.

We will ignore the spin wave part of the partition function, Z_0, since it gives no singular contributions to the free energy. The two-dimensional Coulomb gas has the feature of having an internal energy which consists of contributions that are logarithmic in the separation between the charges of the system. This is a feature which will be of utmost importance in what follows, since it implies that the internal energy and the entropic contribution to the *free energy* will behave precisely in a similar manner as a function of separation of vortices in a vortex–antivortex pair. Since we have that $F = U - TS$, we tentatively conclude that due to the logarithmic interaction, the internal energy U and the entropic term TS compete on an equal footing such that at high enough temperatures T, the entropy wins.

This means that in the Coulomb gas language, a low-temperature dielectric phase of a system of charges consisting of tightly bound dipoles is converted, at some temperature, into a plasma of dissociated dipoles. In other words, the two-dimensional Coulomb gas is capable of sustaining a metal–insulator transition. This would not happen in the one-dimensional Coulomb gas, which is always in the dielectric phase because the internal energy is linear in separation of charges, while the entropy is logarithmic, and hence the former dominates for large separations. Nor would it happen in the three-dimensional Coulomb gas which is always in the metallic phase, due to the fact that the internal energy is inversely proportional to separation between charges, while the entropy is logarithmic, the latter always wins for large separations. So the two-dimensional case is quite special, and it is the logarithmic character of the Coulomb-interaction between charges which is the central point. Coulomb-potentials (in any dimension) also have the special property that they are not screened by dipoles.

A cautionary remark to begin with is absolutely necessary. One-dimensional superfluids *do not* map onto the one-dimensional Coulomb gas, and three-dimensional superfluids *do not* map onto the three-dimensional Coulomb gas! The three-dimensional case will be dealt with in Chapter 10 by similar techniques as in the two-dimensional case, but the result will not be a system of point charges interacting with a Coulombic potential in three dimensions. The topological defects of the three-dimensional superfluid turn out to be *vortex-lines*, not *point-vortices*. Therefore, although a three-dimensional Coulomb gas does not have a finite-temperature phase transition, a three-dimensional superfluid certainly does! When the two-dimensional Coulomb gas transitions from a dielectric phase to a metallic phase, it is the screening properties of the system which are altered in a non-analytic fashion as a function of temperature.

We now follow the pioneering work of Kosterlitz and Thouless [130, 131], with only slight modifications due to Young [132], and José and co-workers [133]. The approach basically amounts to setting up a self-consistent description of a charge-neutral system in terms of scale-dependent electrostatics. The interested reader may also find it useful to consult the excellent review article on the two-dimensional Coulomb gas by Minnhagen [134]. Imagine that we have a system with a low density of charges, such that an asymptotically exact low-density approximation to its dielectric constant is given by the following expression

$$\varepsilon = 1 + 2\pi n_d \alpha \tag{9.46}$$

where α denotes the polarizability of the medium. In the following, we will work in arbitrary dimensions d, and specialize to the $d = 2$ case at the end. The following closely follows the treatment of the problem by Kleinert *et al.* [135], which again builds on the approach of Young to the problem. In arbitrary dimensions, we assume that the point-charges of the system interacts with an interaction of the following form

$$V(x) = \frac{\Gamma\left(\dfrac{d - 2 - \eta_A}{2}\right)}{2^{\eta_A}(4\pi)^{d/2}\Gamma\left(\dfrac{2 + \eta_A}{2}\right)} \left[\left(\frac{|x|}{a}\right)^{2-d+\eta_A} - 1\right] \tag{9.47a}$$

where the Γ-function is defined by

$$\Gamma(x) = \int_0^\infty dt \, t^{x-1}e^{-t} \tag{9.47b}$$

This is a generalized potential which will correspond to precisely a logarithmic interaction when $2 - d + \eta_A = 0$. Here, η_A is some parameter that describes deviations from standard Coulomb interactions in d dimensions. The standard Coulomb case for arbitrary d corresponds to $\eta_A = 0$, and we thus get a logarithmic interaction for the Coulomb-case if and only if $d = 2$.

In this section, we will derive to lowest order in the fugacity the recursion relations for a scale-dependent stiffness parameter $K(l)$ and a scale-dependent fugacity $y(l)$ for a d-dimensional plasma where the bare pair-potential is given by Eq. 9.47a. This potential reduces to a logarithmic interaction when $d = 2$, and $\eta_A = 0$. As will become clear below, the stiffness parameter is basically the inverse of a scale-dependent dielectric constant. The scale-dependent fugacity $y(l)$ basically gives the density of free unbound charges in the system, provided the density is low. Thus, when $\lim_{l\to\infty} y(l) \to 0$, we have no free charges in the system, only tightly bound dipoles. On the other hand, when $\lim_{l \to \infty(l)} y(l) \to \infty$ this is an indication that vortex–antivortex pairs dissociate and free charges appear in the system. Hence, we will look for changes in the scaling of the fugacity as the hallmark of the phase transition in the system.

The starting point will be a low-density approximation for a dielectric constant of this system. Introducing the solid angle in d dimensions $\Omega_d = 2\pi^{d/2}/\Gamma(d/2)$ and the density of dipoles in the fluid by n_d, a low-density approximation for the dielectric constant is given by

$$\varepsilon = 1 + n_d \, \Omega_d \, \alpha \tag{9.48}$$

where a standard linear-response analysis gives $\alpha = 4\pi^2 K \langle s^2 \rangle / d$ for the polarizability, and $\langle s^2 \rangle$ is the mean square of the dipole moment in the system. To compute this, we need the low-density limit of the pair-distribution function $n^{\pm}(r)$ of the plasma, which is readily obtained from the grand canonical partition function Ξ expanded to second order in the bare fugacity ζ, and replacing the thermal de Broglie wavelength by a short-distance cutoff r_0, as follows

$$n^{\pm}(r) = \frac{\zeta^2}{r_0^{2d}} e^{-4\pi^2 K V} \tag{9.49}$$

In this way, we may now go on to express a *scale-dependent* dielectric constant as follows

$$\varepsilon(r) = 1 + \frac{4\pi^2 \Omega_d K}{d} \int_{r_0}^{r} ds \, s^{d+1} n^{\pm}(s) \tag{9.50}$$

Note, however, that in Eq. 9.50, a mean-field approximation is understood to be used by replacing the bare potential V in $n^{\pm}(r)$ by an *effective potential* $U(r)$. This effective screened potential must be selfconsistently determined by demanding that it gives rise to an electric field in the problem given by

$$\frac{\partial U}{\partial r} = E(r) = \frac{K \tilde{f}(d)}{\varepsilon(r) \, r^{1-\rho}} \tag{9.51}$$

where

$$\rho = 2 - d + \eta_A \tag{9.52}$$

and $\tilde{f}(d)$ is defined by

$$\tilde{f}(d) = \frac{(d - 2 - \eta_A)\Gamma\left(\dfrac{d - 2 - \eta_A}{2}\right)}{2^{\eta_A}(4\pi)^{d/2}\Gamma(1 + \eta_A/2)} \tag{9.53}$$

Such a mean-field procedure has been consistently used with success in the two-dimensional case, and the origin of the success lies in the long range of the ln-interaction. In higher dimensions, such a procedure will work even better since the logarithmic potential is felt over even longer distances due to extra volume factors.

Let us introduce a logarithmic length scale $l = \ln(r/r_0)$ along with the new variables

$$\tau(l) = \frac{\varepsilon(r_0 \exp l)}{4\pi^2 K}$$

$$x(l) = 4\pi^2 K U(r_0 \exp l) \tag{9.54}$$

Here, $x(l)$ is determined selfconsistently by integrating the effective field $E(r)$. Then we get from Eqs 9.50 and 9.51

$$\tau(l) = \tau(0) + \frac{\Omega_d \zeta^2}{dr_0^{d-2}} \int_0^l dv\, e^{(d+2)v - x(v)} \tag{9.55}$$

and

$$x(l) = x(0) + \tilde{f}(d) \int_0^l dv\, \frac{r_0^\rho e^{\rho v}}{\tau(v)} \tag{9.56}$$

From Eqs 9.55 and 9.56 we may derive coupled renormalization group equations for $\tau(l)$ and $x(l)$. However, in order to obtain equations that have a form more similar to equations that have appeared in the literature on the d-dimensional Coulomb gas, we introduce a new variable $K(l)$ representing a scale dependent stiffness constant, as follows

$$K^{-1}(l) \equiv \frac{\tau(l)}{r_0^\rho e^{\rho l}} \tag{9.57}$$

Thus, we see that the effect of a non-zero ρ on the stiffness amounts to a scaling change $K(l) \to e^{\rho l} K(l)$. Using Eq. 9.56, we have

$$\frac{\partial x(l)}{\partial l} = 4\pi^2 \tilde{f}(d) K(l) \tag{9.58}$$

Differentiating $K^{-1}(l)$ with respect to l and using Eq. 9.55, we obtain

$$\frac{\partial K^{-1}(l)}{\partial l} = -\rho K^{-1}(l) + \frac{2\Omega_d \zeta^2}{dr_0^{d-2+\rho}} e^{[(d+2-\rho)l - x(l)]} \tag{9.59}$$

From this expression, we define a scale-dependent fugacity $y(l)$ given by

$$y(l) \equiv \frac{\sqrt{2\Omega_d}\, \zeta\, e^{[(d+2-\rho)l - x(l)]/2}}{\sqrt{d}\, r_0^{(d-2+\rho)/2}} \tag{9.60}$$

Thus, we see explicitly that the renormalization of $K(l)$ in principle influences the flow equation for $y(l)$, which is obtained by differentiating with respect to l and using Eq. 9.58. We thus get the two coupled equations

$$\frac{dK^{-1}(l)}{dl} = y^2 - \rho K^{-1}(l)$$

$$\frac{dy(l)}{dl} = [d - \eta_y - 2\pi^2 \tilde{f}(d) K(l)] y(l) \qquad (9.61)$$

where $\eta_y = (d - 2 + \rho)/2$, and ρ is given by Eq. 9.52. It has a physical interpretation as an anomalous scaling dimension of the fugacity. Let us set $\eta_A = 0$, i.e. we consider the standard Coulomb-case (our starting point!), but in arbitrary dimensions. This means that $\eta_y = 0$ as well. We then get the following set of equations for the scale-dependent stiffness parameter $K(l)$, or inverse dielectric constant, and fugacity $y(l)$

$$\frac{dK^{-1}}{dl} = y^2 - (2 - d) K^{-1} \qquad (9.62)$$

$$\frac{dy}{dl} = [d - 2\pi^2 f(d) K] y \qquad (9.63)$$

where $f(d) = (d - 2)\Gamma[(d - 2)/2]/(4\pi)^{d/2}$. These equations describe the flows of the stiffness parameter and the fugacity of the d-dimensional Coulomb gas. Precisely the same results were first obtained by Kosterlitz by a method completely different from the one we have used above, a good sign [136]. Finally, if we now set $d = 2$ we arrive at the celebrated Kosterlitz–Thouless flow equations for the dielectric constant and the fugacity of the two-dimensional Coulomb gas, namely

$$\frac{dK^{-1}}{dl} = y^2 \qquad (9.64)$$

$$\frac{dy}{dl} = [2 - \pi K] y \qquad (9.65)$$

Notice how the second term on the right-hand side of Eq. 9.62 vanishes when $d = 2$. This is of crucial importance. Eqs 9.64 and 9.65 describe how the dielectric constant and the effective fugacity of the Coulomb gas change when the length scale l they are viewed on, varies. These are therefore not standard renormalization group (RG) equations in the sense of Wilson-RG. Rather, they have the status of selfconsistency equations for scale-dependent electrostatics.

Let us now consider what these equations predict. The result of the flows for $K(l)$ and $y(l)$ will depend on what the values of the bare parameter for the dielectric constant and the fugacity are. The bare values in this context refer to what the value of these quantities are at short length scales. Suppose that the bare value of K is larger than $2/\pi$. Then the right-hand side of Eq. 9.65 is negative, and $y(l)$ is reduced as l increases. Consider now Eq. 9.64 under these circumstances. Since $y(l)$ decreases with l, it means that $K(l)$ increases. But then the right-hand side of Eq. 9.65 gets even more negative, thus accelerating the reduction of $y(l)$. Thus, the system flows to a regime where $\lim_{l \to \infty} y(l) \to 0$. Consider now the case where K is smaller than $2/\pi$. Then $y(l)$ increases with increasing l, and K is reduced further, thus accelerating the increase in $y(l)$. Thus, in this case, $\lim_{l \to \infty} y(l) \to \infty$. Hence, the value $K^* = 2/\pi$ is a particular value of the stiffness which separates the regime where $\lim_{l \to \infty} y(l) \to 0$ from the regime where $\lim_{l \to \infty} y(l) \to \infty$.

What does all of this mean, physically? The case $\lim_{l \to \infty} y(l) \to 0$ simply means that in an appropriate temperature regime, there are no free charges in the two-dimensional Coulomb gas, all charges are bound into tight dipoles. Such a system of tightly bound dipolar pairs is incapable of carrying an electric current when it is subjected to an electric field, and is therefore an insulator. The case $\lim_{l \to \infty} y(l) \to \infty$ corresponds to proliferation of free charges in the system, and such a system *is* capable of carrying a current when subjected to an electric field. This is therefore a metallic phase. Hence, we reach the conclusion that the sort of phase transition the two-dimensional Coulomb gas undergoes when it passes through the point $K^* = 2/\pi$, is a metal-insulator transition.

Note how all of this fails when $d \neq 2$. For $d = 1$, the second term on the right-hand side of Eq. 9.62 is finite and negative. This means that no matter how large we initially make the fugacity (for instance by making K initially very large), $K^{-1}(l)$ will decrease, hence increasing $K(l)$ and thus reducing $y(l)$. For all bare values of K, we always end up in a situation where $\lim_{l \to \infty} y(l) \to 0$. Conversely, when $d = 3$ the second term on the right-hand side of Eq. 9.62 is finite and positive. By a similar type of argument as for $d = 1$, we always end up in the situation where $\lim_{l \to \infty} y(l) \to \infty$. The $d = 1$ result is a statement that the one-dimensional Coulomb gas is always in the dielectric phase, never in the metallic phase, whereas the three-dimensional Coulomb gas is never in the dielectric phase, but always in the metallic phase.

It is now very appropriate to go back to the two-dimensional superfluid, and ask ourselves what all of this means in that system. In other words, the metal–insulator transition of the two-dimensional Coulomb gas maps onto precisely what in the two-dimensional superfluid? We have already established in the previous section, that the charges in the two-dimensional Coulomb gas are to be identified with vortices in the phase-field of the superfluid/superconducting ordering field. Thus the metal–insulator transition corresponds to a phase transition in the superfluid from a low-temperature phase with only tightly bound vortex–antivortex pairs into a high-temperature phase where one has dissociated vortex–antivortex pairs. These vortices are the topological defects in the phase-texture of the superfluid. When they proliferate as a result of the dominance of the entropic contribution to the free energy, they convert the powerlaw decay of the phase–phase correlator, (Eq. 9.19), into an exponential decay characteristic of short-range order. The phase-stiffness of the system, which is non-zero in the critical low-temperature regime, has been lost at the phase transition as a result of the appearance of vortices. Now, the phase-stiffness of the XY model is the energy cost of introducing twists in the phase of the superconducting order parameter, i.e. the energy cost of introducing gradients $\nabla\theta$. Superconductivity is a macroscopic quantum phenomenon associated with phase-coherence. The phase stiffness, often called the *helicity modulus* Υ of the XY model, is defined as the cost in free energy of an initial twist $\Delta\theta$ in the phase of the order parameter across the system,

$$\Upsilon \sim \frac{\partial^2 F}{\partial \Delta\theta^2}\Big|_{\Delta\theta=0} \tag{9.66}$$

Going back to the partition function for the XY model, we have (upon normalizing Υ to its value at zero temperature)

$$\frac{\Upsilon}{J} = \frac{1}{N}\left\langle \sum_{\mathbf{r}} \cos(\Delta_\mu\theta)\right\rangle - \frac{J}{k_B T N}\left\langle \left[\sum_{\mathbf{r}} \cos(\Delta_\mu\theta)\right]^2\right\rangle \tag{9.67}$$

This quantity probes the global phase-coherence of the system and can be directly identified with the superfluid density ρ_s of the system. This is a consequence of the London equation

$$\mathbf{J} = -\frac{1}{\mu_0\lambda^2}\mathbf{A} \tag{9.68}$$

as follows. The twist in the phase θ may be viewed as adding a vector potential $\mathbf{A}$ to the argument of the cosine-potential in the XY model (by minimal coupling). This alters the free energy. Now, by standard quantum mechanics, the current given above may be viewed as a linear-response expression for the derivative of the free energy with respect to the added vector potential. The second derivative of the free energy with respect to an added vector potential, *evaluated* at $\mathbf{A} = 0$, is therefore the (negative of the) derivative of the current with respect to the vector potential, which is nothing but $1/\lambda^2 \sim \rho_s$. Hence, we have a major result, namely

$$\Upsilon \sim \rho_s \tag{9.69}$$

Vortices are responsible for destroying the phase stiffness, i.e. Υ, of the XY model, and they are therefore also responsible for destroying the superfluid density in the 2D superfluid/superconductor.

   To put this into perspective, it is worthwhile reminding ourselves that for the two-dimensional Ising model, we are able to consider the magnetization as a useful order parameter. The Ising model for $d = 2$ is capable of sustaining a low-temperature phase with true long range spin–spin correlations because the spins can only point in two directions. The model has a *discrete* symmetry, unlike the *XY* model, and therefore does not sustain gapless spin wave modes that can be excited at arbitrary low temperatures. However, if we preferred, we could detect the phase transition in the two-dimensional Ising model, not by monitoring the magnetization (a local order parameter), but by monitoring the proliferation of lines of overturned spins. These constitute *two-dimensional domain walls* and are the topological objects responsible for destroying the spin-stiffness (a global order parameter) in the two-dimensional Ising model. The Ising-analogue of the spin stiffness is the free-energy cost of forcing a domain wall into the system (since the minimal twist of boundary conditions is an angle π in the Ising case). The free energy of the domain line per unit length is the line tension of the domain-line. When domain-lines proliferate, they do so because their line tension vanishes, and it can be demonstrated that in fact this quantity vanishes precisely at T_c. It was, as a matter of fact, computed *exactly* for the two-dimensional Ising model in 1942 by Onsager [137], in his stellar paper on the exact solution to the two-dimensional Ising model. The line-tension of domain-lines may therefore also serve as a candidate generalized stiffness in this case.

The spirit of the domain-line approach for the two-dimensional Ising model is similar in spirit to monitoring the proliferation of vortices in the two-dimensional *XY* model. The difference is that in the Ising-case, we have the luxury of being able to use a local order parameter, if we so wish, to detect the transition, whereas in the case of the *XY* case in two dimensions, we do not. We are forced to the (conceptually quite charitable!) conclusion that in the two-dimensional *XY* model, the general scheme of identifying topological objects and asking when their proliferation annihilates a generalized stiffness is the *method* of detecting the phase transition. This is basically a by now well accepted realization of Anderson's quite general concept that a phase transition may be viewed as the proliferation of topological defects in some ordering field with an associated breakdown of a generalized rigidity [138].

9.7 Jump in superfluid density

Let us consider in some detail the solution to the flow equations Eqs 9.64 and 9.65. We are primarily interested in solving the equations for low fugacity, since the equations themselves are based on the notion that the density of charge (free vortices) is small. What happens for high densities is beyond the current approach, and will not concern us here. We rewrite Eq. 9.65 slightly as follows using $K^{-1}(l) = \tau(l)$

$$\frac{dy(l)}{dl} = 2 \left[\frac{\frac{2}{\pi}\tau(l) - 1}{\frac{2}{\pi}\tau(l)} \right] y(l)$$

$$\equiv 2T(l)y(l) \tag{9.70}$$

Multiplying on each side by $y(l)$ this may be recast into the form

$$\frac{dy^2(l)}{dl} = 4T(l)y^2(l) \tag{9.71}$$

We will now look for solutions consistent with a low fugacity, and consider therefore the regime close to the point where the scaling of $y(l)$ changes which as we have argued above happens when $\tau = \pi/2$. In this case, we have

$$T(l) \approx \frac{2}{\pi}\tau(l) - 1$$

$$\frac{dT(l)}{dl} \approx \frac{2}{\pi}y^2(l)$$

$$\frac{dT^2(l)}{dl} \approx \frac{4}{\pi}T(l)y^2(l) \tag{9.72}$$

We may now combine Eqs 9.71 and 9.72, upon redefining $y(l) \rightarrow= y/\sqrt{\pi}$, as follows

$$\frac{d[y^2(l)] - d[T^2(l)]}{dl} = 0 \tag{9.73}$$

or equivalently

$$y^2(l) - T^2(l) = \pm\omega^2 \tag{9.74}$$

where ω is taken to be some positive integration constant, and we have deliberately written it in such a way as to distinguish two cases, namely that where it is negative and that where it is positive. We consider the former case first, and for this case, it is furthermore necessary to distinguish between two cases, namely that where $T(l) < 0$ and that where $T(l) > 0$.

We consider therefore first the case where $y^2 - T^2 < 0$, $T < 0$. From Eq. 9.72 it follows that

$$\frac{dT(l)}{dl} = -\frac{2}{\pi}\left[\omega^2 - T^2(l)\right] \tag{9.75}$$

Upon introducing $\tilde{l} = (2\omega l)/\pi$ and $\tilde{T}(l) = T(l)/\omega$, Eq. 9.75 takes the simpler form

$$\frac{d\tilde{T}(l)}{d\tilde{l}} = \tilde{T}^2 - 1 \tag{9.76}$$

or equivalently

$$\frac{d\tilde{T}(l)}{\tilde{T}^2 - 1} = d\tilde{l} \tag{9.77}$$

which immediately may be integrated to yield

$$\coth^{-1}(\tilde{T}(l)) = -(\tilde{l} + \theta) \tag{9.78}$$

where $\theta > 0$ is yet another integration constant, and the choice of sign guarantees that $T(l) < 0$, since this is the case we now consider. Thus, we finally get

$$T(l) = -\omega \coth(u)$$
$$u = \frac{2}{\pi}\omega l + \theta \tag{9.79}$$

Moreover, using Eq. 9.74 and the fact that the fugacity is positive, we also obtain

$$y(l) = \frac{\omega}{\sinh(u)} \tag{9.80}$$

Now, consider the limit $l \to \infty$, i.e. the long distance physics of the problem. It is immediately clear that both $\lim_{l \to \infty} y(l) = 0$ and $\lim_{l \to \infty} T(l) = 0$. Thus, for all initial values such that $y^2 < T^2$, we end up in a dielectric insulating phase of the Coulomb gas. This corresponds to the low-temperature phase of the superfluid, where only spin waves, but no vortices, are thermally excited. The above also implies that in this regime, $\lim_{l \to \infty} \tau(l) = \pi/2$, or equivalently,

$$K^* \equiv \lim_{l \to \infty} K(l) = \frac{2}{\pi} \tag{9.81}$$

Next consider the regime where $y^2 - T^2 < 0, T > 0$. We proceed by integrating the equations precisely as above, but we need to make another choice of sign in the integration constant θ appearing in Eq. 9.78 in order to guarantee the correct sign of T, and we get

$$\coth^{-1}(\tilde{T}(l)) = (\theta - \tilde{l}) \tag{9.82}$$

where $\theta > 0$, whence

$$T(l) = \omega \coth(u)$$
$$u = \theta - \frac{2}{\pi}\omega l \tag{9.83}$$

Consequently, we have

$$y(l) = \frac{\omega}{\sinh(u)} \tag{9.84}$$

where the argument u is given by that in Eq. 9.83. It is clear from the above that the solutions for $y^2 - T^2 < 0, T > 0$ cannot be trusted as l is increased indefinitely, since $y(l)$ in fact diverges, contradicting the assumption that $y(l)$ should be small. However, we *can* say the following. As l increases, starting from an arbitrary, but small value of $y(l)$, the solutions predict that $y(l)$ increases. Thus, instead of $y(l)$ flowing to zero, as in the case $T(l) < 0$, we have an unambiguous indication that tightly bound dipoles dissociate, resulting in a proliferation of free charges, or equivalently a proliferation of free vortices. The consequence of an increasing $y(l)$ for $K(l)$ is obvious, it decreases, thus accelerating the increase of $y(l)$, further reducing $K(l)$. In the regime $y^2 - T^2 < 0$, $T > 0$ we therefore obtain

$$K^* \equiv \lim_{l\to\infty} K(l) = 0 \tag{9.85}$$

Based on Eqs 9.81 and 9.85, we are now ready to state a major result of the whole formalism we have presented in this section and the preceding one. It says that as we go through the metal–insulator transition, the renormalized stiffness parameter $\lim_{l\to\infty} K(l)$ of the system (equivalently the inverse dielectric constant of the Coulomb gas), *exhibits a discontinuity of universal magnitude*, given by $2/\pi$. In a moment, we shall translate this into the language of superfluidity.

First, however, we complete our analysis of the flow equations by considering also the regime $y^2 - T^2 > 0$. In this case, the differential equation corresponding to Eq. 9.75 reads

$$\frac{dT(l)}{dl} = \frac{2}{\pi} \left[\omega^2 + T^2(l) \right] \tag{9.86}$$

which is straightforwardly solved, much as above, to yield the solutions

$$T(l) = \omega \cot(u)$$
$$y(l) = \frac{\omega}{|\sin(u)|}$$
$$u = \theta - \frac{2\omega}{\pi} l \tag{9.87}$$

where again $\theta > 0$ is an integration constant. Since $\cot(u)$ does not have a definite sign, this solution encompasses both cases $T > 0$ and $T < 0$.

The above three cases of the solutions to the flow equations may be summarized in the flowdiagram for $y(l), T(l)$ in Figure 9.4 (which any student of statistical physics and superconductivity should know!), where the arrows indicate in which direction the quantities flow as l is increased

We close the discussion of the flow equations by remarking that to the extent that they exhibit fixed points (it turns out to be an entire line of fixed points), this occurs at zero fugacity, i.e. $y^* = \lim_{l\to\infty} y(l) = 0$. This is a very fortunate circumstance, since the basic starting point for deriving the flow equations was Eq. 9.50 which is a low-density approximation to the dielectric constant. The fact that the fixed point occurs at zero fugacity means that the assumption of low

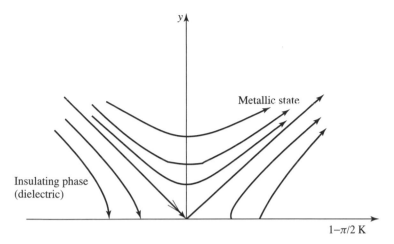

Figure 9.4 Flowdiagram showing variation in renormalized fugacity $y(l)$ and stiffness parameter $K(l)$. Note how $\lim_{l\to\infty} y(l) \to 0$ for *all* $T < 0$. The entire line $Y = 0$, $T < 0$ is therefore a *fixed line*, in contrast to a fixed point, for the flow equations. The two straight lines are given by $y = -T$, $T < 0$ and $y = T$, $T > 0$, and the thick line $y = -T$ is the separatrix separating the two scaling regimes $\lim_{l\to\infty} y(l) = 0$ and $\lim_{l\to\infty} y(l) = \infty$. The thick solid line $y = 0$, $T < 0$ is a line of fixed points.

densities is selfconsistently shown to be correct. Had we encountered a finite-fugacity fixed point, the entire calculation would have been uncontrolled and we would have, at the very least, been forced to include higher order fugacity terms. As it turns out, this need not concern us.

Let us now translate the statement about the jump of $2/\pi$ in K^* at the phase transition into the language of superfluidity. To do this, we need to establish a connection between the superfluid density of the two-dimensional superfluid, and the the scale-dependent inverse dielectric constant $(K(l))$ of the two-dimensional Coulomb-gas. The rather tedious, but straightforward details of this may for instance be found in the textbook of Huang [139], Appendix A. The result is very straightforward, namely that the superfluid density, normalized to its zero-temperature value, may be expressed in terms of the density–density correlation function for vortices as follows

$$\rho_s = \lim_{\mathbf{k}\to 0}\left[1 - \frac{1}{k_B T}\frac{\langle n_{\mathbf{k}} n_{-\mathbf{k}}\rangle}{V k^2}\right] \tag{9.88}$$

where $n_{\mathbf{k}}$ is the Fourier transform of the vortex density, or equivalently the charge density, and V is the volume of the system. The $\mathbf{k} \to 0$ limit of this is nothing but the spatial integral of the pair-distribution function. Hence Eq. 9.88 is nothing but the inverse of the dielectric constant as given in Eq. 9.46, in the low-density limit. Hence, the superfluid density and the stiffness parameter are related as follows,

$$K(T) = \frac{\hbar^2 \rho_s(T)}{m^2 k_B T} \tag{9.89}$$

where m is the mass of the bosons constituting the superfluid, and we have reinstated Planck's constant $\hbar$. (Converting from natural units back to physical units, a density acquires the dimension $(\hbar/m)^2$). Therefore, if we measure $\rho_s(T)$ in a two-dimensional superfluid as a function of temperature approaching T_c from below, we have the quite remarkable prediction that at $T = T_c$, we will observe a discontinuity in $\rho_s(T)$ as a function of temperature as we cross the transition. It is finite up to T_c and vanishes at and above T_c in such a way that

$$\lim_{T \to T_c^-} \frac{\hbar^2 \rho_s(T)}{m^2 k_B T} = \frac{2}{\pi} \tag{9.90}$$

exhibits a universal jump. This is a very powerful prediction of the theory, first made by Nelson and Kosterlitz [140], ranking on par with the prediction of BCS of a universal ratio between the zero-temperature gap and the critical temperature, as we discussed in Chapter 4. The prediction was quickly verified in experiments on superfluid thin He^4 films by Bishop and Reppy [141], and effectively removed any serious skepticism that might have existed about the existence of the Kosterlitz–Thouless transition. A qualitative sketch of the superfluid density as a function of temperature is given in Figure 9.5.

A comment is certainly in order here. Note how the global order parameter ρ_s of the superfluid vanishes discontinuously at the phase-transition. This is reminiscent of a first order phase transition. On the other hand, the correlation length of the superfluid is finite in the high-temperature phase but infinite in the low-temperature phase. This is a situation which is reminiscent of a second order transition. In fact, the Kosterlitz–Thouless transition is neither first nor second order. The above is an indication of the fact that the classification scheme of first and second order phase transitions is not entirely adequate. The Kosterlitz–Thouless transition is in fact an infinite order phase-transition.

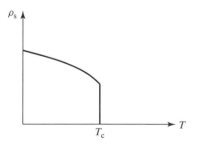

Figure 9.5 Superfluid density $\rho_s(T)$ of a two-dimensional superfluid as a function of temperature. Note the discontinuity at the critical temperature. The jump in ρ_s at T_c, normalized to the critical temperature, is universal, $\lim_{T \to T_c^-} \hbar^2 \rho_s(T)/(m k_B T) = 2/\pi$.

Let us now proceed to see what happens when we apply the duality technology to the three-dimensional case. As we will see in the next chapter, the topological defects of the superfluid/superconductor in three dimensions are vortex lines and not point-vortices. This precludes the possibility of a Kosterlitz–Thouless type of RG approach to this problem.

10

Dual Description of the Superconducting Phase Transition

10.1 Introduction

In this chapter, we will utilize the three-dimensional analogue of the duality transformation introduced in the previous chapter to gain further insight into the character of the superconductor-normal metal transition. Recall that in Chapter 4, where we established and solved the BCS gap equation, it was emphasized that this was a mean-field approach which neglected fluctuations in the critical regime. The BCS approach works extremely well in superconductors like Al and Sn, even at the quantitative level. This is due to the fact that these superconductors are deep into the type I regime and fluctuation effects are tiny. In this chapter, we turn to the case of extreme type II superconductors like the high-T_c cuprates and uncharged superfluids like He^4. The latter can be seen as a limit of the extreme type II case, by letting the Ginzburg–Landau parameter $\kappa \to \infty$. Fluctuation effects are now much more prominent, especially for the high-T_c cuprates. The reason is twofold. First, because of the large values of critical temperatures, thermal fluctuation effects are expected to be larger than for superconductors with low T_c. Another way of saying this is that a high T_c reflects the fact that Cooper-pairs are tightly bound, and hence the correlation length ξ, which is a measure of the spatial extent of a Cooper-pair, i.e. over what distances in real-space two electrons are correlated into opposite-momentum and opposite-spin states, is small compared to its value in unconventional low-temperature superconductors. Hence, the van der Waals type of argument involving large ξ that leads to a justification of the mean-field approach, is no longer expected to work as well. Secondly, extreme type II superconductivity involves $\kappa = \lambda/\xi \gg 1$. Now, $\xi \sim 1/T_c$ and $\lambda \sim 1/\sqrt{\rho_s}$. Hence, $\kappa \sim T_c/\sqrt{\rho_s}$. In high-$T_c$ cuprates, κ is large not only because T_c is

Superconductivity: Physics and Applications   Kristian Fossheim and Asle Sudbø
© 2004 John Wiley & Sons, Ltd   ISBN 0-470-84452-3

large, but also because high-T_c cuprate superconductivity arises out of very poor metals involving low charge carrier density ρ_n. This carrier density is an upper limit on the superfluid density ρ_s, which is therefore also low.

This has important physical ramifications with respect to fluctuation effects. Recall from Chapter 10 that we established a relation between the phase stiffness (helicity modulus Υ) of the superconducting condensate, and ρ_s as follows,

$$\Upsilon \sim \frac{\partial^2 F}{\partial \Delta \theta^2}|_{\Delta\theta=0} \sim \rho_s \tag{10.1}$$

where F is the free energy and $\Delta \theta$ is a phase-twist in the complex scalar superconducting order parameter across the system. This means that a low superfluid density implies the existence of large phase-fluctuations in the system. In superconductors arising out of good metals like Al, ρ_s and hence the phase stiffness, is much larger and these phase fluctuations may usually be ignored except in an immeasurably narrow temperature range around the critical temperature.

We emphasize that the approach to be detailed here is not in contradiction to the BCS theory. Rather, it complements it in the situations where the critical fluctuation effects that are left out in the BCS mean-field approach need to be taken into account. We will base our discussion on the Ginzburg–Landau theory, just as in Chapter 10. The strength of such an approach is that the Ginzburg–Landau theory simply postulates the existence of superconducting order without asking how it has established itself. Therefore, what we shall have to say in this chapter is general and entirely model independent, applying in principle to extreme type II and even moderate type II superconductors with a complex scalar order parameter (spin-singlet superconductivity).

We shall be interested in studying the critical properties of the superconductor and for this purpose the Ginzburg–Landau theory, which is a theory of a bosonic matter field coupled to a fluctuating gauge field, will suffice both for s-wave superconductors and d-wave superconductors. The two differ in one important respect, namely in that the former has no gapless fermionic excitations on the Fermi-surface, whereas the latter does. The nodal fermions in the d-wave superconductors in principle need to be taken into account in the superconducting state, but at T_c we may assume them to be fully thermalized and decoupled from the superconducting condensate. Were we, however, to study *quantum fluctuations* in the superconducting order parameter at very low temperatures, then we would need to consider the nodal fermions seriously, and a d-wave superconductor in this temperature regime has a quite different behavior than an s-wave superconductor. This is a quite involved situation which we will not deal with here, and where much work remains to be done.

We take as our starting point the Ginzburg–Landau theory of the system, introduced in Chapter 4, which when written in dimensionful form is defined by the partition function of the system written as a functional integral over the fluctuating condensate order field ψ and the fluctuating gauge field $\mathbf{A}$ as follows

$$Z = \int D\psi \, D\mathbf{A} e^{-\beta H(\psi, \mathbf{A})}$$

$$H(\psi, \mathbf{A}) = \int d\mathbf{r} \left[\alpha(T)|\psi|^2 + \frac{u}{2}|\psi|^4 + \frac{\hbar^2}{2m^*}|(\nabla - ie^*\mathbf{A})\psi|^2 + \frac{1}{2\mu_0}(\nabla \times \mathbf{A})^2 \right]$$
(10.2)

Here, m^* and e^* are effective masses and charges for the bosonic condensate constituents, $\mathbf{A}$ is a fluctuating gauge field related to a local induction $\mathbf{B}$ via the equation $\mathbf{B} = \nabla \times \mathbf{A}$, and μ_0 is the vacuum permeability. Next, we scale the quantities appearing in Eq. 10.2 as follows

$$\mathbf{A} \to \sqrt{\frac{\mu_0}{\beta}}\mathbf{A}, \, e^* \to \sqrt{\frac{\beta}{\mu_0}}e^*, \, \psi \to \sqrt{\frac{2m^*}{\beta\hbar^2}}\psi, \, \frac{m^*\alpha(T)}{2\beta\hbar^2} \to m^2, \, u \to \frac{\beta\hbar^4}{2(m^*)^2}u$$
(10.3)

Note the parameter m^2 appearing in the above. As $\alpha(T)$ becomes negative for low enough temperatures, we must think of m^2 as some mass *parameter*, not a physical mass. Then, Eq. 10.2 may be written on the form

$$Z = \int D\psi \, D\mathbf{A} e^{-S(\psi, \mathbf{A})}$$

$$S(\psi, \mathbf{A}) = \int d\mathbf{r} \left[m^2|\psi|^2 + u|\psi|^4 + |(\nabla - ie\mathbf{A})\psi|^2 + \tfrac{1}{2}(\nabla \times \mathbf{A})^2 \right] \quad (10.4)$$

where there now appears a dimensionless action S in the exponent of the Boltzmann weight. Since S is dimensionless, all terms under the integral in S carry dimension L^{-d}. Hence, we have the *naive* scaling dimensions of the quantities appearing in Eq. 10.4

$$[\psi] = L^{\frac{1-d}{2}}, \quad [\mathbf{A}] = L^{1-\frac{d}{2}}, \quad [u] = L^{d-2}, \quad [m^2] = L^{-1}, \quad [e] = L^{\frac{d}{2}-1}$$
(10.5)

If we now express all dimensionful quantities as dimensionless via the coupling constant e as follows

$$\psi \to \psi e^{-1}, \quad \mathbf{A} \to \mathbf{A} e^{-1}, \quad \mathbf{r} \to \mathbf{r} e^2, \quad m^2 \to y e^4, \quad u \to x e^2 \quad (10.6)$$

we may write the action in a form which we will use explicitly at the end of this chapter, namely

$$S(\psi, \mathbf{A}) = \int d\mathbf{r} \left[y|\psi|^2 + x|\psi|^4 + |(\nabla - i\mathbf{A})\psi|^2 + \tfrac{1}{2}(\nabla \times \mathbf{A})^2 \right] \quad (10.7)$$

Here, y appears as a temperature-like parameter, while whereas x essentially is like κ^2, such that a large x corresponds to extreme type II, small x corresponds to type I, and the mean-field demarcation line between type I and type II is given by $x = 1/2$. We shall come back to this point in Section 10.8.

Most of this chapter will be concerned with the situation for large $x \gg 1$. Consider for the moment the situation with $x \to \infty$. This means that for a fixed $|\psi|^4$-coupling u, the theory corresponds to that of an uncharged superfluid, and we may simply drop the gauge field $\mathbf{A}$ from the description. It will live its own uneventful life as a free Maxwell field decoupled from the matter field ψ. We now make an approximation that is expected to be excellent in the type II regime, but which we will later show fails in the type I regime. *In the type II regime, we will ignore amplitude fluctuations in* $\psi = |\psi|e^{i\theta}$, at least for the purposes of studying the critical properties of the system. We simply set the magnitude of ψ equal to unity. Thus, as in the two-dimensional case all we are left with are phase-fluctuations in ψ, and as always they come in two varieties, longitudinal and transverse.

10.2   Lattice formulation of the Ginzburg–Landau theory

Our starting point is the continuum Ginzburg–Landau model. In quantum field theory, the GL model is also referred as the scalar QED or the $U(1)$+Higgs model or the Abelian Higgs model. For completeness, we also take anisotropy into account, i.e. the effect that the electron transport exhibits uniaxial anisotropy such that motion in the $[x, y)$-plane is easier than motion along z-axis. The effective Hamiltonian for the Ginzburg–Landau model in an anisotropic system is given by

$$
H_{GL} = \int d^3 r \left[\alpha(T)|\psi|^2 + \frac{g}{2}|\psi|^4 + \sum_{\mu=x,y,z} \frac{\hbar^2}{2m_\mu} \left| \left(\nabla_\mu - i\frac{2\pi}{\Phi_0}A_\mu \right) \psi \right|^2 \right.
$$
$$
\left. + \frac{1}{2\mu_0}(\nabla \times \mathbf{A})^2 \right]
\tag{10.8}
$$

Here, $\psi(\mathbf{r}) = |\psi(\mathbf{r})|e^{i\theta(\mathbf{r})}$ is a complex order field representing the superconducting condensate. In superconductors, the amplitude $|\psi(\mathbf{r})|^2$ should be interpreted as the local Cooper-pair density. Furthermore, m_μ is the effective mass for *one* Cooper-pair when moving along the μ-direction, $\Phi_0 = h/2e$ is the flux quantum, and μ_0 is the vacuum permeability. In Eq. 10.8, the gauge field $\mathbf{A}$ is related to the local magnetic induction, $b(\mathbf{r}) = \nabla \times \mathbf{A}(\mathbf{r})$. Finally, the Ginzburg–Landau parameter g is assumed to be temperature independent, while $\alpha = \alpha(T)$ changes sign at a mean field critical temperature $T_{MF}(B)$, where Cooper-pairs start to form. B is the spatial average of the magnetic induction.

The critical temperature T_c where phase-coherence develops, is in principle always smaller than T_{MF}. Hence, the existence a finite Cooper-pair density does not in itself imply that the system is in a superconducting state.

Later on, we shall recast the Ginzburg–Landau theory in a quite different form that also exhibits a $U(1)$-symmetry, but where the field conjugate to the relevant phase is the number operator for the topological excitations destroying the order of the Ginzburg–Landau theory itself. Although this may seem like an unnecessary complication, it has the advantage of facilitating a detailed discussion of the vortex-liquid phase of the Ginzburg–Landau theory in terms of the ordering of some local field, namely the complex scalar field $\phi(\mathbf{r})$ to be introduced below. This is not possible using the Ginzburg–Landau function, $\psi(\mathbf{r})$, since $\langle\psi(\mathbf{r})\rangle$ is always zero in the vortex liquid phase. In the zero-field low-temperature ordered phase, the system spontaneously chooses a preferred phase angle Θ, and explicitly breaks the $U(1)$ symmetry. The vortex-sector of the Ginzburg–Landau theory also exhibits a $U(1)$-symmetry breaking, but where $U(1)$-symmetry is broken in the high-temperature phase, and restored in the low-temperature phase.

Eq. 10.8 has two intrinsic length scales, the mean-field coherence length

$$\xi_\mu^2(T) = \frac{\hbar^2}{2m_\mu|\alpha(T)|} \tag{10.9}$$

and the magnetic penetration depth

$$\lambda_\mu^2 = \frac{m_\mu\beta}{4\mu_0 e^2|\alpha(T)|} \tag{10.10}$$

ξ_μ is the characteristic length of the variation of $|\psi(\mathbf{r})|$ along the μ-direction, and λ_μ is the characteristic length of the variation of the current flowing along the μ-direction.

In order to carry out Monte Carlo simulations of the Ginzburg–Landau model, the model is discretized by replacing the covariant derivative in the continuum Ginzburg–Landau Hamiltonian (Eq. 10.8), with a covariant lattice derivative,

$$D_\mu\psi = \left(\nabla_\mu - i\frac{2\pi}{\Phi_0}A_\mu\right)\psi$$

$$\rightarrow \mathcal{D}_\mu\psi = \frac{1}{a_\mu}\left(\psi(\mathbf{r}+\hat{\mu})e^{-i\frac{2\pi}{\Phi_0}a_\mu A_\mu(\mathbf{r})} - \psi(\mathbf{r})\right) \tag{10.11}$$

The resulting model has all three spatial directions discretized. The effective Hamiltonian for the lattice Ginzburg–Landau model is given by,

$$H_{\text{LGL}} = a_x a_y a_z \sum_{\mathbf{r}} \left[\alpha |\psi|^2 + \frac{g}{2} |\psi|^4 \right.$$

$$+ \sum_{\mu=x,y,z} \frac{\hbar^2}{2m_\mu a_\mu^2} \left| \psi(\mathbf{r} + \hat{\mu}) e^{-i\frac{2\pi}{\Phi_0} a_\mu A_\mu(\mathbf{r})} - \psi(\mathbf{r}) \right|^2$$

$$+ \left. \sum_{\mu=x,y,z} \frac{1}{2\mu_0 a_\mu^2} (\Delta \times \mathbf{A})_\mu^2 \right] \qquad (10.12)$$

Here, a_μ and $\hat{\mu}$ is a lattice constant and a unit vector along the μ-axis, respectively. Furthermore, the lattice derivative is defined as

$$\Delta_\mu \psi(\mathbf{r}) = \psi(\mathbf{r} + \hat{\mu}) - \psi(\mathbf{r})$$

Taking the continuum limit $a_\mu \to 0$, the effective Hamiltonian for the lattice Ginzburg–Landau model Eq. 10.12 reduces correctly to the Ginzburg–Landau effective Hamiltonian in the continuum (Eq. 10.8). As defined in Eq. 10.12, the lattice Ginzburg–Landau model does not contain vortices. To reintroduce the vortices in the model, we must compactify the gauge-theory by requiring that the gauge invariant phase differences satisfy

$$\left[\theta(\mathbf{x} + \hat{\mu}) - \theta(\mathbf{x}) - \frac{2\pi}{\Phi_0} a_\mu A_\mu(\mathbf{x}) \right] \in [-\pi, \pi) \qquad (10.13)$$

Whenever this constraint is used to bring the gauge invariant phase differences back to its primary interval, we automatically introduce a unit closed vortex loop, and the net vorticity of the system is guaranteed to be conserved at every stage of the Monte-Carlo simulation. From the renormalization group point of view the continuum Ginzburg–Landau model and the lattice Ginzburg–Landau model belong to the same universality class. We therefore expect the lattice Ginzburg–Landau model and the continuum Ginzburg–Landau model to give, qualitatively, the same results.

10.2.1 Lattice Ginzburg–Landau model in a frozen gauge approximation

In extreme type II superconductors, the zero temperature mean-field penetration depth is much greater than the zero temperature coherence length, $\lambda_\mu(T = 0) \gg \xi_\mu(T = 0)$. Thus, fluctuations of the gauge field represented by the last term in Eq. 10.8, around the extremal field configuration are strongly suppressed and can therefore be neglected. The effective Hamiltonian for the frozen gauge (FG) model can be written as

$$
H_{FG} = \frac{|\alpha(0)|^2}{g} a_x a_y a_z \sum_{\mathbf{r}} \left[\frac{\alpha(T)}{\alpha(0)} |\psi'|^2 + \frac{1}{2} |\psi'|^4 \right.
$$

$$
\left. + \sum_{\mu=x,y,z} \frac{\xi_\mu^2}{a_\mu^2} |\psi'(\mathbf{r}+\hat{\mu})||\psi'(\mathbf{r})| \left[2 - 2\cos(\Delta_\mu\theta \quad \mathcal{A}_\mu) \right] \right] \quad (10.14)
$$

Here, we have defined a dimensionless order field and vector potential

$$
\psi' = \frac{\psi}{\sqrt{\frac{|\alpha(0)|}{g}}} \rightarrow |\psi'| \sim [0, 1]
$$

$$
\mathcal{A}_\mu = \frac{2\pi}{\Phi_0} a_\mu A_\mu
$$

The natural energy scale along the μ-direction is,

$$
J_\mu = 2 \frac{|\alpha(0)|^2}{g} a_x a_y a_z \frac{\xi_\mu^2}{a_\mu^2}
$$

Assuming a uniaxial anisotropy along the z-axis, the natural energy scale for the FG model is

$$
J_0 = J_x = \frac{2|\alpha(0)|^2}{g} \xi_{ab}^2 a_z = \frac{\Phi_0^2 d}{4\pi^2 \mu_0 \lambda_{ab}^2} \quad (10.15)
$$

Here, we have put our coordinates (x, y, z)-axis parallel to the crystals (a, b, c)-axis. Furthermore, $\xi_x = \xi_y = \xi_{ab}$ and $\xi_z = \xi_c$ is the coherence length in the CuO-planes and along the crystal's c-axis, respectively. Furthermore, $\lambda_x = \lambda_y = \lambda_{ab}$ and $\lambda_z = \lambda_c$ is the penetration depth in the CuO-planes and along the crystals' c-axis, respectively. In Eq. 10.15, d is the distance between two CuO superconducting planes in adjacent unit cells. The energy scale J_0 is roughly the energy scale of exciting a unit vortex loop.

The ratio between the energy scales J_x and J_z serves as an anisotropy parameter,

$$
\Gamma = \sqrt{\frac{J_x}{J_z}} = \frac{\xi_{ab} a_z}{\xi_c a_x} = \frac{\lambda_c a_z}{\lambda_{ab} a_x} \quad (10.16)
$$

In this model, the lattice constant a_μ should be defined as

$$
a_\mu = \max(d_\mu, C_0 \xi_\mu)
$$

Here, d_μ is an intrinsic length along the μ-direction in the underlying super-conductor to be modelled. Examples of such an intrinsic length are the distance between CuO-planes in adjacent unit cells, the (a, b)-dimension of the unit cell. To be consistent, the constant C_0 should be larger than ~ 4. This requirement $a_\mu/\xi_\mu > 4$ ensures that the amplitude of the order field does not overlap [142]. Such overlap will give rise to a domain wall term $(\nabla|\psi|)$, which is absent in the lattice Ginzburg–Landau model.

Within the frozen gauge approximation, the gauge field serves only as a constraint, fixing the value of the uniform induction. In terms of magnetic induction this approximation is valid when $B \gg B_{c1}(T)$, where the field distribution from individual flux lines overlap strongly, giving uniform induction. Note that $B_{c1}(T)$ also vanishes when the temperature approaches T_c. In zero field, this approximation is valid for all temperatures except an inaccessibly narrow temperature region around T_c.

10.3   Preliminary results

Before we proceed, let us make a few remarks on the neglect of amplitude fluctuations in the type II regime. Is the neglect of such fluctuations in ψ in the above theory defined by Eq. 10.7 really a good approximation? It is well established to be correct in two dimensions. However, as is well known, in four dimensions mean field theory is exact, and in this case amplitude fluctuations alone suffice to correctly describe the phase transition. Three dimensions is between two and four, but is it, in some sense, closer to two dimensions than to four dimensions or vice versa? To answer this, consider the results of large-scale Monte Carlo simulations performed directly on the lattice version of the theory defined by Eq. 10.14 in the absence of a gauge field [29]. All phase fluctuations as well as amplitude fluctuations are taken into account, and we now ask ourselves what these various fluctuations do close to the critical point where the superfluid density vanishes. This is shown in Figure 10.1.

The Cooper-pair density $\langle|\psi|^2\rangle$ behaves quite differently from the superfluid density Υ and the superconducting order parameter $|\langle\psi\rangle|^2$, in that the former still is quite sizeable at and slightly above T_c, whereas the superfluid density vanishes. We thus conclude that if amplitude fluctuations alone had been capable of describing the phase transition from superfluid to normal fluid or from superconductor to metal, then the Cooper-pair density would have vanished at T_c. It does not, and the superfluid density is brought to zero at T_c by phase-fluctuations [29, 31, 32, 143]. We shall now investigate the effect of longitudinal phase-fluctuations first, following the same path as in Chapter 9. As we shall see, longitudinal phase-fluctuations are even more innocuous in three dimensions than in two, whence it follows that the destruction of superfluid density is caused by transverse phase fluctuations.

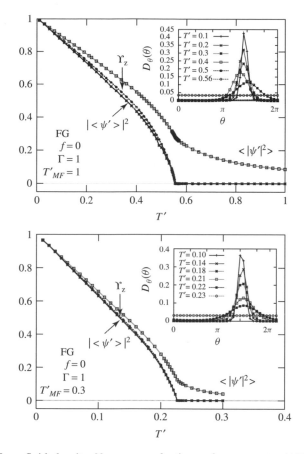

Figure 10.1 Superfluid density Υ, superconducting order parameter $|\langle\psi\rangle|^2$ (often called condensate fraction in the He^4 literature, and Cooper-pair density $\langle|\psi|^2\rangle$). Note how Υ and $|\langle\psi\rangle|^2$, both of which contain phase-fluctuations, are brought to zero at T_c, while $|\langle\psi\rangle|^2$ which does not contain phase-fluctuations, is still finite at T_c. This demonstrates that amplitude fluctuations are not critical at T_c. (adapted from Fig. 1 in Ref. [29]).

As in the two-dimensional case, we may write the effective action when only smooth phase-fluctuations are taken into account, in the form

$$S = \frac{\rho_0}{2} \int d^3r \ (\nabla\theta)^2 \tag{10.17}$$

where ρ_0 formally appears as same bare phase stiffness in the problem (analogous to J in the XY-model) which in the above units is given by $\rho_0 = 1$. However, we will retain it in the following in order to remind ourselves of the presence of this bare stiffness in the problem. In the above, it is assumed that $\nabla\theta$ is non-singular, which means that we ignore the angular (compact) nature of the phase-field θ.

The following discussion closely parallels that of the low-temperature discussion of the two-dimensional XY-model in Chapter 9. Consider the effect of the longitudinal phase fluctuations in computing the correlation function probing superconducting off-diagonal long-range order (ODLRO),

$$\Gamma_2(\mathbf{r}) = \langle \psi^*(\mathbf{r})\psi(0) \rangle = \langle e^{i[\theta(\mathbf{r})-\theta(0)]} \rangle$$
$$= e^{\langle [\theta(\mathbf{r})-\theta(0)]\theta(0) \rangle} = e^{G(\mathbf{r})} \tag{10.18}$$

when only spin-wave fluctuations are included in the Hamiltonian, and where

$$G(\mathbf{r}) = \langle [\theta(\mathbf{r}) - \theta(0)]\theta(0) \rangle \tag{10.19}$$

just as in Chapter 9. By using the equipartition theorem precisely in the same way as in Chapter 9, we obtain in the present units

$$G(\mathbf{r}) = \frac{k_B T}{\rho_0} \int \frac{d^3q}{(2\pi)^3} \frac{e^{i\mathbf{q}\mathbf{r}} - 1}{q^2} \tag{10.20}$$

Of interest for the purposes of establishing ODLRO, is the asymptotic long-distance behaviour of $G(\mathbf{r})$. Compute the above integral by introducing a large momentum cutoff Λ in the problem. It is straightforward to see that $G(\mathbf{r})$ has the following form,

$$G(\mathbf{r}) = -K + \frac{r_0}{r}$$
$$r_0 = \frac{k_B T \, Si(\Lambda)}{2\pi^2 \rho_0}$$
$$K = \frac{k_B T \Lambda}{2\pi^2 \rho_0} \tag{10.21}$$

where r_0 is some characteristic length, and $Si(\Lambda) = \int_0^\Lambda du \sin(u)/u$ is the sine-integral. Note how $G(\mathbf{r})$ again vanishes as $T \to 0$. Now we insert this into $\Gamma_2(\mathbf{r})$ to obtain the long-distance behaviour

$$\Gamma_2(\mathbf{r}) = e^{-K + \frac{r_0}{r}} \tag{10.22}$$

where the expression for $\Gamma_2(\mathbf{r})$ basically applies for $r \geq \Lambda^{-1}$. It is seen that $\Gamma_2(\mathbf{r})$ decreases with r, but it is finite when $r \to \infty$. Moreover, since the entire exponent in $\Gamma_2(\mathbf{r})$ is negative, this implies that the correlation function is suppressed with increasing T at fixed r. Eq. 10.22 is drastically different from the two-dimensional case in that the spinwaves are not capable of destroying ODLRO at all in three dimensions, whereas they are capable of reducing ODLRO to quasi-long range (but not short-range) order in two dimensions.

Hence, as in the two-dimensional case, the fluctuations ultimately responsible for producing short-range order and a transition to a normal fluid/metal, are transverse phase fluctuations. Let us now incorporate them in the description, in much the same way as was done in Chapter 9. There will, however, be crucial differences, since transverse phase fluctuations in three dimensions give rise, not to point-like vortices as in two dimensions, but rather closed vortex-loops, considered recently in great detail in the context of high-temperature superconductors by Tesanovic and Sudbø and co-workers [29, 32].

10.4 Vortex-loops as topological defects of the order parameter

We will proceed as in the two-dimensional case by regularizing the Ginzburg–Landau theory (Eq. 10.4) on a lattice, ignoring amplitude fluctuations, as follows

$$Z = \int d\theta \, d\mathbf{A} e^{-S(\theta, \mathbf{A})}$$

$$S(\theta, \mathbf{A}) = \sum_{\mathbf{x}} \left[\frac{\beta J}{2} |(\Delta - ie\mathbf{A})e^{i\theta}|^2 + \frac{1}{2}(\Delta \times \mathbf{A})^2 \right]$$

$$J = \frac{a|\psi_0|\hbar^2}{m^*} \tag{10.23}$$

where Δ is a lattice derivative, a is a lattice constant of the numerical lattice, and ψ_0 is the amplitude of the ψ-field, which we take to be uniform throughout the system. Finally, we expand the kinetic energy operator as follows

$$|(\Delta_\mu - ie\mathbf{A}_\mu)e^{i\theta}|^2 \rightarrow |e^{i\theta(\mathbf{x}+a\hat{\mu})-iaA_\mu(\mathbf{x})}|^2 = 2\left[1 - \cos(\Delta_\mu \theta(\mathbf{x}) - eA_\mu(\mathbf{x}))\right] \tag{10.24}$$

We ignore the constant term, and set $J = 1$, thus getting

$$Z = \int d\theta \, d\mathbf{A} e^{-S}$$

$$S = \sum_{\mathbf{x}} \left[-\beta \cos(\Delta_\mu \theta(\mathbf{x}) - eA_\mu(\mathbf{x})) + \tfrac{1}{2}(\Delta \times \mathbf{A})^2 \right] \tag{10.25}$$

We now follow steps similar to those in Chapter 9, and replace this cosine-potential by a periodic Gaussian one, thus obtaining the partition function

$$Z = \sum_{\mathbf{n}} \int d\theta \, d\mathbf{A} e^{-S(\psi, \mathbf{A})}$$

$$S = \sum_{\mathbf{x}} \left[\frac{\beta}{2} (\Delta\theta - e\mathbf{A} + 2\pi\mathbf{n})^2 + \frac{1}{2}(\Delta \times \mathbf{A})^2 \right] \qquad (10.26)$$

which is now the Villain approximation to the three dimensional lattice London model (i.e. three dimensional XY model coupled minimally to a fluctuating gauge field). Here, we have introduced an integer valued velocity field $\mathbf{n}$, which we may view as the integer part of the gauge-invariant phase-difference $\Delta\theta - ie\mathbf{A}$. The field $\mathbf{n}$ will figure prominently in our discussion below.

The kinetic term is linearized, as in Chapter 9, by introducing the Hubbard–Stratonovic decoupling

$$e^{-\frac{\beta}{2}\mathbf{u}^2} = \int d\mathbf{v} \, e^{-\frac{\mathbf{v}^2}{2\beta} + i\mathbf{v}\mathbf{u}} \qquad (10.27)$$

which allows us to write the partition function of the form

$$Z = \sum_{\mathbf{n}} \int d\theta \, d\mathbf{A} \, d\mathbf{v} \, e^{-S}$$

$$S = \sum_{\mathbf{x}} \left[\frac{1}{2\beta}\mathbf{v}^2 + i\mathbf{v}(\Delta\theta - e\mathbf{A} + 2\pi\mathbf{n}) + \frac{1}{2}(\Delta \times \mathbf{A})^2 \right] \qquad (10.28)$$

Now, we proceed by performing the θ-integrations. This yields the constraint

$$\Delta_\mu v_{\mu i} = 0 \qquad (10.29)$$

which is solved by introducing the new vector field $h_{\mu i}$ satisfying

$$v_{\mu i} = \varepsilon_{\mu\nu\lambda} \Delta_\nu h_{\lambda i} \qquad (10.30)$$

where $\varepsilon_{\mu\nu\lambda}$ is the three-dimensional Levi–Civita tensor. Now the partition function is given by

$$Z = \sum_{\mathbf{n}} d\mathbf{A} \, d\mathbf{h} \, e^{-S}$$

$$S = \sum_{\mathbf{x}} \left[\frac{1}{2\beta}(\Delta \times \mathbf{h})^2 - i\Delta \times \mathbf{h}(e\mathbf{A} - 2\pi\mathbf{n}) + \frac{1}{2}(\Delta \times \mathbf{A})^2 \right] \qquad (10.31)$$

By a partial integration of the second and third term in this action, it may be written as

$$S = \sum_{\mathbf{x}} \left[\frac{1}{2\beta}(\Delta \times \mathbf{h})^2 + ie\mathbf{h}(\Delta \times \mathbf{A}) - i2\pi\mathbf{h}(\Delta \times \mathbf{n}) + \frac{1}{2}(\Delta \times \mathbf{A})^2 \right]$$

$$(10.32)$$

Next, we perform the sum over **n** in Eq. 10.31. By the Poisson summation formula, this converts the **v**-field into an integer valued field, equivalently **h** becomes integer valued. We will come back in a moment to exactly what the physical interpretation of this new vector field is. As a result of the above, we may write the partition function as

$$Z = \sum_{\{\mathbf{h}\}} \int d\mathbf{A}\, e^{-S}$$

$$S = \sum_{\mathbf{x}} \left[\frac{1}{2\beta}(\Delta \times \mathbf{h})^2 - ie\Delta \times \mathbf{h}\mathbf{A} + \frac{1}{2}(\Delta \times \mathbf{A})^2 \right] \qquad (10.33)$$

Now, it is a somewhat awkward fact that while the gauge field **A** is real valued, the vector field **h** is integer valued. This may be remedied by again using the Poisson summation formula to lift the field **h** from $\mathbb{Z}$ to $\mathbb{R}$ by introducing another integer valued field **m** in the partition function, thus writing it on the form

$$Z = \sum_{\{\mathbf{m}\}} \int d\mathbf{h}\, d\mathbf{A} e^{-S}$$

$$S = \sum_{\mathbf{x}} \left[\frac{1}{2\beta}(\Delta \times \mathbf{h})^2 - ie\Delta \times \mathbf{h}\mathbf{A} + \frac{1}{2}(\Delta \times \mathbf{A})^2 - i2\pi\mathbf{m}\mathbf{h} \right] \quad (10.34)$$

By comparing the action in Eq. 10.34 with the action Eq. 10.32, we arrive at the following physical interpretation of the new vector field **m**. We see that

$$\mathbf{m} = \Delta \times \mathbf{n} \qquad (10.35)$$

Since **n** is the integer part of the gauge-invariant phase difference we started with, it is a current. Since **m** is the curl of this, it is the vorticity of this phase difference. In other words, the fields **m** are segments of vortex excitations in the superconducting ordering field. Finally, from Eq. 10.35, we see that

$$\Delta_\mu m_{\mu i} = 0 \qquad (10.36)$$

This means that the vortex segments are divergence free fields, and form closed loops. Thus, we have identified the topological defects in the Ginzburg–Landau model, they are precisely these vortex loops. (Note how the use of the phrase *vortex ring* is meticulously avoided!) The analogue of the Kosterlitz–Thouless vortex-pair unbinding taking place in the two-dimensional case will now be a *vortex-loop* unbinding characterized by a low-temperature phase of vortex loops with finite linetension and finite perimeter into a high-temperature phase characterized by vortex tangle with no line tension of the vortex loops, which therefore exist on all length scales up to and including the system size.

Note how there is one feature which distinguishes the three-dimensional situation from the two-dimensional. Vortex–antivortex pairs have one characteristic feature, and that is the separation between the vortex and the antivortex. A vortex loop is an infinitely more complicated object, in that it also has *shape*. This is the basic feature which precludes the possibility of constructing, in a consistent manner, a Kosterlitz–Thouless style of renormalization group approach to the problem in three dimensions.

Next, let us integrate out the electromagnetic gauge field **A**. We rewrite the action introducing Fourier-modes of the field **A** as follows

$$S = \sum_q \left[\frac{1}{2\beta} |\mathbf{Q_q}|^2 + i\pi (\mathbf{m_q h_{-q}} + \mathbf{m_{-q} h_q}) - \frac{ie}{2} \left[A_q^\mu (\mathbf{Q_q} \times \mathbf{h_{-q}})_\mu \right. \right.$$
$$\left. \left. + A_{-q}^\mu (\mathbf{Q_{-q}} \times \mathbf{h_q})_\mu \right] + \frac{1}{2} |\mathbf{Q_{-q}} A_{-q}^\mu|^2 \right] \tag{10.37}$$

Completion of squares is accomplished by introducing the auxiliary quantities

$$A_q^+ = \frac{1}{2} (\mathbf{A_q} + \mathbf{A_{-q}})$$

$$A_q^+ = \frac{1}{2i} (\mathbf{A_q} - \mathbf{A_{-q}}) \tag{10.38}$$

and

$$\Gamma_0^+ = \frac{ie}{2} (\mathbf{Q_q} \times \mathbf{h_{-q}} + \mathbf{Q_q} \times \mathbf{h_{-q}})$$

$$\Gamma_0^- = \frac{e}{2} (\mathbf{Q_{-q}} \times \mathbf{h_q} - \mathbf{Q_{-q}} \times \mathbf{h_q}) \tag{10.39}$$

As a result of this, the **A**-part of the action may be written in the form

$$S = \frac{1}{2} \sum_q \left[|\mathbf{Q_q Q_{-q}}| [(\tilde{A}^+)^2 + (\tilde{A}^-)^2] - \underbrace{\frac{1}{|\mathbf{Q_q Q_{-q}}|} [(\Gamma_0^+)^2 + (\Gamma_0^-)^2]}_{e^2 \mathbf{h_q h_{-q}}} \right]$$

$$\tag{10.40}$$

The integrals over the shifted gauge-fields $\tilde{A}^\pm$ in the partition function are Gaussian and may be performed. Hence, the net result of integrating out these fields is to produce a term in the remaining part of the action which is simply a mass-term for the **h**-field. Note how the presence of the mass term depends on the existence of a charge e in the problem. In order to interpret this mass, we proceed by integrating out the **h**-field. The remaining part of the action prior to integrating out the **h**-field now reads

$$S = \sum_{\mathbf{q}} \left[\pi i (\mathbf{m_q h_{-q}} + \mathbf{m_{-q} h_q}) + \frac{1}{2\beta} Q_q Q_{-q} \mathbf{h_q h_{-q}} + \frac{e^2}{2} \mathbf{h_q h_{-q}} \right] \quad (10.41)$$

We introduce

$$\mathbf{h_q^+} = \frac{1}{2} (\mathbf{h_q} + \mathbf{h_{-q}})$$

$$\mathbf{h_q^+} = \frac{1}{2i} (\mathbf{h_q} - \mathbf{h_{-q}}) \quad (10.42)$$

along with

$$\Omega_0^+ = \frac{i\pi}{2} (\mathbf{m_{-q}} + \mathbf{m_{-q}})$$

$$\Omega_0^- = \frac{e}{2} (\mathbf{m_q} - \mathbf{m_{-q}})$$

$$\xi^2 = \frac{1}{2} \left(e^2 + \frac{Q_q Q_{-q}}{\beta} \right) \quad (10.43)$$

Introducing these variables into the action and completing squares, we get

$$S = \frac{1}{2} \sum_{\mathbf{q}} \left[\xi^2 \left(\mathbf{h_q^+} + \frac{\Omega_0^+}{\xi^2} \right)^2 + \left(\mathbf{h_q^-} - \frac{\Omega_0^-}{\xi^2} \right)^2 - \frac{1}{\xi^2} ((\Omega_0^+)^2 + (\Omega_0^-)^2) \right] \quad (10.44)$$

The shifted **h** field may now be integrated over, to give the final form of the action in terms of an interacting gas of closed vortex loops as follows

$$S = \beta \sum_{\mathbf{q}} \frac{\mathbf{m_q m_{-q}}}{e^2 \beta + Q_q Q_{-q}} \quad (10.45)$$

This may be written in term of real-space vortex segments interacting with a potential in terms of the following Hamiltonian, to be used in the partition function

$$H = -2\pi^2 J_0 \sum_{\mathbf{x_1, x_2}} \mathbf{m(x_1)} V(\mathbf{x_1} - \mathbf{x_2}) \mathbf{m(x_1)}$$

$$V(\mathbf{x}) = \sum_{\mathbf{q}} \frac{e^{-i\mathbf{qr}}}{4 \sum_{\mu} \sin^2(q_\mu/2) + \lambda^{-2}} \quad (10.46)$$

where $\lambda^{-2} = e^2 \beta$.

We now see that the mass e^2 of the gauge-field $\mathbf{h}$ shows up as a screening length in the interaction between vortex segments. The charge in the original problem, i.e. the charge of the superconducting condensate, leads to short-range interactions between the vortex-loop segments.

Setting the charge equal to zero in the above, we obtain an equivalent description of a neutral superfluid, such as He^4. The gauge field decouples from the order parameter in Eq. 10.25, and can simply be ignored as it gives rise to an additive analytic contribution to the free energy. The remaining model is the three-dimensional XY model. It may be written in terms of vortex-loop segments following the steps we have just gone through, and the result is Eq. 10.46 with screening length set to infinity. This means that the potential $V(\mathbf{x})$ is the unscreened Coulomb potential in three dimension. Note that such a potential satisfies the Laplace equation $-\partial^2 V(\mathbf{x}x) = \delta(\mathbf{x})$ in the continuum.

Let us rederive the above result for the neutral case in an alternative manner following what we did in Chapter 9. In the continuum limit the three-dimensional XY model may be written as

$$
\begin{aligned}
H &= \frac{J_0}{2} \int d^3 r (\partial \theta)^2 \\
&= \frac{J_0}{2} \int d^3 r \left[(\partial \theta_L)^2 + (\partial \theta_T)^2 \right]
\end{aligned}
\tag{10.47}
$$

Again, we have split the gradient of the phase into a regular spin-wave part $\partial \theta_L$ and a part $\partial \theta_T$ giving rise to vortex segments. The latter part satisfies the equation

$$
\varepsilon_{\mu\nu\lambda} \partial_\nu (\partial_\lambda \theta_T) = 2\pi n_\mu
$$

where $n_\mu(r)$ is a local vorticity in the μ-direction, and $\varepsilon_{\mu\nu\lambda}$ is the three-dimensional Levi–Civita tensor. A representation of $(\partial_\mu \theta_T)$ that satisfies this equation is the following

$$
(\partial_\mu \theta_T) = -2\pi \varepsilon_{\mu\nu\lambda} \partial_\nu \int d^3 r' G(r - r') n_\lambda(r')
\tag{10.48}
$$

where $G(r)$ is a solution to the three-dimensional Laplace equation, $-\partial^2 G(r) = \delta(r)$. The verification of this follows the lines given for the two-dimensional case in Chapter 9. If we now insert this representation into Eq. 10.47, using the Laplace equation for $G(r)$ and the fact that $\varepsilon_{\nu\lambda\rho} \varepsilon_{\nu\eta\sigma} = \delta_{\lambda\eta} \delta_{\rho\sigma} - \delta_{\lambda\sigma} \delta_{\rho\eta}$, we find

$$
H = \frac{J_0}{2} \int d^3 r (\partial \theta_L)^2 + 2\pi^2 J_0 \int d^3 r \int d^3 r' n_\mu(r) G(r - r') n_\mu(r')
\tag{10.49}
$$

Here, we have used a partial integration in the second term along with the fact that $\partial_\mu n_\mu(r) = 0$. The above describes a system of spin waves decoupled from a system of vortex segments forming closed loops and interacting with a three-dimensional unscreened Coulomb potential (anti Biot–Savart law). Eq. 10.49 is therefore the continuum version of Eq. 10.46 with no screening, $\lambda = \infty$, when the spin-wave part is ignored. Coupling between spin waves and vortices are seen to be higher-order gradient terms and are therefore, as in the two-dimensional case, irrelevant in renormalization group sense.

10.5 Superconductor–superfluid duality in $d = 3$

In this section, we will formulate a Ginzburg–Landau theory for the above action for the system of interacting vortex segments forming closed loops. This Ginzburg–Landau theory is therefore a field theory of the ensemble of topological defects of the *original* Ginzburg–Landau theory of the superconductor, and hence a *dual* theory to the original system. Incidentally, we note in passing that we can also describe superfluidity in He^4 simply by setting the charge $e = 0$. We shall now demonstrate that the dual theory to the Ginzburg–Landau theory of a superconductor is the Ginzburg–Landau theory of a charge-neutral superfluid, and vice versa. The details are as follows.

The key point is to compare two equations derived above from the original Ginzburg–Landau theory. The two equations we compare are Eqs 10.28 and 10.32, written out in real space as follows

$$
Z(\beta) = \int \mathcal{D}\mathbf{A} \sum_{\mathbf{v}}' \exp\left[-\sum_{\mathbf{x}} \left(\frac{\mathbf{v}^2}{2\beta} - i\mathbf{v}\mathbf{A} + \frac{1}{2}(\nabla \times \mathbf{A})^2 \right) \right]
$$

$$
Z(\beta) = \int \mathcal{D}\mathbf{h} \sum_{\mathbf{m}}' \exp\left[-\sum_{\mathbf{x}} \left(\frac{\Gamma \mathbf{m}^2}{2} + 2\pi i\mathbf{m}\mathbf{h} + \frac{e^2}{2}\mathbf{h}^2 + \frac{1}{2\beta}(\nabla \times \mathbf{h})^2 \right) \right]
$$

(10.50)

The first of these is Eq. 10.28 after having performed the θ-integrations and **n**-sum. The second equation is Eq. 10.32 written in real space, with a term $\Gamma \mathbf{h}^2/2$ added by hand. The crucial observation now is that Eq. 10.28 is basically a version of the original Ginzburg–Landau model! Eq. 10.32 on the other hand, is a version of the vortex-loop gas formulation of the model. The point to note is that the action in these two partition functions are basically identical, provided we make the identification

$$
\frac{1}{2\beta}\mathbf{v}^2 \to \frac{\Gamma}{2}\mathbf{h}^2
$$

$$
\frac{1}{2}(\nabla \times \mathbf{A})^2 \to \frac{1}{2\beta}(\nabla \times \mathbf{h})^2
$$

$$
-\mathbf{b}\mathbf{A} \to 2\pi\mathbf{m}\mathbf{h}
$$

(10.51)

Adding the extra term $\Gamma \mathbf{h}^2/2$ in the second of Eq. 10.50 is like adding a *short-range* interaction between vortex segments. This is an irrelevant perturbation, in renormalization group sense, to the already existing short-range piece of the interaction between vortex segments, and is not expected to alter the long-distance physics. Now, since we know that the first of the two equations Eq. 10.50 corresponds to a Ginzburg–Landau theory, it follows that the second one also does so. In the continuum limit, the field theory description of the second of Eq. 10.50 is therefore of a form [29, 32, 144, 145]

$$S(\mathbf{h}, \phi) = \int d^3x \left[\frac{1}{2\beta} (\nabla \times \mathbf{h})^2 + \frac{e^2}{2} \mathbf{h}^2 + m_\phi^2 |\phi|^2 + u_\phi |\phi|^4 + |(\nabla - i e_d)\phi|^2 \right]$$

$$(10.52)$$

The field ϕ is a complex matter field describing the vortex segments of the theory, in the same way that the field Ψ in the original Ginzburg–Landau theory was a a matter field describing Cooper-pairs giving rise to segments of supercurrents. Moreover, $e_d = 2\pi$ in the units we are working with here. It has the interpretation of being a *dual charge* coupling the vortex segments, or equivalently the matter field ϕ, to the gauge-field $\mathbf{h}$ which mediates the (anti) Biot–Savart interaction between vortex segments. Finally, the term $m_\pi^2 |\phi|^2$ is a local chemical potential term for the vortex segments, while the term $u_\phi |\phi|^4$ represents a steric repulsion between vortex segments. Eq. 10.52 is of the same form as Eq. 10.4 except for the mass term $e^2 \mathbf{h}^2/2$ of the gauge-field $\mathbf{h}$.

Note that if we had started out with a neutral theory in Eq. 10.4, $e = 0$, we would have ended up with a dual theory Eq. 10.52 of the same form as Eq. 10.4 with *finite charge e*. Hence, the dual theory of a superfluid in $d = 3$ is immediately seen to be the Ginzburg–Landau theory of a superconductor.

To see the converse, namely that the dual of a superconductor is a neutral superfluid, is only slightly more involved. A physically intuitive argument proceeds as follows (more involved arguments involving renormalization group calculations merely confirm the correctness of the physical picture we shall give). In the presence of a finite charge in Eq. 10.4, the gauge field $\mathbf{h}$ in Eq. 10.52 is massive. This means (in complete analogy with, for instance, the massive meson-field mediating short-range nuclear interactions) that when $e^2 \neq 0$, the gauge field mediates a short-range interaction between vortex segments, with range given by e^{-1}. Now, by the dimensional analysis leading to Eq. 10.5, charge has a positive scaling dimension when $d = 3$, whence

$$\frac{\partial e^2}{\partial l} > 0 \qquad (10.53)$$

Thus, for the purposes of describing the long-distance physics, we may set the renormalized effective charge $\lim_{l \to \infty} e^2(l) = \infty$ once the bare charge is non-zero. This means that on long length scales the gauge-field is suppressed completely in the action and may be discarded. The physics of this is that the short-range Biot–Savart interaction is an irrelevant perturbation to the steric repulsion represented by the term $u_\phi |\phi|^4$. Hence, the long-wavelength physics is described correctly by Eq. 10.52, but with **h** removed from the action, namely

$$S(\phi) = \int d^3 x \left[m_\phi^2 |\phi|^2 + u_\phi |\phi|^4 + |\nabla \phi|^2 \right] \tag{10.54}$$

This is nothing but the Ginzburg–Landau theory of a neutral superfluid. Thus, the dual of a three-dimensional superconductor is a three dimensional superfluid. Another point to note here, as a consistency check, is the following. Suppose we start out with a superconductor and dualize it once. The resulting dual action is that of a neutral superfluid. Suppose now that we dualize once more. This means that we are out to obtain the dual of the dual of a superconductor, which is the dual of a neutral superfluid. This, as we have seen, is just a superconductor. Hence, dualizing twice brings us, quite correctly, back to the starting point, $\hat{D}^2 = 1$, where $\hat{D}$ is the duality operator.

Based on the above, we can immediately infer that a vortex-loop tangle consisting of non-interacting vortex segments (a very peculiar case, admittedly, but as we shall see not entirely academic), is simply described by the Gaussian field theory

$$S(\phi) = \int d^3 x \left[m_\phi^2 |\phi|^2 + |\nabla \phi|^2 \right] \tag{10.55}$$

The essential content of the above is shown in Figure 10.2. Observe the isomorphism along both the diagonals. Of course what is *dual*, and what is *original* theory eventually becomes a *relative* notion.

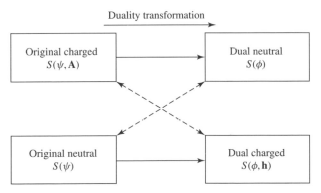

Figure 10.2 The relations between charged and neutral versions of the original and dual theory. The dashed lines indicate isomorphism.

10.6   Zero-field vortex-loop blowout

Consider now for the moment the case of zero magnetic field. The main advantage of the above formulation is that the probability of finding a connected path of vortex segments, starting at $\mathbf{x}$ and ending at $\mathbf{y}$, $G(\mathbf{x}, \mathbf{y})$, is given by the two-point correlation function of the ϕ-field [145]

$$G(\mathbf{x}, \mathbf{y}) = \langle \phi^*(\mathbf{x})\phi(\mathbf{y}) \rangle \tag{10.56}$$

A vortex-loop unbinding will lead to a finite $G(\mathbf{x}, \mathbf{y})$ when $|\mathbf{x} - \mathbf{y}| \to \infty$, because infinitely large loops will connect opposite sides of the vortex system. On the other hand, if $\lim_{|\mathbf{x}-\mathbf{y}| \to \infty} G(\mathbf{x}, \mathbf{y}) \neq 0$, this implies that $\langle \phi^*(x) \rangle \neq 0$, corresponding to a broken $U(1)$-symmetry. Therefore, the dual field $\phi(\mathbf{r})$ is an order parameter of a vortex-loop unbinding transition. The broken $U(1)$-symmetry is associated with the loss of number conservation of connected vortex-paths threading the system in any direction (including direction perpendicular to an applied magnetic field, if that is present). This limit of the two-point correlation function is closely related to a quantity O_L we will introduce below, which probes the connectivity of the vortex tangle in an extreme type II superconductor. The above connection makes it at the very least plausible that an abrupt change in this connectivity, as probed by the change in O_L, is associated with breaking a $U(1)$-symmetry of the vortex-sector of the GL-theory, equivalently an onset $\langle \phi^* \rangle$ or $\langle \phi \rangle$. Since this only happens above a critical temperature, we may view ϕ as a disorder-field, in contrast to the order-parameter field ψ of the original GL-theory. We will make explicit use of this connection later on.

In zero magnetic field, we will show that the loss of superfluid density, specific heat anomaly, change in vortex-loop distribution, loss of long-wavelength vortex-line tension, and abrupt change in vortex tangle connectivity abruptly occurs precisely at the same temperature both for the three-dimensional XY-model, also when amplitude fluctuations are included. Thus we may associate the the phase transition from superconductor to normal metal or neutral superfluid to normal fluid (such as the λ-transition in He^4) as a vortex-loop blowout.

The idea of a vortex-loop blowout being responsible for the λ-transition in He^4 (also denoted helium II) must be credited to Onsager [146]. Ref. [146] is a discussion remark from as early as 1949, and the two last sentences of the remark are well worth quoting: 'Finally, we can have vortex rings in the liquid, and the thermal excitation of Helium II, apart from the phonons, is presumably due to vortex rings of molecular size. As a possible interpretation of the λ-transition, we can understand that when the concentration of vortices reaches the point where they form a connected tangle throughout the liquid, then the liquid becomes normal'. This is qualitatively the right idea. Note, however, how we have meticulously avoided the use of the phrase 'vortex rings' in our treatment, and used the phrase 'vortex-loops' instead. Vortex rings are precisely that, namely rings, while vortex-loops can have much more complicated shapes, being crinkled to a degree that one cannot speak even crudely of ring-like shapes, the hole in the doughnut is in fact completely filled in. This is an essential feature, since it endows the vortex loops with enough configurational entropy to be able to overcome the internal energy such that the line tension of the vortices, which is the free energy per unit length, may be driven to zero. An object that on large length scales looks like a ring, will have a Hausdorff dimension of 1. As we shall see in this chapter, however, the vortex-loop tangle arising out of a neutral superfluid has Hausdorff dimension larger than 2. For a superconductor, the Hausdorff dimension of the vortex-loop tangle is slightly smaller than 2. In any case, it is clear that one cannot speak of ring-like objects. The truly remarkable thing about Onsager's vision from 1949, is that he correctly identified the nature of the topological defects in the order parameter which are responsible for driving the phase transition. The idea (as so many of Onsager's ideas) was far ahead of its time, and there was no hope of putting them on a quantitative level back in 1949. Only with the advent of the Kosterlitz–Thouless theory in 1973 [130, 131] was the much simpler problem of vortex–antivortex unbinding tackled in two dimensions! Note that if one takes a vortex loop and slices through it with a plane, the loop cuts the plane in two places, which are the locations of a vortex and an antivortex. Hence, the two-dimensional counterpart of Onsager's vortex-loop blowout is the vortex-antivortex unbinding of Kosterlitz and Thouless.

10.6.1   Definitions

Specific heat C

To calculate the specific heat per site C, we use the fluctuation formula,

$$\frac{C}{k_B} = \frac{1}{\mathcal{V}} \frac{\langle H^2 \rangle - \langle H \rangle^2}{(k_B T)^2}. \tag{10.57}$$

Here, the dimensionless volume $\mathcal{V} = L_x L_y L_z$, and L_μ is the dimensionless linear dimension of the system along the μ-direction. L_μ is measured in units of the lattice constant a_μ. As a check of consistency, we may also calculate the specific heat per site using the numerical derivation of the internal energy U,

$$C_U = \frac{1}{\mathcal{V}} \frac{\partial U}{\partial T} \tag{10.58}$$

Note that for the FG model, where the effective Hamiltonian depends explicitly on the temperature, there strictly speaking is an additional term in the expression for the internal energy,

$$U = -\frac{\partial \ln Z}{\partial \beta} = \langle H \rangle - T \left\langle \frac{\partial H}{\partial T} \right\rangle$$

For the three-dimensional XY model, if the simulations are properly done, $C \cong C_U$. For the FG model, where the effective Hamiltonian explicitly depends on the temperature, $C \neq C_U$. Note however that $\langle \partial H / \partial T \rangle$ arises out of introducing a temperature dependence of the coefficients of the Ginzburg–Landau theory. The temperature dependence of these coefficients always has a temperature dependence set by the *mean-field* critical temperature. Thus, close to the true T_c these corrections, arising from integrating out the fermions of the underlying microscopic description of the superconductor, are always negligible compared to the contribution coming from the true critical fluctuations of the order parameter, i.e. the transverse phase-fluctuations. In terms of Eq. 10.15 the term $-T \langle \partial H / \partial T \rangle$ originates from the temperature dependence of the *amplitude* of the order-parameter. Were this to actually vanish at $T = T_c$, substantial corrections to the specific heat and entropy would result. As we will see later, at $T = T_c$, the ensemble average of the amplitude of the order parameter is far from renormalized to zero by vortex-loop fluctuations. Hence, at the critical point the correction term $-T \langle \partial H / \partial T \rangle$ in the internal energy, with its resulting corrections to entropy, is negligible. There is now ample experimental evidence from the group of Ong at Princeton [147], that critical fluctuations are indeed important over a sizeable temperature window in the high-T_c cuprates of order several Kelvin below T_c a window which is consistent with a coherence length of order $10 \, \text{Å}$, about two orders of magnitude shorter than in conventional superconductors.

Local Cooper-pair density $\langle |\psi'|^2 \rangle$

As a probe for the local Cooper-pair density, we calculate

$$\langle |\psi'|^2 \rangle \equiv \frac{1}{V} \sum_{\mathbf{r}} \langle |\psi'(\mathbf{r})|^2 \rangle \tag{10.59}$$

We see in Eq. 10.59 that $\langle |\psi'|^2 \rangle$ involves both thermal and space average. Recall that $\psi' \equiv \psi / \sqrt{|\alpha(0)|/g}$. At the mean field level, we expect $\langle |\psi'|^2 \rangle$ to develop an expectation value below the mean field critical temperature $T_{MF}(B)$.

Superfluid condensate density $|\langle \psi' \rangle|^2$

As a probe for the local condensate density (density of Cooper-pairs participating in the superconducting condensate), we calculate

$$|\langle \psi' \rangle|^2 \equiv \frac{1}{V} \sum_{\mathbf{r}} |\langle \psi'(\mathbf{r}) \rangle|^2 \tag{10.60}$$

Note the difference between $\langle |\psi'|^2 \rangle$ and $|\langle \psi' \rangle|^2$. The former describes local Cooper-pair density, while the latter describes what is commonly known as the condensate density in ^{4}He-physics. In zero field, we expect $|\langle \psi' \rangle|^2$ to develop an expectation value below the critical temperature T_c.

Distribution of the order field phase angle

To probe the distribution of the phase angle in $\psi'(\mathbf{r}) = |\psi'(\mathbf{r})|e^{i\theta(\mathbf{r})}$, we define the distribution function

$$D_\theta(\theta') = \frac{1}{\mathcal{V}} \left\langle \sum_{\mathbf{r}} \delta_{\theta(\mathbf{r}), \theta'} \right\rangle \qquad (10.61)$$

Here, $\delta_{i,j}$ is the Kronecker delta function. In the simulations, we have chosen to work with a discrete set of phase angles, $\theta', \theta(\mathbf{r}) = 2\pi n/N_\theta$. Here, $n \in [0, N_\theta]$ is an integer, and N_θ is the number of allowed discrete phase angles. In our experience, the simulation results do not depend on N_θ, provided $N_\theta \stackrel{>}{\sim} 16$. In zero field, when the phase is disordered, we expect $D_\theta(\theta)$ to be uniformly distributed, $D(\theta) = 1/N_\theta$. In the ordered phase, we expect $D_\theta(\theta)$ to show a peak around a preferred phase angle.

Helicity modulus Υ_μ

To probe the global superconducting phase coherence across the system, we consider the helicity modulus Υ_μ, defined as the second derivate of the free energy with respect to an infinitesimal phase twist in the μ-direction. Finite Υ_μ means that the system can carry a supercurrent along the μ-direction. Within the three-dimensional *XY*-model, the helicity modulus along the μ-direction becomes,

$$\frac{\Upsilon_\mu}{J_\mu} = \frac{1}{\mathcal{V}} \left\langle \sum_{\mathbf{r}} \cos[\Delta_\mu \theta - \mathcal{A}_\mu] \right\rangle$$

$$- \frac{J_\mu}{k_B T \mathcal{V}} \left\langle \left[\sum_{\mathbf{r}} \sin[\Delta_\mu \theta - \mathcal{A}_\mu] \right]^2 \right\rangle .$$

For the FG case,

$$\frac{\Upsilon_\mu}{J_\mu} = \frac{1}{\mathcal{V}} \left\langle \sum_{\mathbf{r}} |\psi'(\mathbf{r})||\psi'(\mathbf{r} + \hat{\mu})| \cos[\Delta_\mu \theta - \mathcal{A}_\mu] \right\rangle$$

$$- \frac{J_\mu}{k_B T \mathcal{V}} \times \left\langle \left[\sum_{\mathbf{r}} |\psi'(\mathbf{r})||\psi'(\mathbf{r} + \hat{\mu})| \sin[\Delta_\mu \theta - \mathcal{A}_\mu] \right]^2 \right\rangle .$$

Note the difference between $|\langle \psi' \rangle|^2$ and Υ_μ, they are not identical. The former quantity probes the superfluid condensate density, which is a locally defined quantity, while the latter quantity probes a global phase coherence along a given direction μ. Since $\langle \psi' \rangle$ is the order parameter of the Ginzburg–Landau theory, close to the critical point we have

$$|\langle \psi' \rangle|^2 \sim |\tau|^{2\beta} \tag{10.62}$$

where $\tau = (T - T_c)/T_c$. On the other hand, $\Upsilon_\mu \propto \rho_{s\mu}$, where $\rho_{s\mu}$ is the superfluid density in the μ-direction. Using the Josephson scaling relation $\rho_{s\mu} \sim \xi^{2-d} \sim |\tau|^{\nu(d-2)}$ along with the scaling laws $\gamma = \nu(2 - \eta)$ and $2\beta = 2 - \alpha - \gamma$, we find

$$\Upsilon_\mu \sim |\tau|^{2\beta - \eta\nu} \tag{10.63}$$

Here, d is the dimensionality of the system, ν is the correlation length exponent of the system, β is the order parameter exponent, γ is the order parameter susceptibility exponent, and η is the anomalous dimension of the order parameter two-point correlation function at the critical point. Therefore, although $|\langle \psi' \rangle|^2$ and Υ_μ are in principle different, they may *appear* to be very close if the anomalous dimension η of the ψ-field is small, as indeed is the case for the Ginzburg–Landau theory, where $\eta \approx 0.04$. Note that for $\eta > 0$, the curve for Υ_μ should bend more sharply towards zero at the critical point than $|\langle \psi' \rangle|^2$. We will explicitly show by direct calculations within the Ginzburg–Landau theory that $|\langle \psi' \rangle|^2$ is very close to Υ_μ both in zero field and finite magnetic field. In zero magnetic field this is precisely what one would expect based on the above, when $\eta \ll 1$. For the special case of $d = 3$, we have $2\beta - \eta\nu = \nu < 2\beta$. To high precision, we have for the three-dimensional XY-model, that $\nu = 0.673$ and $\eta = 0.038$.

Vortex loop distribution $D(p)$

To probe the typical perimeter $L_0(T)$ and the effective long-wavelength *vortex-line tension* $\varepsilon(T)$ (not to be confused with the *flux-line tension*, which is always zero when gauge-fluctuations are completely suppressed due to the absence of tubes of confined magnetic flux), we define a vortex-loop distribution function $D(p)$, which measures the ensemble-averaged number of vortex-loops in the system having a perimeter p. In order to compare results from different system sizes, we normalize $D(p)$ with respect to the system size.

We search for a vortex-loop using the following procedure. Given a vortex configuration, we start with a randomly chosen unit cell with vortex segments penetrating its plaquettes. We follow the directed vortex path and record the trace. When the directed vortex path encounters a unit cell containing more than one outgoing direction, we choose the outgoing direction *randomly*. When the vortex path encounters a previously visited unit cell, i.e. when it crosses its own trace, we have a closed vortex loop, its perimeter being p. We now delete the vortex-loop from the vortex configuration, to prevent double counting, and continue the search. The search is continued until all vortex segments are deleted from the system.

Using a three-dimensional non-interacting boson analogy to the vortex system, it can be shown that the distribution function $D(p)$ can be fitted to the form

$$D(p) = A p^{-\alpha} \exp\left[-\frac{\varepsilon(T) p}{k_B T} \right] \qquad (10.64)$$

Here, A is a temperature-independent constant, and the exponent $\alpha \approx 5/2$ to a first approximation. When $\varepsilon(T)$ is finite, there exists a typical length scale $L_0 = k_B T / \varepsilon$ for the thermally excited vortex loops. The probability of finding vortex loops with much larger perimeter than L_0 is exponentially suppressed, according to Eq. 10.64. When $\varepsilon = 0$, $D(p)$ decays algebraically, and the length scale of the problem $L_0 = k_B T / \varepsilon(T)$, has diverged. As a consequence, configurational entropy associated with topological phase-fluctuations is gained without penalty in free energy. In zero field, there is only one critical point, and in this case L_0 must be some power of the superconducting coherence length $\xi(T)$.

Probe of vortex-connectivity, O_L

For probing the connectivity of a vortex tangle in a type II superconductor, in zero as well as finite magnetic field, we introduce a quantity O_L, defined in zero magnetic field as the probability of finding a vortex configuration that *can* have at least one connected vortex path threading the entire system in any direction. In the presence of a finite magnetic field, O_L is defined as the probability of finding a similarly connected vortex path in a direction transverse to the field direction, without using the PBC along the field direction. In zero field, we use the same procedure as in the finite-field, namely searching for connected vortex paths perpendicular to the z-direction, although in this case we could just as well have used any direction. Note that O_L is very different from the winding number W in the two-dimensional boson analogy. There, W is proportional to the number of vortex paths percolating the system transverse to the field direction. However, in the calculation of W, the PBC along the field direction is used many times.

This is entirely consistent with a number of other Monte Carlo simulation results on the three-dimensional XY-model which all show the loss of longitudinal phase-coherence and onset of longitudinal dissipation precisely at the vortex lattice melting transition. This is measured simply by the helicity modulus Υ_z, which is quite different from O_L. To the contrary, in our calculation of O_L, we do not allow for the use of periodic boundary conditions in the z-direction to measure vortex-tangle connectivity in the x- or y-directions. We have

$$O_L = \frac{N_c}{N_{total}} \tag{10.65}$$

N_{total} is the total number of independent vortex configurations provided by the Monte Carlo simulation. Furthermore, N_0 is the number of vortex configurations containing *at least one* directed vortex path that traverses the entire system perpendicular to the direction, without using the PBC along the field direction. For convenience, we treat the zero field case as the limit $\lim_{B \to 0}$ keeping the 'field direction' intact.

We search for the possibility of finding a vortex path such as described above by using the following procedure. Assume that the magnetic induction points along the z-axis. We follow all paths of directed vortex segments starting from all four boundary surfaces with surface normal perpendicular to $\hat{z}$, and check whether at least one of these vortex paths percolates the system and reaches the opposite surface, without applying the PBC in the z-direction. Note that when crossing vortex segments are encountered, the procedure is to attempt to continue in a direction that will bring the path closer to the opposite side of the system, rather than randomly resolving the intersection. O_L is therefore a *necessary*, but not *sufficient* condition for finding an actual vortex path crossing the system. However, in zero field this procedure does not make a difference to that of resolving the intersections randomly. This is demonstrated by the correlations of the change in O_L and $D(p)$, to be detailed in the next section.

If a vortex path is actually found crossing the system in any direction in zero field, or without using PBC in the field-direction when a field is present, one may safely conclude that the vortex-line tension has vanished. If it were finite, it would not be possible to find such a path at all, either because all vortex-lines form closed confined loops in zero field, or because the vortex-line fluctuations along the field direction would be *diffusive* in finite field. In zero-field, this is clear by the above mentioned correlation between the change in O_L and $D(p)$, cf. the results of the next section. In this paper, we also investigate this in detail for the finite-field case, by considering the position of the lowest temperature T_L where we have $O_L = 1$ both as a function of system size and aspect ratio $L_x/L_z = L_y/L_z$. If vortex-line physics remains intact, T_L should move monotonically up with system size, and should scale with L_x/L_z. Instead, we will find that T_L moves *down* slightly, and saturates with increasing system size at fixed aspect ratio. In addition, we find that T_L is virtually independent of aspect ratio for large enough systems.

In the VLL phase $O_L = 0$, since the field induced flux lines are well defined and do not 'touch' each other, and the thermally excited vortex-loops are confined to sizes smaller than the magnetic length. $O_L = 1$ in the normal phase above the crossover region where the remnant of the zero field vortex-loop blowout takes place. Needless to say, it is a matter of interest to investigate precisely where O_L changes value from zero to one.

Note that O_L itself is not a genuine thermodynamic order parameter, although it may be said to *probe* an order-disorder transition [29]. However, by the transcription of the vortex-content of the Ginzburg–Landau theory to the form of Eq. 10.52, it is brought out that probing the vortex-tangle connectivity by considering O_L is closely connected to probing the two-point correlator of a local complex field $\phi(\mathbf{r})$, the dual field of the local vorticity-field $m_\mu(\mathbf{r})$ of the Villain approximation and London approximation to the Ginzburg–Landau theory (Eq. 10.26). The two-point correlator $\langle \phi^*(\mathbf{r})\phi(\mathbf{r}')\rangle$ is ultimately the probe of whether or not the ϕ-theory (Eq. 10.52) exhibits off-diagonal long-range order and a broken $U(1)$-symmetry. An entirely equivalent interpretation of the change in O_L does not involve a local field $\phi(\mathbf{r})$, but rather number conservation of vortex-lines threading the entire superconductor. This number is conjugate to the phase-field of the local complex field $\phi(\mathbf{r})$. An advantage of the present formulation involving Eq. 10.52 is that it directly relates he change in O_L to the long-distance part of a correlation for a local field, and hence to a local order parameter $\langle \phi(\mathbf{r})\rangle$. This connection makes it at least plausible that the change in vortex-tangle connectivity, i.e. a change in the geometry of the vortex-tangle, may be related to a thermodynamic phase-transition. This is illustrated in Figure 10.3, where it is shown that the anomaly in the specific heat, the loss of superfluidity and the change in O_L all occur at the same temperature.

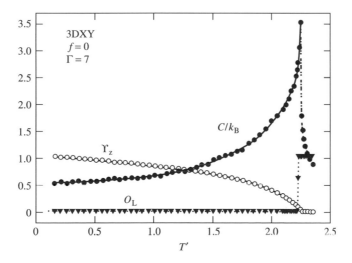

Figure 10.3 Specific heat capacity C, helicity modulus Υ_z and vortex-connectivity probe O_L as functions of temperature. Note how the the anomaly in C, the vanishing of Υ_z, and the onset of O_L all occur at the same temperature.

10.7 Fractal dimension of a vortex-loop tangle

One great advantage of the dual field theory formulation of the previous section is that it allows us to obtain information about critical exponents of the (dual) field theory by investigating *geometric* properties of the vortex-loop tangle. To see this, we first introduce the quantity $P(\mathbf{x}, \mathbf{y}, N)$ which is the probability of being able to perform a walk from the point $\mathbf{x}$ to the point $\mathbf{y}$ via a connected path of vortex segments in precisely N steps. Now, from this we can can derive several useful quantities. The one we shall focus on here is the probability distribution $D(N)$ of finding vortex loops of perimeter N. This is nothing but the probability of starting in point $\mathbf{x}$ and ending up in the same point in N steps, summed over all possible starting points $\mathbf{x}$ along the closed loop, normalizing to the number of starting points along the loop, which is just N. That is, we have

$$D(N) \propto \frac{1}{N} \sum_{\mathbf{x}} P(\mathbf{x}, \mathbf{x}, N) \propto N^{-\alpha} \tag{10.66}$$

This relation *defines* the geometric exponent α describing the vortex-loop tangle. (It should not be confused with the specific heat exponent.) Note also that all contributions in the above sum are equal, by translational invariance. Hence, we may pick out an arbitrary contribution to the sum, and invert to obtain

$$P(\mathbf{x}, \mathbf{x}, N) \propto N^{1-\alpha} \tag{10.67}$$

Note also that we have so far been careful to denote the configurations of vortex segments closing on themselves as vortex-loops, not vortex-rings. In fact the loops we are describing turn out to be highly crinkled line-objects which do not resemble rings at all. They are instead *fractal* objects with a Hausdorff dimension D_H that can be computed from α. Not only that, but we may compute the *anomalous scaling dimension* η_ϕ for the dual matter field ϕ, from α. Let us demonstrate explicitly how this close connection between geometry and criticality in superconductors and superfluids can be found.

Let us start with the simplest case, that of a tangle of non-interacting vortex segments. Then an analytic exact expression for $P(\mathbf{x}, \mathbf{x}, N)$ would be given by

$$P(\mathbf{x}, \mathbf{y}, N) \propto \frac{1}{N^{d\Delta}} \exp\left(-\frac{(\mathbf{x} - \mathbf{y})^2}{N^{2\Delta}}\right) \tag{10.68}$$

with $\Delta = 1/2$. In analogy with this, we write down a scaling Ansatz for $P(\mathbf{x}, \mathbf{y}, N)$ for the general case of interacting loops as follows

$$P(\mathbf{x}, \mathbf{y}, N) \propto \frac{1}{N^{d\Delta}} F\left(\frac{|\mathbf{x} - \mathbf{y}|}{N^{2\Delta}}\right) \tag{10.69}$$

where F is some scaling function and Δ is some exponent, often called the *wandering exponent*, to be determined. This exponent is defined via the relation between the linear extent L of a walk and the number of steps N in the same walk, namely

$$L \sim N^{\Delta} \tag{10.70}$$

Conversely, the number of links in a path of linear extent L is given by

$$N \sim L^{1/\Delta} \sim L^{D_H} \tag{10.71}$$

where D_H is the so-called Hausdorff dimension, or fractal dimension, of the vortex-loop tangle in the thermodynamic limit. For the non-interacting case, it can be shown that $D_H = 2$. This follows immediately from what we have stated above, $D_H = 1/\Delta = 2$. If we merely add a steric repulsion to this, passing from Eq. 10.55 to Eq. 10.54, the effect must surely be to push vortex segments apart, thus reducing the Hausdorff-dimension, hence for this case we have $D_H < 2$. Physically, this corresponds to the dual case of a charged three-dimensional superconductor. On the other hand, if we start out with a neutral superfluid, then the correct dual description is given by Eq. 10.52 with $e^2 = 0$. In this case, in addition to having a steric repulsion between vortex segments we also have a long-range anti-Biot–Savart interaction between vortex segments. This latter interaction tends to produce a long-range *attraction* between vortex segments that are oppositely directed. This interaction will tend to compress the vortex tangle compared to the case with only steric repulsion. Moreover, we expect the long-range interaction to overcompensate the dilution of the tangle produced by the local steric repulsion such that in fact the tangle is even more compact than the non-interacting case. Thus, we expect that for the dual of a three-dimensional neutral superfluid, $D_H > 2$. The various cases described above are illustrated in Figure 10.4.

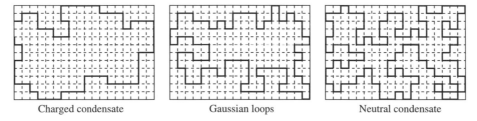

Charged condensate   Gaussian loops   Neutral condensate

Figure 10.4   A schematic illustration of vortex loops in three cases. The three fractal dimensions are ordered according to: $D_H(e \neq 0) < D_H^G < D_H(e = 0)$, where D_H^G is the fractal dimension corresponding to a Gaussian field theory, i.e. random loops. Qualitively the three loops are *self-avoiding, random* and *self-seeking*.

We will be interested in the scaling function for $|\mathbf{x} - \mathbf{y}| = 0$. By combining Eqs 10.67 and 10.69 we find by simple inspection the scaling relation

$$d\Delta = 1 - \alpha \tag{10.72}$$

Let us now connect these geometrical properties of the vortex-loop tangle to statistical physics and critical properties. To do this, we consider the two point correlation function for the dual field $\phi(\mathbf{x})$. We have

$$G(\mathbf{x}, \mathbf{y}) = \langle \phi(\mathbf{x})\phi^*(\mathbf{y}) \rangle \tag{10.73}$$

This correlation function (propagator) is nothing but the probability amplitude of going from the point $\mathbf{x}$ to the point $\mathbf{y}$ via any connected vortex path. Hence, we have

$$G(\mathbf{x}, \mathbf{y}) \propto \sum_N P(\mathbf{x}, \mathbf{y}, N) \tag{10.74}$$

The sum over N in Eq. 10.74 cannot be performed exactly; however we will now focus on long loops such that the discrete character of the excursions of the vortex segments on the lattice can be neglected. We therefore replace the sum by an integral such that

$$G(\mathbf{x}, \mathbf{y}) \sim \int dN \frac{1}{N^{d\Delta}} F\left(\frac{(|\mathbf{x} - \mathbf{y}|)}{N^{2\Delta}}\right) = \frac{1}{|\mathbf{x} - \mathbf{y}|^{d - 1/\Delta}} \tag{10.75}$$

Close to criticality, on the other hand, we have the standard scaling Ansatz for the two-point correlation function

$$G(\mathbf{x}, \mathbf{y}) \sim \frac{1}{|\mathbf{x} - \mathbf{y}|^{d - 2 + \eta_\phi}} \tag{10.76}$$

which defines the anomalous scaling dimension η_ϕ. By comparing Eqs 10.72, 10.71, 10.75, and 10.76 we obtain the following scaling relations connecting geometric exponents of the vortex tangle to critical exponents of the dual field theory

$$\frac{d}{D_{\mathrm{H}}} = \alpha - 1$$
$$\eta_\phi + D_{\mathrm{H}} = 2 \tag{10.77}$$

For computational purposes it is most convenient to compute the geometric exponent α in simulations. The reason is that this quantity appears to have less finite-size effects than for instance those observed in direct computations of D_{H}. However, once α is determined, D_{H} and η_ϕ follow.

Table 10.1   Value of the loop-size distribution exponent, as determined from simulations

Exponent	Gaussian	$e = 0$	$e \neq 0$	Limit
α	5/2	**2.312 ± 0.003**	**2.56 ± 0.03**	$\alpha > 2$
D_H	2	2.287 ± 0.004	1.92 ± 0.04	$D_H < 3$
Δ	1/2	0.437 ± 0.001	0.52 ± 0.01	$\Delta > 1/3$
η_ϕ	0	−0.287 ± 0.004	0.08 ± 0.04	$\eta_\phi > -1$

The remaining exponents have been calculated using scaling relations Eq. 10.77. Gaussian results are exact, and confirmed for the different exponents independently.

In Table 10.1 we show results of large-scale Monte Carlo simulations [148] where the vortex-loop distribution exponent α has been computed, and scaling relations used to derive D_H and η_ϕ.

The qualitative picture described is Eq. 10.71 is confirmed by the results shown in Table 10.1. For the original charged case $e \neq 0$, corresponding to a dual theory which only features a steric repulsion, we indeed find $D_H \langle 2, \eta_\phi \rangle 0$, while for the original neutral case corresponding to the dual charged case, we find $D_H \langle 2, \eta_\phi \rangle 0$.

10.8   Type I versus type II, briefly revisited

In this section, we will revisit the issue of type I versus type II superconductivity [45, 149]. As we have noted earlier, the distinction between the two is given in terms of the Ginzburg–Landau parameter $\kappa = \lambda/\xi$. At the mean-field level, the value of κ separating type I from type II behaviour is given by

$$\kappa = \frac{1}{\sqrt{2}} \qquad (10.78)$$

Note also that when we speak of stable vortices as excitations in a superconductor, we are implicitly talking about type II superconductors. The reason is that in type I superconductors, vortices are not stable. This is due to the fact that for too small values of κ, the vortex-attraction of range ξ, overwhelms the vortex-repulsion due to circulating currents around each individual vortex, a repulsion of range λ. Alternatively, therefore, type I and type II superconductors can be viewed as superconductors where vortices interact attractively and repulsively, respectively.

In this section, we start out with the Ginzburg–Landau model once again, but unlike the type II case studied in previous sections of this chapter, we do not ignore amplitude fluctuations of the superconducting order parameter. What we shall find is that the value of κ that separates type I from type II behaviour is not precisely given by the mean-field estimate, but rather by [45, 149]

$$\kappa = \frac{0.76 \pm 0.04}{\sqrt{2}} \tag{10.79}$$

This may be seen by reasoning as follows. We start out by performing large-scale Monte Carlo simulations on the full Ginzburg–Landau model, including amplitude fluctuations and phase-fluctuations in Ψ, as well as gauge-field fluctuations. As the system is warmed up, it passes from a low-temperature superconducting state to a high-temperature disordered state with no Meissner effect, which we obviously identify as the normal state. Now, we may compute the latent heat of this phase-transition as a function of κ. A first order phase transition has a non-zero latent heat, i.e. a jump in the entropy of the systems across the transition, whereas a second order phase transition does not. What is found is that the latent heat is finite deep in the type I regime, $\kappa \ll 1$. The latent heat is zero deep in the type II regime, $\kappa \gg 1$. The latent heat is found to vanish at a critical value of κ

$$\kappa = \kappa_{\text{tri}} \tag{10.80}$$

where the numerical value is given in Eq. 10.79. We have denoted this critical value of κ by κ_{tri}, since it is the value that separates a first order phase transition from a second order one. A critical endpoint terminating a first order phase-transition line and where a second order phase-transition line commences, is a *tricritical* point. The above numerical result is in rather remarkable agreement with an early analytical calculation [144], which gave

$$\kappa_{\text{tri}} = \frac{3\sqrt{3}}{2\pi} \sqrt{1 - \frac{4}{9} \left(\frac{\pi}{3}\right)^4} = \frac{0.798}{\sqrt{2}} \tag{10.81}$$

Now, let us use the results of the previous sections to connect this change in character of the phase transition, from a first order to a second order on the one hand, to the notion of type I and type II superconductivity on the other. We will follow the discussions of Refs. [45, 149].

Because the critical fluctuations in a superconductor are vortex-loops, criticality at the phase transition from the superconducting state to the normal metallic state requires stable vortices to exist. I. e. they must interact repulsively. By the previous discussion, this requires the vortex tangle to have Hausdorff dimension $D_{\text{H}} \leq 3$ [148, 45]. A Hausdorff dimension equal to 3 would imply that the vortex tangle fills space in a compact manner, which is just another way of stating that the vortex system has collapsed on itself. Hence, if we were to monitor the vortex-loop distribution function in the first order regime, we would find $D_{\text{H}} = 3$. By the scaling relation $D_{\text{H}} + \eta_\phi = 2$, we would conclude that $\eta_\phi = -1$. Now, we may use the well-known Rushbrooke scaling law

$2\beta = \nu(d - 2 + \eta_\phi)$ to conclude that the order parameter exponent $\beta = 0$. That is, the order parameter vanishes discontinuously, the hallmark of a first order phase transition. On the other hand, for repulsive vortex interactions and a stable vortex tangle, $D_H < 3$, $\eta_\phi < 1$, and $\beta > 0$. This means that the superconducting order parameter vanishes in a continuous fashion, which is the hallmark of a second-order phase transition. Thus, via the connection between geometry and criticality, we find a first-order phase transition when vortices attract each other, and a second-order phase transition when vortices repel each other.

However, as noted above, attractive and repulsive vortex interactions are precisely the characteristics of type I and type II superconductivity, respectively. Hence, we conclude that the distinction between type I and type II superconductivity is conceptually the same as the distinction between superconductors undergoing first order and second order phase transitions, respectively. The transition from one type of behaviour to the other happens at κ_{tri}, which is slightly renormalized from the mean field value $1/\sqrt{2}$. It should be noted that the latent heat even deep in the type II regime is tiny, and hence extremely precise measurements close to the critical temperature would be necessary in order to see the first order character. However, liquid crystals are described by a similar Ginzburg–Landau theory, as Eq. 10.7. In that case the Ginzburg–Landau parameter κ can be varied by varying the chemical compositions of the liquid crystal compounds and the latent heat involved in the smectic-A to nematic phase transition is sizeable. For such systems it has indeed been found experimentally that both first and second order phase transitions take place, depending on the value of κ [150].

We may summarize the discussion on effects of fluctuations in the Ginzburg–Landau model as follows. If we perform a mean-field analysis of the Ginzburg–Landau model, we arrive at the conclusion that the superconductor-normal metal transition is a second order phase transition with classical critical exponents. We may include effects of fluctuations both in the order parameter and the gauge-field $\mathbf{A}$, arriving at different conclusions depending on what fluctuations we take into account.

Historically, the first attempt at seriously considering fluctuation effects in the Ginzburg–Landau model were made by Halperin and co-workers who froze the superconducting order parameter ψ and considered only gauge-field fluctuations [43]. This is a procedure valid for extreme type I superconductors. The result is that this generates a term $|\psi|^3$ in the free energy, with a negative coefficient. Since this now contains a cubic invariant, a first order phase-transition is inevitable. For the superconductors present at the beginning 1980s the critical region is extremely narrow, however the predicted jump in the entropy across the first order transition is found to be small, of order $0.01k$ per degree of freedom, where k is Boltzmann's constant. Consequently the question was difficult to settle experimentally. However there is an isomorphism between superconductors and liquid crystals first discussed by Halperin and co-workers [151, 152], and on this system experiments can be done [153, 150].

Table 10.2   A schematic view of some important results regarding the *order* of the phase transition in the GL model

Description	Fluctuations	Order
1 At the mean field level the model reduces to the $\|\psi\|^4$ theory, i.e. the effect of the gauge field vanishes completely.	None	2.
2 $\|\psi\|^4$ theory is a well known theory, which is much studied [154].	$\|\psi\|$, arg ψ	2.
3 To study the effect of the gauge field Halperin, Lubensky and Ma integrated out the gauge field exactly, this gives a cubic term in the remaining effective ϕ theory. This is a sound approach in the $\kappa \ll 1$, i.e. strongly type I regime.	**A**	1.
4 Dasgupta and Halperin studied the combined effects of phase and gauge field fluctuations, *excluding* amplitude fluctuations [155]. Various duality relations were derived, and the concept of an *inverted* XY transition was introduced. Excluding amplitude fluctuations is a valid approximation in the $\kappa \gg 1$, i.e. strongly type II regime.	arg ψ, **A**	2.
5 For the superconductors present at the beginning 1980s the critical region is extremely narrow, and the predicted jump across the first order transition very small. Consequently the question was difficult to settle experimentally. However there is an isomorphism between superconductors and liquid crystals [151, 152], and on this system experiments can be done [153, 150].	$\|\psi\|$, arg ψ, **A** (Experiment)	1. **and** 2.

Especially from cases 3 and 4 one can conclude that there must be a κ_{tri} separating first order and second order transitions. It is also important to realize that a correct description of this feature of the GL model requires a *full description*, including fluctuations in $\|\psi\|$, arg ψ and **A**. Adapted from Ref. [156].

The next serious step at considering fluctuation effects in superconductors was taken by Dasgupta and Halperin [155]. Here, they took the opposite approach, and ignored the gauge-field fluctuations altogether, focusing on the fluctuations in ψ of a specific type, namely phase fluctuations. As we have seen, this is a sensible procedure in extreme type II superconductors. A second order phase transition in the *inverted* three-dimensional XY universality class is found, i.e. a three-dimensional XY phase transition with inverted temperature axis (and a negative anomalous scaling exponent). In fact, we may also include the gauge-field fluctuations with the phase fluctuations, and the result is always a second-order phase transition, as we have seen above [45, 149, 148]. Moreover, we may include amplitude fluctuations with phase fluctuations only, the result is still a permanent second order phase transition [154]. Thus, as long as we

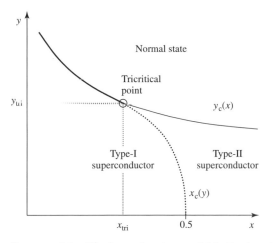

Figure 10.5 Phase diagram of the Ginzburg–Landau model in the (x, y) plane (parameters as in Eq. 10.7). The value $x_{tri} = 0.295 \pm 0.025$, corresponding to a tricritical value of the Ginzburg–Landau parameter given by $\kappa_{tri} = \frac{0.76 \pm 0.04}{\sqrt{2}}$. The solid line is the phase transition line from superconducting to normal state. The dashed line $x_c(y)$ is a line separating type I from type II superconductivity. The tricritical point separates a first-order phase transition (thick solid line) from a second-order phase transition (thin solid line), and also separates type I from type II superconductivity. The mean field value separating type I from type II superconductivity is given by $x = 1/2$. Note how it is possible to cross over from type I to type II behaviour in one and the same compound, for moderate values of κ, as the temperature is increased!.

ignore amplitude fluctuations in the order parameter ψ, a second-order phase transition always is obtained, either in the three-dimensional XY universality class (gauge-field fluctuations not included) or in the *inverted* three-dimensional XY universality class (gauge-field fluctuations included).

Kleinert [135] then considered an attempt at combining gauge-field fluctuations and phase fluctuations with amplitude fluctuations by mimicking the amplitude fluctuations using a fluctuating temperature in the Ginzburg–Landau model. He then found that the superconductor–normal metal phase transition changes from first to second order as κ passes through the tricritical value given by Eq. 10.81.

The next step was taken by Bartholomew who considered the Ginzburg–Landau model in Monte Carlo simulations, including all fluctuations in phase, amplitude, and gauge fields, finding a second order phase transition when κ is larger than some tricritical value $\kappa_{tri} = 0.4/\sqrt{2}$ [44]. This value was improved on in Refs [45, 149] using very large scale Monte Carlo simulations in conjunction with sophisticated finite-size scaling techniques, finding a first order phase transition below $\kappa = 0.76/\sqrt{2}$ and a second order phase transition above it. This is in rather remarkable agreement with Eq. 10.81.

All of the above may be put into a table showing various descriptions of the Ginzburg–Landau theory, including which fluctuations are taken into account at the superconductor–normal metal transition, and what the character of the resulting phase-transition is. This is shown in Table 10.2.

In terms of the parameters used in Eq. 10.7, the phase diagram of the Ginzburg–Landau model is given in Figure 10.5.

PART III

Selected Applications

11

Small Scale Applications

As a common introductory for the two chapters on applications we should moti-
vate the contents as follows: Superconductor technology divides naturally into
two main categories: small scale, usually electronic components or devices,
and large scale where magnets and energy applications are of most importance.
The applied part of this book intends to review some selected examples of
superconductor applications. The entire spectrum of implemented and potential
applications is so great today, that a comprehensive review is far beyond the
scope of any physics text. However, there is a need to see the physics of superconductivity in the context of applications which is a major driving force in
research on superconductors today. To this end, some examples of applications
are given, and in important cases the underlying theory is worked out in some
detail. A characteristic for superconductor-based technology is often that it is
either *superior* to alternative methods, or *only possible* using superconductors.
It should be emphasized that the potential for future applications other than
those discussed in Chapters 11 and 12, or for improvements of existing ones,
is enormous. Usually the question whether a demonstrated new superconductor
technology is finally brought to the market is one of economy. This is always
the case when alternative technologies are available, for instance in electric
energy transport and storage, and in magnet technology. Long-term reliability
and safety is another important consideration. The situation is somewhat different in areas where superconductor technology is unique, i.e. when no real
comparable alternative exists. This is clearly the case with regard to some of
the technologies based the Josephson junction. Although only a few examples
have been chosen here, they suffice to demonstrate existing and future superconductor technology on scales ranging from micrometres in the Josephson
junction to tens of kilometres in electric energy transport and maglev passenger
movement.

Superconductivity: Physics and Applications   Kristian Fossheim and Asle Sudbø
© 2004 John Wiley & Sons, Ltd   ISBN 0-470-84452-3

11.1   More JJ-junction and SQUID basics

11.1.1   Introductory remarks

In the present chapter we will discuss some further basic properties and how these are important for applications of superconductivity on a small scale, as well as giving outstanding examples of present day use.[1] The most important small scale electronic device applications so far are based on Josephson junctions, a topic we already encountered in Chapter 5. Among these, the SQUID has reached by far the widest range of useful implementations due to its unique property as the most sensitive and versatile device for detection and measurement of magnetic flux.

The SQUID has been under continuous development and improvements during several decades. It owes its sensitivity to the properties of the superconducting wavefunction phase which provides the basis for both flux quantization and Josephson tunnelling. The broad versatility is due to the possibility it offers for high resolution measurements of any physical quantity which can be converted into magnetic flux. This is the case with quantities like magnetic field, magnetic susceptibility, magnetic field gradient, electric current, voltage, and displacement. Therefore, a wide range of uses are open to the use of SQUID technology. A better understanding of the SQUID can only be reached through a good understanding of the Josephson effect. We therefore proceed to discuss the Josephson junction further, and then again turn to the SQUID and other applications.

11.1.2   RSJ – the resistively shunted Josephson junction

We showed in Chapter 5 that the current trough an unbiased Josephson junction can be expressed as

$$I = I_0 \sin \gamma, \tag{11.1}$$

where I_0 is the maximum supercurrent the junction can sustain, usually referred to as the critical current, and $\gamma = \phi_1 - \phi_2$ is the difference in superconducting phase across the junction. We also discovered that while the current I could be expected to flow at zero voltage, another new phenomenon would appear when the current was forced to exceed I_0: A voltage V developed across the junction, and the phase γ became time dependent, giving an AC current with so-called Josephson frequency ω_J according to $\omega_J = \partial \gamma / \partial t = (2e/\hbar)V = 2\pi V/\Phi_0$. The current across the junction should now be written

$$I = I_0 \sin[\gamma_0 - (2eV/\hbar)t] \tag{11.2}$$

---

[1] Suggested additional reading: Säppä H, Ryhänen T, Ilmoniemi R and Knuutila J, in Fossheim K., (ed) *Superconducting Technology. 10 case studies*, World Scientific Publishing, 1991.

where γ_0 in the phase across the junction without the voltage V present. In junctions of high structural perfection the $I-V$ characteristic is often found to be hysteretic as shown in graphs (b) and (c) of Figure 11.1. We will return to the details of this figure later on. For a junction to be practical in a SQUID it cannot be allowed to have such characteristics. Experience and theoretical analysis have taught us that the problem can be solved by shunting the junction, i.e. by adding an external resistance in parallel. This modification is commonly referred to as the RSJ-model (resistively shunted junction), as shown in Figure 11.2. One identifies from left to right: The current source, the resistive shunt R, the Josephson junction, the self-capacitance C bound to be present in the layered SIS junction structure, and finally the voltage V.

We take the current source to be in a current controlled mode. This forces the phase γ in the equation $I = I_0 \sin \gamma$ to adjust itself accordingly. The current now determines the phase difference across the junction, and not vice versa.

We might further take into account the possibility that there is a fluctuating noise current $I_n(t)$ associated with the components of the circuit, as is commonly true. The equation of motion for the circuit is then

$$C\frac{dV}{dt} + I_0 \sin \gamma + \frac{V}{R} = I + I_n(t) \tag{11.3}$$

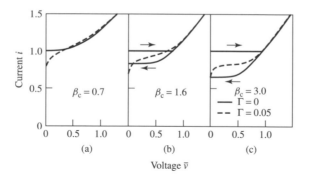

(a)   (b)   (c)

Voltage $\bar{v}$

Figure 11.1 Numerically calculated current-voltage characteristics of Josephson junctions with $\beta_c = 0.7$, 1.6 and 3.0 at zero temperature, i.e. $\Gamma = 0$ (solid lines), and with $\Gamma = 0.05$ (dashed lines). (For reference, see the footnote at the foot of the opposite page). Parameters β_c and Γ are defined below.

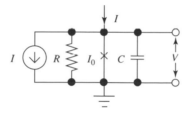

Figure 11.2 The RSJ model for a Josephson junction.

an equation first studied by Stewart [157] and McCumber [158]. We choose first to analyze the differential equation (Eq. 11.3) under the assumption that $I_n(t)$ can be neglected. We use the previously found relationship $\partial \gamma / \partial t = (2e/\hbar)V$ and write Eq. 11.1 as

$$\frac{\hbar C}{2e} \ddot{\gamma} + \frac{\hbar}{2eR} \dot{\gamma} = I - I_0 \sin \gamma. \tag{11.4}$$

Upon defining the quantity

$$U = -\frac{\Phi_0}{2\pi}(I\gamma + I_0 \cos \gamma) \tag{11.5}$$

Eq. 11.4 may be written as

$$\frac{\hbar C}{2e} \ddot{\gamma} + \frac{\hbar}{2eR} \dot{\gamma} = -\frac{2e}{\hbar} \frac{\partial U}{\partial \gamma} \tag{11.6}$$

This is recognized as the equation of motion for a mass point moving down a corrugated surface, often referred to as a 'washboard potential', where U is a function of γ.

Two characteristic situations are illustrated in Figure 11.3, $I < I_0$, and $I > I_0$ respectively. In the first case the point mass gets trapped in one of the potential minima, and oscillates locally there. In the other case it goes on sliding. The first of these situations corresponds to zero time average phase difference, i.e. $\langle \dot{\gamma} \rangle = 0$. This corresponds to the average voltage across the junction being zero.

The second type of motion arises when the tilt is increased to make $I > I_0$. In this case the time average is $\langle \dot{\gamma} \rangle > 0$, in other words, the point mass keeps sliding down the washboard. It is helpful also to note that in the mechanical analogue the capacitance C represents the particle mass, and $1/R$ is proportional to the friction coefficient, or damping of the motion.

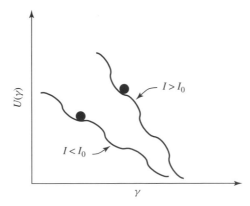

Figure 11.3   The washboard potential with a sliding point mass. Here I represents the tilt of the washboard as can be seen from Eq. 11.5.

Numerical analysis of Eqs 11.3–11.6 shows that hysteresis is controlled by the parameter β_c, called the Stewart–McCumber parameter

$$\beta_c = \frac{2\pi I_0 R^2 C}{\Phi_0} \tag{11.7}$$

Hysteresis is found for $\beta_c > 1$. Calculations of current versus voltage is shown in Figure 11.1 for different values of β_c. Distinct hysteresis appears for $\beta_c > 1$. Further properties of Figure 11.1 will be discussed later. It requires highly qualified laboratory expertise to create junctions which meet the particular requirement $\beta_c < 1$ for hysteresis-free junctions.

Until now we neglected the current noise $I_n(t)$. For a better understanding of the Josephson contact and SQUID physics the $I_n(t)$ term should be reintroduced, and its consequences analysed. The full equation of motion is now:

$$\frac{\hbar C}{2e}\ddot{\gamma} + \frac{\hbar}{2eR}\dot{\gamma} + I_0 \sin \gamma = I + I_n(t) \tag{11.8}$$

The effects of noise are usually discussed in terms of noise spectral density $S_I(f)$ of $I_n(t)$. The meaning of $S_I(f)$ is 'noise power per bandwidth due to current noise $I_n(t)$'. Theoretical analysis shows that the noise spectral density of $I_n(t)$ is given by the Nyquist result:

$$S_I(f) = 4k_B T / R \tag{11.9}$$

Since the current I represents the tilt of the washboard potential, the physical effect of $I_n(t)$ is to cause fluctuations in that tilt. If these fluctuations become large enough they will cause the particle to escape from is local trapping site. This creates a condition for $\langle \dot{\gamma} \rangle$ to differ from zero, which again means that a voltage will appear, in this case corresponding to the appearance of voltage noise with increasing temperature. This will cause a rounding of the $I–V$ curve before the steeper onset of current driven by the external voltage V. The noise corresponds to an additional voltage dV_n and a total effective voltage $V + dV_n$, causing the precursor "noise- rounding" effect in the $I–V$ characteristic observed in the calculations of Figure 11.1.

This current noise has an additional effect of lowering the effective critical current value, a potentially harmful consequence of noise since the critical current I_0 must be above a certain value for proper operation of the junction. A criterion for I_0 to be large enough to maintain a coherent supercurrent or 'coupling of the junction' is that the coupling energy exceeds the available thermal energy, expressed by the relation

$$I_0 \Phi_0 / 2\pi \gg kT \tag{11.10}$$

The quantity $I_0\Phi_0/2\pi$ is what we referred to as the coupling energy of the junction. It is a measure of the energy associated with the current pattern for each flux quantum in the junction. A typical value of I_0 is in the μA range at low temperatures, a small but easily measured current.

11.1.3   Further modelling of the Josephson junction

Having discussed the Josephson junction in rather simple terms above, let us examine it a little further. Before we rewrite the equation of motion, we remark first on the presence of the resistance R. Going back to the Josephson result for the current, we should, to be accurate, add a term coming from quasiparticle current in the presence of a voltage. The total current is now:

$$I = I_0 \sin \gamma + V/R_q \tag{11.11}$$

R_q is a voltage dependent quasiparticle resistance. The term $I_0 \sin \gamma$ still determines the main properties of the junction. Next, we rewrite the equation of motion by defining new, normalized variables:

$$t^* = (2\pi I_0 R/\Phi_0)t; \quad i_n(t^*) = I_n(t)/I_0 \tag{11.12}$$

We assume further that the total resistance R, which is formed by R_q in parallel with the shunt resistance, satisfies the condition $R \ll R_q$. For such cases the equation of motion (Eq. 11.8) describes the junction accurately.

The thermal noise current I_n is the current in the resistance R. Its autocorrelation function is $\langle I_n(t + \tau)I_n(t)\rangle = 2kT\delta(\tau)/R$, where $\delta(\tau)$ is the Dirac delta function. With a little effort one finds the following reduced form of the equation of motion:

$$\beta_c \frac{\mathrm{d}^2\gamma}{\mathrm{d}t^{*2}} + \frac{\mathrm{d}\gamma}{\mathrm{d}t^*} = -\frac{\mathrm{d}}{\mathrm{d}\gamma}(-i\gamma - \cos\gamma) + i_n(t^*) \tag{11.13}$$

The current $i_n(t^*) = I_n/I_0$ is expressed as a Langevin function describing the thermal noise, and according to standard treatment obeys the relation

$$\langle i_n(t^* + \tau^*)i_n(t^*)\rangle = 2\Gamma\delta(\tau^*) \tag{11.14}$$

where $\Gamma = 2\pi kT/I_0\Phi_0$ is the normalized thermal energy. The parameter β_c is now analogous to mass, and inversely proportional to the damping of the system. Numerical analysis of Eq. 11.13 yields further insight into the current–voltage characteristics, as seen in Figure 11.1. What was computed there were the quantities $i = I/I_0$, and corresponding values of $\bar{v} = \langle v \rangle = \bar{V}/RI_0$, resulting in the function $\bar{v}(i)$. $\bar{V}$ is the time averaged voltage over the current-biased SQUID.

The graphs demonstrate the dependence of the $I-V$ characteristics, here expressed as i versus $\bar{v}$, on the two parameters Γ and β_c. First, we observe that Γ embodies the rounding effect at finite temperatures mentioned above. It is therefore referred to as the noise rounding parameter, and the phenomenon itself as noise rounding. Second, as pointed out before, at $\beta_c \leq 1$ the characteristics is hysteresis free when $\Gamma = 0$, but at higher values of β_c there is distinct hysteresis even when $\Gamma = 0$. This result defines one of the main technical challenges in applications of the Josephson effect: To control the parameter β_c and thereby the characteristics of the junction.

The use of Josephson contacts in the SQUID, requires that hysteresis is absent, as mentioned. This provides the important design criterion $2\pi I_0 R^2 C / \Phi_0 < 1$. Fortunately, this nontrivial requirement on the $I_0 R^2 C$ product can be met in practice.

11.1.4   The autonomous DC SQUID

Next, we proceed to describe the structure, properties, and operational principles of the double junction DC SQUID, consisting of a superconducting loop interrupted by two identical junctions, both resistively shunted to avoid hysteresis. In the absence of hysteresis the state of the junctions and the DC SQUID are unambiguously determined by the applied voltage. Furthermore, since the flow patterns of supercurrent are characteristically different when the ring encloses an integral number n of flux quanta as opposed to $(n + \frac{1}{2})$ quanta, the corresponding two $I-V$ characteristics are also distinctly different. This fact provides the basis for practical use of the DC SQUID.

We will now analyse the double-junction DC SQUID by extending the RSJ model to this case. This procedure neglects some technical aspects. But the DC SQUID is in reality such a complex device that one should make some initial simplifications to gain insight into the basic properties. Next, further aspects can be discussed on that basis. For instance, when it comes to practical implementation one cannot avoid considering the coupling of the SQUID to the external world in order to fully understand and control the properties. But it is not a good idea to include all such problems in the first basic analysis. Hence we start the analysis by using the RSJ model equation for both junctions simultaneously. First we note that in each junction, hereafter numbered $i = 1, 2$, the current I_i is

$$I_i = \frac{\hbar}{2e} C \frac{\mathrm{d}^2 \gamma_i}{\mathrm{d}t^2} + \frac{\hbar}{2eR} \frac{\mathrm{d}\gamma_i}{\mathrm{d}t} + I_0 \sin \gamma_i + I_{\mathrm{n},i} \qquad (11.15)$$

This provides us with two equations. First we observe that the total current through the DC SQUID is $I = I_1 + I_2$, and next that the total magnetic flux threading the ring is

$$\Phi = \frac{\Phi_0}{2\pi}(\gamma_1 - \gamma_2) = \Phi_a + L(I_1 - I_2)/2 \tag{11.16}$$

In other words, the flux in the ring it is the sum of the applied flux and the flux generated by the (difference of) currents I_1 and I_2. Φ_a is as before, the external applied flux. The quantities C, R and I_0 take on the same meaning as discussed in the RSJ model discussed before. L is the inductance of the ring, and $I_{n,i}$ is the thermal noise current in the shunt resistance R_i. The junctions are characterized by the phase differences γ_i.

For further analysis we combine the equations of motion for the two junctions by once adding them, and once subtracting one from the other. This allows us to study the behaviour of the quantities $\gamma_1 + \gamma_2$ and $\gamma_1 - \gamma_2$, which turns out to be a useful procedure. To this end we define the quantities

$$\nu = (\gamma_1 + \gamma_2)/2; \phi = (\gamma_1 - \gamma_2)/2 \tag{11.17}$$

The instantaneous voltage across the two junctions is now $V = RI_0 d\nu/dt^*$, using the previously defined rescaled, dimensionless time variable $t^* = 2\pi RI_0 t / \Phi_0$, and the total flux is $\Phi = \Phi_0 \phi/\pi$.

Using the previously defined parameter β_c, and introducing a new parameter

$$\beta_L = 2\pi L I_0/\Phi_0 \tag{11.18}$$

we combine the equations as described, to obtain two new equations of motion. Looking at the case where we add the equations for junction 1 and junction 2 we find

$$I = I_1 + I_2 = \frac{\hbar}{2e}C\frac{d^2}{dt^2}(\gamma_1 + \gamma_2) + \frac{\hbar}{2eR}\frac{d}{dt}(\gamma_1 + \gamma_2)$$
$$+ I_0(\sin\gamma_1 + \sin\gamma_2) + I_{n,1} + I_{n,2}. \tag{11.19}$$

Upon introducing $\nu = (\gamma_1 + \gamma_2)/2$ and $\phi = (\gamma_1 - \gamma_2)/2$ we find

$$I = \frac{\hbar}{e}C\frac{d^2\nu}{dt^2} + \frac{\hbar}{eR}\frac{d\nu}{dt} + 2I_0\sin\left(\frac{\gamma_1 + \gamma_2}{2}\right)$$
$$\times \cos\left(\frac{\gamma_1 - \gamma_2}{2}\right) + I_{n,1} + I_{n,2} \tag{11.20}$$

$$\frac{I}{2I_0} \equiv i = \frac{\hbar}{2eI_0}C\frac{d^2\nu}{dt^2} + \frac{\hbar}{2eRI_0}\frac{d\nu}{dt} + \sin\nu\cos\phi + i_{n,\nu} \tag{11.21}$$

with $i_{n,\nu} = (I_{n,1} + I_{n,2})/2I_0$. By rescaling the time variable and introducing t^* as explained we finally obtain

$$\beta_c\frac{d^2\nu}{dt^{*2}} + \frac{d\nu}{dt^*} + \sin\nu\cos\phi = i + i_{n,\nu}(t^*) \tag{11.22}$$

(The change of sign for $i_{n,v}(t^*)$ is unimportant since this is a quantity which fluctuates around zero.) Proceeding in a similar manner when taking the difference
$I_1 - I_2$ we find

$$\beta_c \frac{d^2\phi}{dt^{*2}} + \frac{d\phi}{dt^*} + \cos v \sin \phi + \frac{2}{\beta_L}(\phi - \phi_a) = i_{n,\phi}(t^*) \tag{11.23}$$

where we introduced the quantity $\phi_a = \pi(\Phi/\Phi_0)$. The Langevin functions for
$i_{n,v}$, and $i_{n,\phi}$ represent uncorrelated thermal noise sources, and each of them
are characterized by the correlation property $\langle i_{n,v}(t^* + \tau)i_{n,v}(t^*)\rangle = \langle i_{n,\phi}(t^* + \tau)i_{n,\phi}(t^*)\rangle = \Gamma\delta(\tau)$.

Γ is again the noise rounding parameter defined previously. The equations
developed here describe what is called the 'autonomous DC SQUID', i.e. one
that is operating in a magnetic field with an external current I applied, but
without other coupling to external devices, like a signal coil or control circuit,
feedback circuit etc.

11.1.5 Simplified model of the DC SQUID

Numerical analysis of Eqs 11.22 and 11.23 has given much insight into the
properties of the DC SQUID. For instance, under the assumption that the loop
inductance L is very small, corresponding to $\beta_c \ll 0$, the equations can be
expanded as a series, from which the voltage, the circulating current, and the
transfer function $\partial \bar{V}/\partial \Phi_a$ can be estimated.

Under the condition that L is negligible so that $\beta_L \ll 1$ and $\beta_c \approx 0$, the total
flux in the ring is $\Phi \approx \Phi_a$. According to Eq. 11.22 the voltage can now be
written as

$$V \approx RI_0(i - \cos \phi_a \sin v) \tag{11.24}$$

By integration of V over the period T of the Josephson oscillation, the average
voltage across the current-biased DC SQUID is found [53],

$$\bar{V} = \frac{1}{T}\int_0^T V \, dt \approx \frac{RI}{2}\left[1 - \left(\frac{2I_0}{I}\cos\frac{\pi\Phi_a}{\Phi_0}\right)^2\right]^{\frac{1}{2}} \tag{11.25}$$

In this limit the double junction DC SQUID behaves like a single Joseph-
son junction with resistance $R/2$, and with an effective critical current $I_{c,\text{eff}} = 2I_0\cos(\pi\Phi_a/\Phi_0)$, in other words a flux-dependent critical current, with a modulation depth $\Delta I_{0,\text{eff}} = 2I_0$ when $\beta_L \ll 1$ and $\beta_L \approx 0$. This is seen by comparison
with the I–V characteristics found for a single junction.

The transfer function $\partial \bar{V}/\partial \Phi_a$ can be estimated on the basis of the above analysis by differentiation of Eq. 11.25 at $\Phi_a = \Phi_0/2$ and $I = 2I_0$ which approximates the usual point of operation in a feedback circuit. To obtain realistic estimates one needs to relax the condition $\beta_L \ll 1$. RSJ model simulations indicate that at $\beta_L = \pi$ the modulation depth of critical current is approximately equal to I_0. This leads to the conclusion that $\partial \bar{V}/\partial \Phi_a \propto \Delta I_{0,\text{eff}}$ is reduced by approximately a factor of 2 for $\beta_L = \pi$. The result is that the transfer function can be approximated as

$$\frac{\partial \bar{V}}{\partial \Phi_a} \approx \frac{R}{2L} \tag{11.26}$$

Similarly, the dynamic resistance can be estimated as $R_{\text{dyn}} = \partial \bar{V}/\partial I \approx R/\sqrt{2}$

Operational properties

The most important aspects of the DC SQUID with regard to operational properties are sketched in Figure 11.4. From left to right is shown the resistive shunting of each junction, and with a capacitance C associated with each, followed by a sketch of the two typical $I-V$ characteristics, once with a flux $\Phi_a = n\Phi_0$ and once with $\Phi_a = (n + \frac{1}{2})\Phi_0$ applied. This leads to a periodically varying voltage vs. applied flux, with a period Φ_0 as shown in the right figure. These insights can be gained through numerical calculations, using the equations of motion discussed above.

Looking through the mathematics of these equations, it is not easy to see from them why the I–V characteristics behave as shown in Figure 11.4. Clarke [159] has given a 'simplistic' analysis with a minimum amount of mathematics. We refer here to Figure 11.5, and also to Figure 11.4: we take the two junctions in the SQUID to be arranged symmetrically, one on each side of the loop, as is always done in a DC SQUID. The bias current I is then divided equally

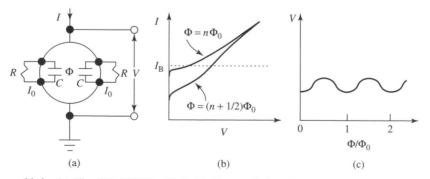

(a)   (b)   (c)

Figure 11.4   (a) The DC SQUID; (b) $I-V$ characteristics; (c) V vs. Φ/Φ_0 at constant bias current I_B. After Clarke [159].

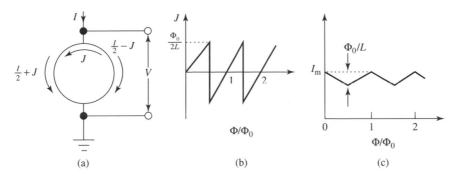

Figure 11.5 Simplistic view of the DC SQUID: (a) a magnetic flux Φ generates a circulating current J that is periodic in Φ as shown in (b); as a result (c), the maximum supercurrent I_m is also periodic in Φ. After Clarke [159].

between the two paths, with $I/2$ in each. As long as the current in a junction is below the critical current I_0, a supercurrent tunnels through the junction at zero potential difference. If we allow an arbitrary amount of flux Φ_a to be applied to the ring, an additional supercurrent I_{circ} is set up, creating a self-generated flux LI_{circ} in the ring. Recalling that the total flux Φ in a superconducting ring or loop is quantized as $n\Phi_0$, the following relation now applies:

$$\Phi = \Phi_a + LI_{\text{circ}} = n\Phi_0 \qquad (11.27)$$

Taking the starting values of the applied flux as $\Phi_a = 0$ or $\Phi_a = n\Phi_0$, the circulating current according to Eq. 11.27 is initially zero. On increasing the applied flux, a finite circulating current $JI_{\text{circ}} = (\Phi_a - n\Phi_0)/L$ develops in order to keep the total flux through the loop at its starting, quantized value.

This internally generated, circulating current adds to the externally applied bias current $I/2$ in junction 1, and subtracts in the other, junction 2. The individual junction currents are related as $I_1 = I_2 + 2I_{\text{circ}}$ and $I_2 = I_1 - 2I_{\text{circ}}$. Since the circulating current flows in opposite sense relative to the bias current in the two junctions, the junctions will respond differently to the same amount of externally applied flux. Junction 1 switches to the voltage state, i.e. a voltage develops across it, when it carries a total current equal to the critical current (see Figure 11.1). This happens when $I_1 = I/2 + I_{\text{circ}} = I_0$. At this point the device current is $I = 2I_0 - 2I_{\text{circ}}$. This is to be regarded as the device critical current at this applied flux. The maximum value clearly is $2I_0$. What is important for use of the device is how it further responds to applied flux: At the same time as Φ_a increases from zero to $\Phi_0/2$, I_{circ} increases from zero to $\Phi_0/2L$ according to Eq. 11.27, and the device critical current diminishes to $2I_0 - \Phi_0/L$. We see from this that the critical current of the device as a whole has decreased by the amount Φ_0/L. Next, as we increase the applied flux beyond $\Phi_0/2$ the SQUID makes a transition from the flux state $n = 0$ to $n = 1$, with reversal of the sign of I_{circ}. As Φ_a is increased to Φ_0, the critical current again increases to its maximum value $2I_0$. The sequence discussed is depicted in Figure 11.5. It repeats periodically with increasing flux, and thereby results in the behavior of the DC SQUID illustrated in Figure 11.4. As Figure 11.4 also indicates, a practical device is biased at a constant current I_B. This bias is above the critical current and therefore puts the entire DC SQUID in the voltage state, which is what we need in order to make it a flux to voltage transformer. It is the resulting voltage response to applied flux, shown in the right figure, which is so extensively exploited in practical applications of the device.

An important question is how this device which is so sensitive, can be used both in low and high magnetic fields. The answer is that a combination of several factors make this possible: The sensitivity of the the transfer function $\partial \bar{V}/\partial \Phi$ at the bias current I_B, combined with compensating feedback circuitry, as well as the use of a flux transformer that amplifies the flux detected at the observation point and transfers it to the SQUID, usually located some distance away. The DC SQUID is typically operated in socalled flux-locked loop. A simplified example is shown in Figure 11.6. The operating point of the SQUID is chosen at the steepest point on one of the periodic oscillations of the $V - \Phi$-curve seen in Figure 11.4. In this design the SQUID is coupled to an external compensation inductor controlled by feedback from the SQUID output, thus causing the flux felt by the SQUID to be constant during operation even in a strongly varying external field. The feedback circuit also produces a voltage proportional to the flux that the circuit compensates. When calibrated, this voltage then measures the field detected by the device. In many uses of the SQUID other types of detector coils are used, such as first-order or second-order gradiometers. Gradiometers are used to eliminate spatially homogeneous external noise superimposed on the signal, like electromagnetic noise in the environment. A simplified but realistic and detailed sketch of a DC SQUID with signal coil attached is shown in Figure 11.7. The development of the SQUID has been a long battle against noise of many kinds. Noise may be environmental type, it may come from the components of the SQUID itself, and from the detecting circuitry and amplifier. Since the SQUID is a non-linear device, thermal noise has a dramatic effect on the dynamics. Figure 11.1 already gave some

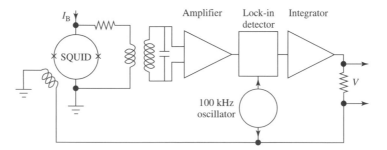

Figure 11.6   Modulation and feedback circuit for the DC SQUID. After Clarke [159].

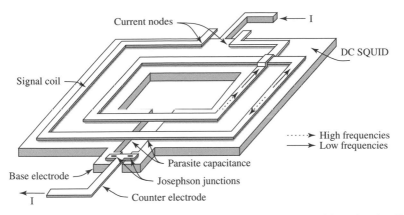

Figure 11.7   Simplified structure of a planar DC SQUID together with a signal coil. The sites for the current nodes of the microstrip resonance and for the effective capacitance coupling are indicated. The directions of the current in the different signal-coil turns at frequencies lower and higher than the microstrip line resonance frequency are shown. After Säppä *et al.* [160].

information on the influence of noise through the parameter Γ as an example. According to careful studies [160] the autonomous DC SQUID discussed above has an optimal energy resolution $\epsilon \approx 12 k_B T \sqrt{LC}$ at $\beta_c \approx 0.7$. This expression can be interpreted as the ratio of thermal energy $\frac{1}{2} k_B T$ divided by the so-called effective noise bandwidth of the SQUID, $\frac{1}{4} R/L$. We conclude that a low-noise DC SQUID should have small junction capacitance, small loop inductance, and proper damping of the junctions. Examples of typical values are: $C = 1\,\mathrm{pF}$ and $L = 20\text{--}40\,\mathrm{pH}$. In Figure 11.7 it is seen that the design attempts to meet these requirements. Obviously, the junction capacitance is made very small.

DC SQUIDS may be shunted in several ways. We have only discussed one of these, the resistively shunted DC SQUID. In addition both capacitive and inductive shunting are possible, and have been well characterized.

11.2   SQUID applications

11.2.1   Biomagnetism: neuromagnetic applications

Introductory remarks[2]

Magnetic fields produced by living organisms has become one of the most wide-spread applications of the SQUID. Especially important is magnetoencephalography (MEG) which is a method for studying brain functions by mapping of the magnetic field produced by neural currents in the human brain.

---

[2]Suggested additional reading: Ahonen A.I, *et al.*, in Fossheim, K. ed. *Superconducting Technology. 10 case studies*, World Scientific Publishing, 1991 [161].

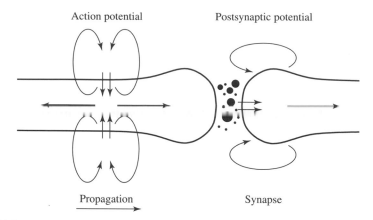

Figure 11.8 Schematic illustration of an action potential, essentially consisting of two nearby opposite current dipoles propagating towards a synapse and the resulting postsynaptic potential corresponding to a single stationary current dipole. After Ahonen *et al.* [161].

Such magnetism is referred to as neuromagnetism. The function of the brain involves activity on the millisecond timescale. For such study there are two main windows: Electroencephalography (EEG) and magnetoencephalography (MEG). In EEG the potential distribution over the head is measured by electrodes attached to the scalp. In MEG the weak electromagnetic fields of typical amplitude 50–500 fT generated by the human brain are measured in order to locate the current sources generating the magnetic fields associated with cortical activity. Magnetoencephalography is a noninvasive tool for such study, offering spatial resolution on the millimetre scale and time resolution on the millisecond scale. A particular advantage of the MEG is its ability to locate cortical activity at deeper levels and spatially more accurately than EEG. The EEG is strongly influenced by currents in the scalp, which the MEG is not. Hence MEG offers superior quality with regard to locating accuracy. There are two types of signals from the brain: spontaneous activity, and evoked response to external stimuli.

The principle of neuromagnetism

The basis for MEG lies in concepts of *action potential* and *postsynaptic potential*. Information is carried from one neuron to another by the action potential via the synapse, mediated by the release of transmitter molecules which increase the permeability of the postsynaptic membrane as illustrated in Figure 11.8. Due to an interaction between chemical and electrical gradients there is a net current flow through the membrane and a change in voltage, called the postsynaptic potential (PSP).

To a first approximation PSP is an intracellular current dipole oriented along the neuron, while the action potential may be described as a current quadrupole. In order for a resulting magnetic field to be detectable outside the head, a large number of neurons need to be activated, and to act in unison. The strength of a PSP current dipole is only 10^{-14} A. The observed magnetic field is believed to be mostly due to PSP cells which are perpendicular to the cortex. The field due to the action potential is very short range due to its quadrupolar nature. In addition, the dipolar PSP lasts an order of magnitude longer. The observed equivalent dipole is of the order of 10^{-8} A, which means that tens of thousands of neurons are activated at the same time. Dipoles which are observed in spontaneous activity, like in elliptic foci, may be an order of magnitude larger still.

In order to calculate the field outside the head, resulting from cortical activity, some model assumptions have to be made. The head is often described as spherical symmetric, filled by a conductive medium with conductivity $\sigma(r)$. The magnetic field outside such a body has important simplifying properties: (1) radial dipoles and their volume currents cancel each other and thus do not generate a magnetic field outside the body, (2) the radial component of magnetic field is produced by the primary current density only, and (3) both radial and tangential field components are independent of the conductivity profile.

During measurement, the evoked response signal is usually assumed to be composed of the true source signal plus gaussian noise. By repeated sampling, say $N = 100$ times, the signal part is integrated up proportional to N, the total number of samples, while the noise integrates up proportional to $\sqrt{N}$. Thus an acceptable signal to noise ratio can be achieved.

Instrumentation

The SQUID is the only instrument with sufficient sensitivity to perform neuromagnetic studies. Usually, the external field is coupled into the SQUID by a a superconducting flux transformer, i.e. a pickup loop coil and a signal coil in series, shown in principle in Figure 11.9. Normally the signal is orders of magnitude smaller than the environmental noise. Therefore, the transformer usually has a gradiometer coil on the detecting side, sensitive only to the derivative of the magnetic field, $\partial B_z / \partial z$, and insensitive to environmental noise with electromagnetic wavelength much longer than the size of the coil. An illustration of such a coil is shown in Figure 11.10. It consist of two identical coils in series, but oppositcly wound. The first part may be regarded as the signal detector, and the second part as a compensator coil. In a case when spatially homogeneous noise is detected by the first coil, it is compensated by the second one, and the resulting signal picked up by the gradiometer is zero. If on the other hand an inhomogeneous field is present, the gradiomctcr coil dctccts a signal proportional to the spatial derivative along the coil axis. Depending on the design one may choose to make gradiometers that measure tangential derivatives $\partial B_z / \partial x$ and $\partial B_z / \partial y$.

Usually the whole device is placed in a magnetically shielded room. However, more and more sophisticated designs have made it possible to operate the

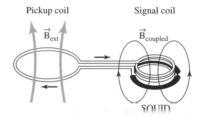

Figure 11.9   Transformer. After Ahonen *et al.* [161].

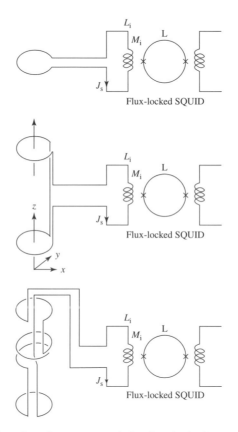

Figure 11.10   Examples of gradiometers coupled to flux-locked SQUID. After Clarke [159].

SQUID in unshielded environment. For a more widespread use of SQUIDs in diagnostics, it is essential to avoid the use of shielded rooms for the sake of cost reduction. A potentially interesting area for the future is in cardiology. The capability to obtain high quality signals from the heart cycle is well established even using high-T_c superconductor SQUIDs. However, sensitivity to the positioning of the device is one of the problems with this method.

During the 1990s a new development took place. Multichannel SQUID systems, containing well over 100 SQUIDs were built to be placed over the head, and

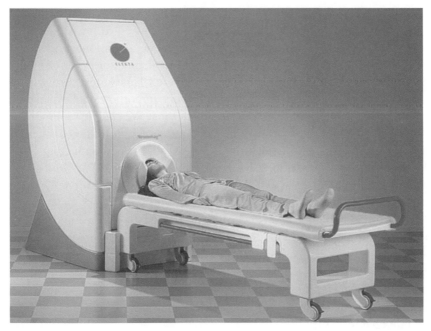

Figure 11.11 Multichannel SQUID system for neuromagnetic studies of the brain. With permission from Neuromag.

allowing a very detailed recording of brain activity in the entire brain. These systems are commercially available, and becoming increasingly important as a tool for brain diagnostics, and as a research tool for the study of brain function more generally. Figure 11.11 shows an example of multichannel SQUID system for neuromagnetic studies of the brain. Determining the precise location of epileptic foci before surgical treatment is an established clinical application of such systems.[3]

11.3 Superconducting electrodynamics in the two-fluid model

11.3.1 Frequency dependent conductivity in the two fluid model

Electrical current in the Meissner phase is confined to the penetration depth λ. This obviously is the DC limit of response to an electromagnetic field. Therefore, any theory for the low frequency response must have Meissner screening

---

[3]Recommended further reading: See two review articles by Clarke J. (1) In Weinstock, H. ed. *SQUID Sensors: Fundamentals, Fabrication and Applications*. NATO ASI Series. Kluwer Academic Publishers, Dordrecht, The Netherlands 1996, and (2) In Weinstock, H. ed. *Applications of Superconductivity*. NATO ASI Series. Kluwer Academic Publishers, Dordrecht, The Netherlands 2000. These reference treat the rf SQUID.

as its limiting low frequency behaviour. We will now treat electrodynamics in the situation where a superconductor surface is exposed to an AC electromagnetic field. Relevant cases may be electromagnetic radiation incident on the superconductor surface, for instance at normal incidence, or an electromagnetic wave traveling along the superconductor surface. This latter case will correspond to the situation in the technologically very important planar wave-guides and filters used in high frequency communications. We treat a superconducting half-space first as the basic entity. In the course of this treatment penetration depths for electromagnetic field in both normal and superconducting materials will be introduced. Furthermore, we need a model description of the superconducting electron gas. It turns out that the so-called two-fluid model is adequate for this purpose over a wide frequency range.

The two-fluid model was historically introduced by Gorter and Casimir in 1934 [67]. It proposes to divide the mobile charges in the superconducting state of a metal into two distinct entities: A normal component of concentration n_n with properties like those of the electron in the normal state, and a superfluid component of concentration n_s whose properties are responsible for the observed screening of a static magnetic field in the Meissner phase. The total concentration of mobile charge is then

$$n = n_n + n_s \tag{11.28}$$

where both n_n and n_s have to be temperature dependent in order to account for the temperature dependence of the penetration depth λ. Corresponding to this division one also foresees that currents are composed of two parts, a normal part J_n associated with n_n, and a superconducting part, J_s, associated with n_s. For the total current then

$$J = J_n + J_s \tag{11.29}$$

Experimentally one finds in low-T_c superconductors that the London penetration depth λ_L obeys the following simple relation for its temperature dependence

$$\lambda_L(T) = \lambda_L(0) \left[1 - (T/T_c)^4\right]^{-\frac{1}{2}} \tag{11.30}$$

This can be interpreted as a consequence of a density of normal component $n_n/n = (T/T_c)^4$, and a corresponding density of superfluid component

$$n_s/n = 1 - (T/T_c)^4 \tag{11.31}$$

Since $\lambda_L = \left(m_s/\mu_0 n_s(T)q_s^2\right)^{\frac{1}{2}}$, substituting n_s from Eq. 11.31 reproduces Eq. 11.30. We define $\lambda_L(0)$, the London penetration depth at 0 K, as

$$\lambda_L(0) = (m_s/\mu_0 n_s(0)q_s^2)^{\frac{1}{2}} \tag{11.32}$$

How does this relate to conductivity? Using Newton's second equation for the case of zero friction we have

$$m_s \frac{d\mathbf{v}}{dt} = q_s \mathbf{E} \tag{11.33}$$

The current density J_s may be expressed as

$$\mathbf{J}_s = n_s q_s \mathbf{v} \tag{11.34}$$

Hence we can rewrite Eq. 11.33 as

$$\Lambda \frac{d\mathbf{J}_s}{dt} - \mathbf{E} = 0 \tag{11.35}$$

which is the first London equation where $\Lambda = m_s/n_s q_s{}^2$. In case of a sinusoidal field of frequency ω we expect a resulting sinusoidal current. After performing the derivation in Eq. 11.35 we obtain

$$\mathbf{J}_s = \frac{1}{i\omega\Lambda} \mathbf{E} = \sigma_s \mathbf{E} \tag{11.36}$$

where we have defined a conductivity σ_s:

$$\sigma_s = \frac{1}{i\omega\Lambda} = \frac{1}{i\mu_0\omega\lambda_L^2} \tag{11.37}$$

Since σ_s is expressed for a condition of zero loss (without a damping term in Newton's equation of motion) it is a purely imaginary quantity. Also, if we assume low frequency, as when $\omega\tau \ll 1$, we can take the usual DC expression for the normal state conductivity to apply for the normal component of the electron gas, with the normal current being in phase with the driving field:

$$J_n = \sigma_n E = \frac{n_n q_n^2 \tau}{m_n} E \tag{11.38}$$

and the total current

$$J = J_n + J_s = \left(\frac{n_n q_n^2 \tau}{m_n} - i\frac{1}{\mu_0\omega\lambda_L^2} \right) E \tag{11.39}$$

The conductivity is now a complex expression:

$$\sigma = \sigma_1 - i\sigma_2 = \frac{n_n q_n^2 \tau}{m_n} - i\frac{1}{\mu_0\omega\lambda_L^2} \tag{11.40}$$

The response to an AC field in a superconductor clearly is out of phase with the applied E-field.

11.3.2   Surface impedance and AC loss

Because of the altered screening effects resulting from the foregoing analysis there is a need to apply a different concept, adequate for this situation, in addition to Meissner screening. This is the concept of a (complex) surface impedance Z, measured in units of Ω, defined by

$$ Z = \frac{E_{\|}}{\int J \, dz} \tag{11.41} $$

where $E_{\|}$ is the electric field in the surface, driving the current J along the surface, and integration is from the surface ($z = 0$) to (in principle) infinity inside the conductor. Using Maxwell's equation one derives for a normal metal, as we will show later,

$$ \begin{aligned} Z_n &= R_n + i X_n \\ R_n &= X_n = (\mu_0 \omega / 2\sigma_n)^{\frac{1}{2}} \end{aligned} \tag{11.42} $$

Consequently, the resistive and reactive parts of the surface impedance in a normal conductor are equal, and, since σ is frequency independent in the range we are considering, both go like $\omega^{1/2}$. These results have been well verified experimentally.

In the superconducting state matters are different due to the difference in AC conductivity discussed in the previous section. Since we already used London's first equation to establish an expression for σ in the superconducting state, and in addition invoked the two-fluid model, superconductivity is built into the σ_s we derived above. We therefore now only need to use Maxwell's equations, and the surface impedance takes the same form both in the superconducting and normal states when expressed in terms of σ_s. We have for the superconductor:

$$ Z_s = R_s + i X_s = \left(\frac{i \mu_0 \omega}{\sigma_s} \right)^{\frac{1}{2}} \tag{11.43} $$

We will show later that

$$ R_s = \frac{1}{2} \mu_0^2 \omega^2 \lambda_L^2 \sigma_n \tag{11.44} $$

$$ X_s = \mu_0 \omega \lambda_L \tag{11.45} $$

where we used the symbol σ_n for the conductivity of the superconductor in its normal state, corresponding to $n_n = n$. We also note that the reactive part X_s of the surface impedance Z_s is purely inductive. The usual expression for such impedance is $X = \omega L$. In the present case therefore, we can conclude that $L_s = \mu_0 \lambda_L$, usually called the equivalent inductance, known also as the kinetic inductance. We now proceed to deduce the results briefly mentioned above, first for the normal state. We want to find the intrinsic impedance Z. This quantity may be defined from Maxwell's equations

$$\nabla \times \boldsymbol{H}_n^s = \boldsymbol{J}_n^s + i\omega\varepsilon\boldsymbol{E}_n^s \tag{11.46}$$

where the second term on the right-hand side is the displacement current due to bound charges, or

$$\nabla \times \boldsymbol{H}_n^s = (\sigma_n + i\omega\varepsilon)\boldsymbol{E}_n^s \tag{11.47}$$

and by use of

$$\nabla \times \boldsymbol{E}_n^s = -i\omega\mu_0\boldsymbol{H}_n^s \tag{11.48}$$

In all cases the superscript s here refers to 'surface' rather than 'superconducting' while subscript n refers to the normal state. Combining Eqs 11.47 and 11.48 by taking the curl of the second one and inserting the result in the first, and further writing $E_n^s = E_{n,o}^s e^{-(\alpha+i\beta)z} = E_{n,o}^s e^{-\gamma z}$ one finds

$$\gamma_n^2 = (\sigma_n + i\omega\varepsilon)\,i\omega\mu_0 \tag{11.49}$$

which can be written in the form

$$\gamma_n = i\omega\,(\mu_0\varepsilon)^{\frac{1}{2}} \left(1 - i\frac{\sigma_n}{\omega\varepsilon}\right)^{\frac{1}{2}} \tag{11.50}$$

In a good metal the displacement current density is very small compared to that caused by free charges. Numerically, therefore, one finds even for normal metals that $\sigma_n/\omega\varepsilon \gg 1$. Hence, the general expression for γ_n can be approximated by

$$\gamma_n = i\omega\,(\mu_0\varepsilon)^{\frac{1}{2}} \left(-i\frac{\sigma_n}{\omega\varepsilon}\right)^{\frac{1}{2}} = i\,(-i\omega\mu_0\sigma_n)^{\frac{1}{2}}$$

$$= (1 + i)\,(\omega\mu_0\sigma_n/2)^{\frac{1}{2}} = (1 + i)\,/\delta_n \tag{11.51}$$

where δ_n is the classical normal state skin depth, defined here as

$$\delta_n = \left(\frac{2}{\omega\mu_0\sigma_n}\right)^{\frac{1}{2}} \tag{11.52}$$

We also note that under this approximation, since $\gamma_n = \alpha_n + i\beta_n$ is a complex quantity, we have for α_n and β_n

$$\alpha_n = \beta_n = \left(\frac{1}{2}\omega\mu_0\sigma_n\right)^{\frac{1}{2}} = \frac{1}{\delta_n} \tag{11.53}$$

These quantities are again closely related to the intrinsic impedance, a quantity we will deduce next. The complex surface impedance $Z \equiv R + iX$ defined as indicated previously (Eq. 11.41), or alternatively by the relation

$$Z \equiv \frac{E^s_{x,o} e^{-\gamma z}}{H^s_{yz}} \tag{11.54}$$

here specifically referring to an electromagnetic AC field applied at normal incidence to the surface whose normal is along z. Using $\nabla \times E^s = -i\omega\mu_0 H^s$ one finds

$$Z = \frac{(i\omega\mu_0)^{\frac{1}{2}}}{(\sigma + i\omega\varepsilon)^{\frac{1}{2}}} = \left(\frac{\mu_0}{\varepsilon}\right)^{\frac{1}{2}} \frac{1}{(1 - i\,(\sigma/\omega\varepsilon))^{\frac{1}{2}}} = \left(\frac{\mu_0\omega}{2\sigma}\right)^{\frac{1}{2}} (1 + i) \tag{11.55}$$

We can at this point easily distinguish between the normal and superconducting states. In the normal state the conductivity is a real quantity σ_n, and we find in this case for $Z_n = R_n + iX_n$, the corresponding terms in

$$Z_n = \left(\frac{\omega\mu_0}{2\sigma_n}\right)^{\frac{1}{2}} + i\left(\frac{\omega\mu_0}{2\sigma_n}\right)^{\frac{1}{2}} \tag{11.56}$$

Next, introduce σ in the superconducting state as a complex quantity $\sigma_s = \sigma_1 - i\sigma_2$ as we did before. Inserting this, we have:

$$Z_s = R_s + iX_s = \left(\frac{i\omega\mu_0}{\sigma_1 - i\sigma_2}\right)^{\frac{1}{2}} \tag{11.57}$$

Since in the superconductor $\sigma_1 \ll \sigma_2$ we can expand the root, and find, using the expressions for σ_1 and σ_2 from Eq. 11.40

$$Z_s = R_s + iX_s = \frac{\mu_0^2\omega^2\lambda_L^3 n_n q_n^2\tau}{2m_n} + i\mu_0\omega\lambda_L \tag{11.58}$$

$$R_s = \frac{1}{2}\mu_0^2\omega^2\sigma_n\lambda_L^3 \tag{11.59}$$

$$X_s = \mu_0\omega\lambda_L \tag{11.60}$$

We notice in particular that the surface resistance predicted by this model increases as the square of the frequency, compared to the square root dependence in the normal case. Observations confirm these predictions with good accuracy in, respectively, normal metals, and in high quality LTS superconductors. In HTS the frequency dependence is sometimes influenced by non-ideal structures. The agreement with predictions is therefore not as good in such cases. Examples of observations are given in Figure 11.12.

11.3.3 Surface resistance measurement

The most crucial information needed in order to evaluate the quality of superconducting surfaces or thin films for use in high frequency communication systems is
the surface resistance R_s. This quantity will determine the loss properties, and this
is the essential information to obtain. Several methods have been established for
such purposes. As far as high-T_c materials are concerned, due to the anisotropic
structure of these materials, measurements must be done on planar surfaces. This
is contrary to well-established methods for low-T_c materials, which are often measured in cylindrical cavities including the walls. Since the planar sample can very
well be used in both cases we describe briefly the so-called endwall replacement
method. Figure 11.13 shows a sketch of a vertical cut through a cylindrical TE_{011}
mode cavity with input and output signal attached as indicated. The short centre
lead sticking into the cylinder are antennas for the excitation of the electromagnetic cavity, and for measurement of the response on the 'out' side. While the
cylindrical walls and the bottom part are made of high quality copper, the top
flange can be exchanged, placing here in one case a copper plate, and in the other
case a high-T_c film deposited on some appropriate substrate.

By sweeping across a certain frequency range around the resonance frequency
f_0 of the cavity, the detecting antenna will register a varying field response due

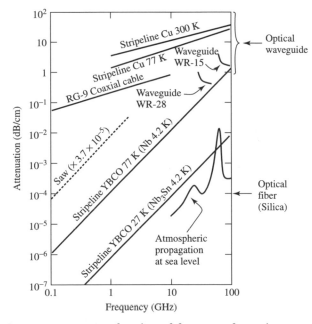

Figure 11.12 Loss per meter as a function of frequency for various transmission media.
(The surface acoustic wave propagation losses on $LiNbO_3$ have been scaled to make a
realistic comparison and represent dB/µs of delay for an equivalent electromagnetic wave in
a media with $\varepsilon_r = 10$). After Lyons and Withers [162].

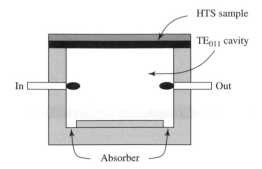

Figure 11.13   Cross-section of a TE_{011} mode cylindrical cavity for measurements of surface resistance. After Shen [163].

to the finite width Δf of the resonance. This width is a measure of the quality of the cavity as a resonator, called its Q-value, defined by

$$\frac{\omega_0}{\Delta \omega} = \frac{f_0}{\Delta f} \equiv Q = \frac{2\pi f_0 W}{P} = \frac{\omega_0 W}{P} \qquad (11.61)$$

We can regard this as a measure of how much energy is dissipated per second (P) relative to how much energy is put into the cavity per second. Here W is the stored energy, which, after multiplication by frequency essentially measures the rate of energy put into the cavity. Therefore, a measurement of the Q-value defined here is an appropriate way to characterize the quality of the cavity. These aspects are determined fully by the loss properties of the material.

The next question is now how a measurement of the Q-value as described above gives access to the surface resistance R_s. First we have to replace the upper Cu flange by the superconductor. The change of Q will tell us how the materials differ. A relative measure of the surface resistance r of the superconductor compared to that of copper is often expressed as

$$r = \frac{R_s(Cu)}{R_s(SC)} = \frac{k}{k + Q_0(Cu)/Q_0(SC) - 1} \qquad (11.62)$$

Here, k is a geometrical constant of the cavity.

$$k = \frac{1}{2 + P_{Sw}(Cu)/P_{ew}(Cu)} \qquad (11.63)$$

where P_{Sw} (Cu) is the power lost in the sidewall of the copper cavity while P_{ew} (Cu) is the power lost in the end wall when the Cu end wall is in place. We did not prove these results. They are referred to here only to show that the analysis can be completed. The important relation is that expressed by r. Clearly, if all that is needed to determine the surface resistance of the superconductor, relative to that of copper, is a measurement of quality factors as explained, with the

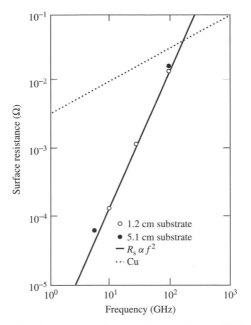

Figure 11.14 Relationship between surface resistance R_s at 77 K and frequency in a $Tl_2Ba_2CaCu_2O_8$ film and in Cu. After Holstein *et al.* [164].

upper wall in place, once as a superconductor and once as a copper wall. Since copper is the standard by which metallic resistance is gauged, this does the job.

What is found, is that although copper is a very good conductor, with low resistance, a superconductor is often 1000 or more times better, i.e. R_s is 10^3 to 10^4 times lower than that of copper at room temperature. This is a drawback for the method described, making it one of relatively low sensitivity. Other techniques have been devised to allow the superconductor to play a larger role in determining the total Q of the cavity. We refer to more specialized literature on this subject.

Figure 11.14 shows results of measurement of surface resistance carried out on $Tl_2Ba_2CaCu_2O_8$ film. We notice in particular that the square law for surface resistance discussed before is verified, even using different substrates. The validity of the result of the two-fluid theory is confirmed even in this high frequency range.

11.4 High-frequency radio technology

11.4.1 Microstrip filters and delay lines

Introduction

Superconductors can have a dramatic impact on selected passive microwave device applications because of two properties that differ greatly from those of normal metals at high frequencies. First, much lower surface resistances is available using superconductors, a fact that transforms into much lower loss and much higher Q-values in superconducting microwave system components. Second, superconductors have a practically frequency independent penetration depth in the microwave frequency range. This has the important consequence that superconductors introduce no dispersion into a microwave device up 1 THz frequency in low-T_c superconductors, and well above this in high-T_c superconductors due to their larger gap frequency. The improvement over normal conducting performance is substantial. Superconductor passive devices may work quite well in cases where normal conductor devices would function very poorly or not at all. A normal metal component would, in many cases, have several orders of magnitude higher conductor loss than the superconducting ones. Important passive microwave device components are bandpass filters, which are becoming essential components in mobile phone base stations. Other uses are in chirp filters which can provide improved resolution in radar images, and delay lines used in various contexts.

As we discussed previously in this chapter, the two-fluid model is adequate for an engineering representation of the physical properties of superconductors in the present context. Figure 11.15 shows how the surface impedance can be represented graphically in a circuit analysis. The surface resistance, which we previously found in Eq. 11.59, can be expressed as follows:

$$R_s = \frac{1}{2}\mu_0^2\omega^2\lambda_L^3(0)\sigma_n \left(\frac{T}{T_c}\right)^4 \left[1 - \left(\frac{T}{T_c}\right)^4\right]^{-\frac{3}{2}} \qquad (11.64)$$

where the temperature dependencies of n_n/n and λ_L, respectively, have been inserted in the two parentheses. According to this expression the surface resistance goes to zero at $T = 0$. In reality this does not happen. Instead a limiting value is reached at low temperatures, independent of temperature but dependent on frequency. This effect is typically caused by small inclusions of normal phase. For niobium the residual resistance falls below $10^{-8}\ \Omega$ in the GHz range at about 1.2 K. High-T_c superconductors have a much shorter coherence length than low-T_c ones. The residual resistance is therefore a more serious problem

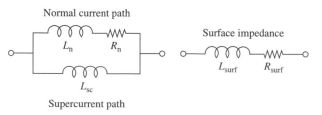

Figure 11.15   Two-fluid model for the bulk impedance of superconductors and its equivalent surface impedance.

in high-T_c superconductors. We refer again to Figure 11.12 for examples of the improved surface resistance properties in superconductors compared to copper, in particular in high-T_c, in the 1–100 GHz range. For YBCO film the calculated curves are based on the two-fluid model. The potential improvement compared to Cu at 77 K is enormous. Although part of the improvement is lost when patterning is performed, compared to the unpatterened continuous film, a substantial advantage remains. Another important aspect is the loss per meter during signal transmission. Figure 11.12 leaves no doubt about the advantage of superconducting transmission media over surface acoustic wave signal transport, normal coaxial cable, and Cu stripline. This figure does not take into account dielectric losses, nor does it take into account transducer and conversion losses in the SAW device. Comparison to Nb and Nb_3Sn at 4.2 K is made.

Implementation: filters and delay lines

Figure 11.16 shows the design of a 4-pole so-called Chebyshev bandpass filter with 3% bandwidth, 0.05 dB passband ripple, and 4 GHz centre frequency. The filter was made by deposition of YBCO film on 425 μm thick $LaAlO_3$ substrate. Data in Figure 11.17 shows a passband insertion loss of only 0.3 dB at 77 K and 0.1 dB at 13 K. Nb filters showed similarly 0.1 dB insertion loss at 4.2 K. Gold filters had an insertion loss of 2.8 dB at 77 K. The superconductor filters have far steeper skirts than gold filters, a very important feature to prevent cross-talk between channels, in particular when used as filters for mobile phone communication.

Another important application of superconductor passive devices is in analogue signal processing where the high frequencies and wide bandwidths required place a premium on conductor loss. The tapped delay line architecture developed at Lincoln Laboratory is illustrated in Figure 11.18. A tapped delay line provides temporary storage and sampling of the input signal at intervals τ_i. The samples are multiplied by weights w_i and coherently combined by spatial summation or time integration. Such delay lines are used to implement chirp

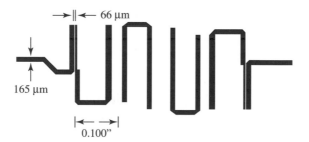

Figure 11.16 Four-pole superconductive microstrip filter layout. The filters were fabricated on $LaAlO_3$ substrates using gold, niobium and YBCO signal lines. After Lyons and Withers [162].

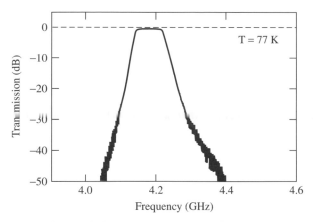

Figure 11.17   Measured transmission response at 77 K of a filter fabricated with a postan-nealed YBCO signal line and ground plane on a 425 μm thick LaAlO$_3$ substrate. Passband insertion loss is 0.3 dB. After Lyons and Withers [162].

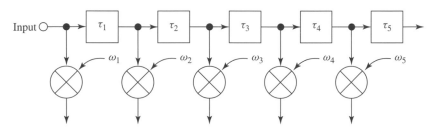

Figure 11.18   Architecture of a transverse filter, where τ_i are time delays and w_i are fixed or programmable tap weights. After Lyons and Withers [162].

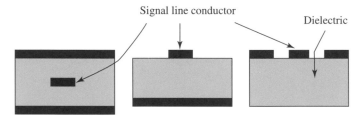

Figure 11.19   Various transmission line geometries considered for the implementation of YBCO delay lines. After Lyons and Withers [162].

filters, matched filtering, correlation and Fourier transformation. The number of information cycles of the waveform gathered coherently in the filter determines the signal processing gain, i.e. the time-bandwidth product. Delay line transmission geometries are illustrated in Figure 11.19, in various combinations of signal line, dielectric and ground. Generally, the various structures perform very much in agreement with predictions.

11.4.2   Superconducting high-frequency devices

Introductory remarks

Superconducting tunnel junctions are well suited for use in low-noise high frequency detectors. The noise level in the these quantum mechanical based detectors can be well below those of competing state-of-the-art detectors. The most demanding applications for mm- and sub-mm observations in radio astronomy already take advantage of these low-noise receivers. According to Dicke's famous radiometer formula [165], the signal to noise ratio for a given bandwidth (Δf) and a given integration time (Δt) is

$$\frac{S}{N} = \frac{T_s}{T_R}\sqrt{\Delta f \Delta t} \tag{11.65}$$

where T_s is the equivalent blackbody temperature of the signal, and T_R is the receiver noise temperature. From this formula we observe that for a given signal to noise ratio the integration time needed to observe a particular source is reduced by a factor of 4 if the receiver noise temperature is reduced by a factor of 2. Since the integration time sets the time scale of the telescope observation time, this transforms into far more efficient use of an expensive satellite telescope with moderate improvement of the detector. Superconducting device components are therefore of great interest. It turns out that the greatest advantage is obtained at several hundred GHz, i.e. in the mm and sub-mm wavelength regime, an interesting range for radio astronomy in space or on the ground.

Phase sensitive heterodyne detectors normally use a mixer to down-convert a high frequency signal to an intermediate frequency (*if*) range where low noise semiconductor amplifiers are available. To perform the down-conversion a high frequency local oscillator (*lo*) is needed. For semiconductor mixers which need about 1 mW power above 200 GHz, these oscillators are rather heavy equipment. In contrast, a superconducting detector needs only low *lo* power. Furthermore, a large number of receivers can use the same *lo*. It is therefore possible to construct an array of detectors for radio astronomy used in satellite telescopes, that can scan the sky simultaneously in several directions. This gives another factor of 10 in efficiency. Lightweight low power *lo* solutions are also available for SIS mixers. For space use the ideal technology will be to build integrated all superconductor receivers, i.e. local oscillator, mixer and preamplifier on the same chip.

The particular property of SIS junctions which makes them useful in the context outlined above, is their extremely sharp non-linear $I-V$ characteristics. According to the physical properties we have discussed in Chapters 4 and 5, there are two distinct possibilities, either the quasiparticle tunnelling $I-V$ characteristic near the gap V_g, or the Josephson effect pair tunneling characteristic at zero voltage, $V = 0$. Both may be used.

Mixer principle

It is the lack of broadband low-noise amplifiers for signals above $\approx 100\,\text{GHz}$ that drives the need for superconductor technology in the mm to sub-mm range of radiowaves. When a local oscillator voltage is applied to a nonlinear element, in our case an SIS tunnel junction, a periodic variation of the small signal conductance dI/dV is produced, giving rise to a number of current components at the fundamental and harmonic frequencies of the *lo*. Combining the received signal voltage at frequency f_s with the *lo* signal at f_{lo} the non-linear characteristic of the detector gives a product response at Fourier component frequencies $f_m = |mf_{lo} + f_{if}|$ where $f_{if} = |f_s - f_{lo}|$ is the so-called intermediate frequency signal, or *if* signal for short. This means that power is transferred to sum and difference frequencies of the two signals that are mixed in the SIS non-linear device. The situation is illustrated in Figure 11.20. From this new spectrum of frequencies the important signal is the one whose frequency has been downconverted to the range accessible to conventional low-noise amplifiers, the last stage before viewing the resulting telescope output. We will briefly outline a simplified mathematical description of the mixing process. The current through the nonlinear element can be written as a power expansion in the voltage V:

$$I(V) = a_1 V + a_2 V^2 + a_3 V^3 + \cdots \qquad (11.66)$$

The input signal consist of three parts: (1) the DC bias voltage V_0, which places the device at the correct operating point on the $I-V$ curve, here the gap voltage V_g, (2) the *lo*-voltage $v_{lo} = V_{lo}\cos(\omega_{lo}t)$, and (3) the weak signal we want to retrieve, $v_s = V_s\cos(\omega_s t)$ where $\omega = 2\pi f$. When these voltages are inserted into the expansion, Eq. 11.66, the resulting terms can be gathered according to their common coefficients a_i. The sum of terms with the coefficient a_1 is a scaled replica of the sum of all inputs. This is not what we are searching. Rather, the term we are interested in is found in the sum with the a_2 coefficient:

$$2V_{lo}V_s\cos\omega_{lo}t\cos\omega_s t = V_{lo}V_s\{\cos(\omega_{lo} - \omega_s)t + \cos(\omega_{lo} + \omega)t\} \qquad (11.67)$$

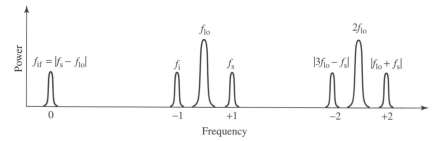

Figure 11.20 Power spectrum for a mixer with the local oscillator (*lo*) and the signal frequencies applied. Note also the intermediate and image frequencies.

Here, the down-converted *if* voltage at $f_{if} = |f_s - f_{lo}|$ is present, as well as a correspondingly up-converted one. By adjusting the local oscillator frequency ω_{lo} various received signals can now all be converted to the same *if* frequency and amplified at the same frequency. The point is here that a single state-of-the art *if* amplifier optimized just for one frequency can handle a wide range of signal frequencies. A sweep of *lo* frequencies will give access to a range of received signal frequencies.

The SIS junction has a pronounced microwave response, as illustrated in Figure 11.21, where the characteristics are shown both with and without *lo*-power applied. At low-frequency radiation, the $I-V$ curve is smeared at the gap singularity, the current increases at sub-gap voltages, and the DC Josephson current is depressed. Photon induced steps appear as the rf frequency is increased. These are of two kinds. Figure 11.21 (b) indicates photon-induced steps in the quasiparticle current at voltages $2\Delta/e \pm mhf_{lo}/e$. They occur as photons are absorbed (or emitted) in the tunnelling process, thus allowing electrons to tunnel into the high density of states above the energy gap. The photon-induced structure increases in strength with microwave power until superconductivity is quenched. The other kind of steps (Figure 11.21(d)), is due to Josephson mixing products falling at zero intermediate frequency for bias voltages corresponding to multiples of the applied frequency ($2eV_k/h = f_J = kf_{lo}$). The amplitude of

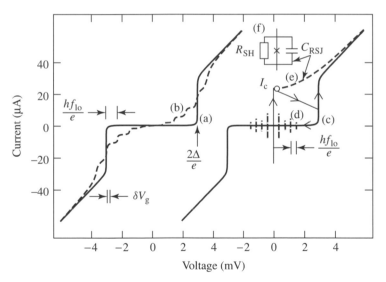

Figure 11.21 Typical $I-V$ curves for SIS tunnel junctions. For a low critical current junction (or with an applied magnetic field) the left-hand curves show the $I-V$ curve without (a) and with (b) *lo* applied ($\sim$240 GHz). The right hand curves show how the Josephson effect changes the $I-V$ curves without (c) and with (d) *lo* applied ($\sim$240 GHz). Outlined in the figure is the $I-V$ curve of a resistively shunted tunnel junction (RSJ), and the corresponding electrical circuit (e resp. f). A Josephson mixer could also be biased at the open circle. After Winkler *et al.* [166].

these Shapiro steps vary as Bessel functions, $J_n(\alpha_1)$, with normalized microwave amplitude $\alpha_1 = 2eV_{lo}/hf_{lo}$.

Josephson junctions can be used in a number of applications, based on the effects that relate current and voltage via the phase difference across the junction. Some prominent examples are high frequency oscillators, voltage standards, parametric amplifiers, Josephson video detectors and mixers.[4]

---

[4]Further recommended reading may be found in the review by Winkler *et al.* [166].

12

Superconducting Wire and Cable Technology

12.1 Low-T_c wire and cable

12.1.1 Introductory remarks

Technical superconductors for energy and magnet applications[1] consist of composite wires of superconducting and normal conducting material, both in high-T_c (HTS) or low-T_c (LTS) applications. To stabilize the superconducting state the superconductor is subdivided into a multi-filamentary structure embedded in low resistivity normal conducting matrix. As far as LTS applications is concerned, only the intermetallic stoichiometric compound Nb_3Sn and the alloy NbTi are important. In both of these the matrix consists of low resistance Cu matrix. Filament diameters range from a few μm to about 100 μm, and the filament number can be as high as 100 000. Wire diameters are between 0.1 mm and 2 mm, and the current capacity range is from a few ampere to 1000 A in fields up to 20 T in the liquid helium range at or below 4.2 K. The fabrication process includes bundling of components into a billet, with subsequent extrusion and drawing. Conductors for high current use are produced in the form of fully transposed cables. Depending on the design, additional stabilizer material of copper and aluminium, as well as strengthening members and cooling channels may be added.

Wires of ceramic high-T_c cuprates are mostly manufactured by the powder-in-tube method. Here the leading first generation technology is based on the development and use of BiSrCaCuO/Ag wires. These can tolerate magnetic

---

[1]Based largely on the review by Krauth in Fossheim: *Superconducting Technology. 10 case studies.* [167] For further reading we recommend Buckel: *Supraleitung* [168], Wilson *Superconducting magnets* [169], R. P. Reed and Clark *Materials at Low Temperature* [170], Collings *Applied Superconductivity, Metallurgy and Physics of Titanium Alloys* [171] and Foner and Schwartz: *Superconductor Material Science* [172].

Superconductivity: Physics and Applications   Kristian Fossheim and Asle Sudbø
© 2004 John Wiley & Sons, Ltd   ISBN 0-470-84452-3

fields up to more than 20 T at 4.2 K. Even at 77 K the quality of these conductors is becoming interesting for some applications, as we will give examples of in the following. A second generation of superconductor tapes, manufactured by film deposition on a suitable substrate is emerging, and should be expected to take over the market.

12.1.2 General design considerations

The primary characteristics in the design and fabrication of technical superconductors may be broken down into three groups:

1. Superconductor properties

    ☐ Critical temperature T_c
    ☐ Upper critical field B_{c2}
    ☐ Critical current density J_c

2. Design of composite wire: Superconductor filaments in normal matrix

    ☐ Stability against transition to normal state (quenching)
    ☐ AC losses
    ☐ Mechanical characteristics

3. Fabrication technology

    ☐ Fabrication of multi-filamentary composites
    ☐ Fabrication of high current conductor cables with additional stabilization, mechanical reinforcement and cooling channels
    ☐ Compatibility of conductor properties with magnet fabrication technologies.

In the following we will discuss these three areas of wire and cable technology in turn, and give examples of results obtained.

12.1.3 Basic superconductor properties

The transition from normal-conducting to superconducting state is determined by three critical parameters, the critical temperature T_c, the upper critical field B_{c2}, and the critical current density J_c. The first two are determined by the electronic properties, including the pairing mechanism for Cooper pair formation. In typical low-T_c materials of technical importance these parameters cannot be manipulated to any important extent. In high-T_c cuprate materials they can be manipulated first by choice of material composition, of which there are now a great number, and next by doping of the chosen material. However, in reality only optimally

Table 12.1   Development of the highest known T_c as a function of time. T_c and B_{c2} values of some prominent materials

Superconductor	T_c	B_{c2} (4.2 K)
Nb	9.2 K	0.27 T
NbZr	10.8 K	11 T
NbTi	10.2 K	11 T
V_3Ga	16.5 K	22 T
Nb_3Sn	18 K	22 T
$(Nb, Ta)_3Sn$	18 K	25 T
Nb_3Ge	23 K	30 T
$PbMo_6S_8$	14 K	45 T
$Y_1Ba_2Cu_3O_7$	90 K	Highly anisotropic
$Bi_2Sr_2Ca_1Cu_2O_8$	90 K	$\geq$50 T, $B\|$Cu-O-layers
$Bi_2Sr_2Ca_2Cu_3O_{10}$	110 K	$\geq$200 T, $B \perp$Cu-O-layers
$Tl_2Ba_2Ca_2Cu_3O_{10}$	125 K	

doped materials, to the highest achievable T_c are technically interesting. For values of T_c and B_{c2} in important materials, we refer to Table 12.1.

The critical current density J_c is strongly dependent on temperature and magnetic field in all classes of materials. In high-T_c materials J_c is in addition, strongly anisotropic, and depends critically on grain boundary alignment (texturing) and on successful introduction of pinning centres. Clearly, these aspects bring processing technology to the forefront as the key to achieving successful material control during manufacture of wires and cables. Optimization of J_c is a demanding task in all bulk superconductor applications, but specially so in the high-T_c cuprates. The great potential advantages of high-T_c materials are their high critical temperature and high upper critical fields. These advantages are easily lost unless extremely careful processing methods are developed to meet the challenges mentioned. In recent years much progress has been made in making multi-filamentary wires even of high-T_c superconductors, especially Bi2223 and Bi2212, by the powder-in-tube method. A new generation of high-T_c wires and cables is under development by superconductor film deposition on non-superconducting substrates with matching lattice constant. The ultimate success of high-T_c applications in the area of wires and cables will depend crucially on the outcome of this extremely demanding processing development.

Only the solid solution alloy NbTi- and the Nb_3Sn-based binary intermetallic compound have so far reached the stage of an industrially produced commercial superconductor. The optimum composition of NbTi contains about 47 to 50 wt. per cent Ti. The exact critical parameters of Nb_3Sn depend on the composition and the perfection of the superconducting A15 phase. In ternary compounds like $(NbTa)_3Sn$ or $(NbTi)_3Sn$ B_{c2} can be enhanced without sacrificing too much in T_c and workability. Adding 7.5 wt. per cent Ta to the Nb results in a B_{c2} shift of about 3 T. According to the B_{c2} values NbTi conductors can be used up to field levels of 10 T and Nb_3Sn-based conductors up to about 20 T. Possible candidates for applications beyond 20 T are Chevrel phases, like $PbMo_6S_8$, or HTS like $Bi_2Sr_2Ca_2Cu_3O_{10}$ and $Bi_2Sr_2Ca_1Cu_2O_8$ when operated at low temperature. Due to their crystal structure with the superconductivity mainly in the CuO_2 planes, the behaviour of HTS in a magnetic field is highly anisotropic, leading to much higher B_{c2} values when the magnetic field is parallel to the CuO_2 planes as compared with the field perpendicular to these planes.

A high critical current density J_c at operational field is a precondition for most applications. Usually J_c values of 10^5 to 10^6 A/cm^2 are desirable. A lower limit is 10^4 A/cm^2. A type II superconductor is penetrated by the magnetic field in form of quantized flux lines. Only the volume outside the flux line remains superconducting. B_{c2} is reached when the flux lines fill up the total volume. An ideal type II superconductor can not carry a superconducting current because the flux lines move by the Lorentz force. This movements leads to electric fields and therefore to losses. The superconductor must therefore be hardened by microstructural features acting as pinning centres for flux lines. Effective pinnings centers can be for instance precipitates of normal conducting phase, like in NbTi, or grain boundaries as in Nb_3Sn. But grain boundaries may at the same time act as weak links, thus impeding the current flow from one grain to its neighbour. Thus it is possible to have high intragrain J_c, yet at the same time low intergrain J_c in the same material, giving low overall transport critical current density. High-T_c materials are well known for such problems, among them YBCO.

As we have discussed in Chapter 8 the critical current density J_c is defined at the point where the Lorenz force begins to exceed the pinning force, i.e. at the onset of *flux flow*. In this regime, above J_c, current transport is no longer lossless. But even at low current density another loss mechanism is active. This is the *flux creep* process. The presence of such effects is seen in the decay of magnetization with time. The creep effect is usually very small in the liquid helium range, but becomes more and more significant as the temperature is increased. This is a well-known problem in all high-T_c materials operating at elevated temperatures up to 77 K or higher. We refer to the Topical Contribution by Muralidhar and Murakami (Section 13.4) showing how a sophisticated interplay between composition and processing can improve pinning up to 77 K in YBCO-type high-T_c materials. Another important phenomenon related to flux penetration and flux pinning is so-called *flux jump*. This phenomenon, which can lead to severe instabilities, may be suppressed by appropriate design.

Due to inhomogeneities of the material the onset of flux flow is not a sharp criterion for determination of critical current. A practical criterion is therefore used: J_c is defined as the current density at the point where a 1 µV/cm electric field is observed across the sample in the transport direction. In small laboratory scale experiments the criterion is often set simply as 1 µV. A corresponding criterion using resistivity measurements is to identify the current where the resistivity reaches 10^{-14} to 10^{-13} Ω m. Often the transition from superconducting to

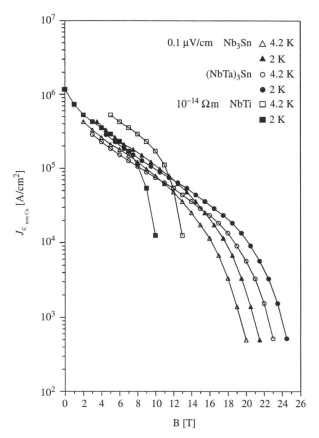

Figure 12.1 Critical current density as a function of magnetic field and temperature for metallic superconductors: NbTi, Nb_3Sn, $(NbTa)_3Sn$ at 4.2 K and 1.8 K, respectively. In case of Nb_3Sn-based conductors the bronze area is included in the calculation of J_c.

normal transport can be described by a power law, as discussed in Chapter 8, i.e. $E \propto I^n$. The higher the exponent n, the higher quality of a technical superconductor. A high value of the exponent n reflects uniform pinning and generally uniform material conditions. In Figure 12.1 J_c values of commercial grade NbTi, Nb_3Sn, and $(NbTa)_3Sn$ are shown as a function of magnetic field at 4.2 K and 1.8 K. In Nb_3Sn the conductors were produced by the bronze route, as will be explained below. In this case the bronze area is included in the calculation of J_c. An estimate of J_c in the superconductor is obtained by multiplying the J_c value by a factor of 3.

12.1.4 Design of technical superconductors

The starting point for the mechanical design of a technical superconductor is its behaviour in a magnetic field. This field may be of external or of internal origin (self-field). According to the Bean model the full penetration field B_p, which is the maximum field that can be shielded is, as shown in Chapter 8

$$B_p = 2\mu_0 J_c a \tag{12.1}$$

in a slab of thickness $2a$. We have referred to this condition as the critical state. Figure 12.2 and the discussion of the Bean model in Chapter 8 should be consulted for a better understanding of how this model describes flux penetration under different external fields B_m. In the Bean model, as we have

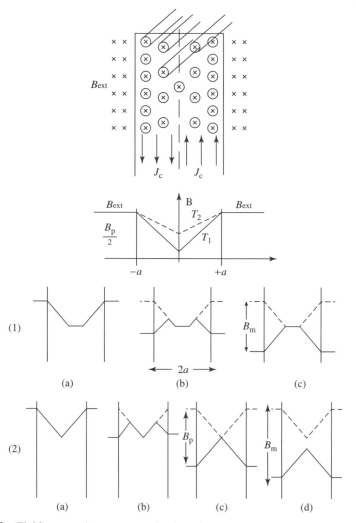

Figure 12.2 Field pattern in a superconducting slab according to the Bean model. Upper: field and current directions and field patterns for two temperatures $T_2 > T_1$, $(J_c(T_2) < J(T_1))$. Lower: Field pattern for a time varying field with amplitude $B_m/2$: (1) $B_m < B_p$, (2) $B_m > B_p$.

seen, J_c is defined by the field gradient, which is taken to be constant, i.e.

$$\frac{\mathrm{d}B}{\mathrm{d}x} = \mu_0 J_c \qquad (12.2)$$

12.1.5 Stabilization

The Bean model does not say anything about stability. In reality there is always a decay of currents. This means that the field profile will flatten out, or 'thermalize', given sufficient time. The time taken for this process can vary by many orders of magnitude depending on the controlling factors for flux creep: temperature and pinning barriers. In addition, strong external disturbances can cause flux jumps with large energy dissipation, which can lead to quench of the superconductor. Such severe disturbances can be of magnetic, thermal, or mechanical nature leading to a positive feedback. Functional breakdown events due to such effects may be avoided by appropriate prevention measures. (i) Reduction of available magnetic energy by fine subdivision of the superconductor volume so that the heat capacity of the superconductor and its nearest surrounding matrix is high enough to limit the temperature increase ΔT following the flux jump event. The technical term for this is *adiabatic stabilization*. (ii) Damping of the rate of flux change by highly conducting surrounding normal conducting matrix, Cu or Al. This ensures both a reduced heat generation and efficient heat removal to the coolant. The technical term for this is *dynamic stabilization*.

The subdivision required to achieve these goals can be calculated by accounting for both energy and power balance, respectively. In case a magnetic field is perpendicular to the wire axis, the adiabatic stabilization requirement on the superconducting filament diameter d is

$$d < \frac{2}{J_c} \left(\frac{C \Delta T}{\mu_0} \right)^{\frac{1}{2}} \qquad (12.3)$$

where C is the volumetric specific heat, and $\Delta T = \frac{-J_c}{\mathrm{d}J_c/\mathrm{d}T}$. Usually J_c decreases linearly with temperature so that $\Delta T = T_c(B_{op}) - T_{op}$, where B_{op} and T_{op} are operational field and temperature, respectively. For typical parameter values of NbTi at 4.2 K this results in $d < 100\,\mu$m filament diameter. When this criterion is used at 77 K for HTS systems one finds that much larger diameters are allowed due to the far higher heat capacity of both the superconductor and the matrix. Obviously, such thin filaments have a quite low current carrying capacity. Therefore a large number of them may be needed, bundled together and embedded in a normal conducting matrix. If this matrix is electrically and thermally highly conductive, one can achieve dynamic stabilization according to the above analysis. To avoid an unacceptable temperature increase inside a filament, the following criterion must be fulfilled to prevent flux jumps:

$$d < \left(\frac{8k_c \Delta T \alpha}{\rho_{st}} \right)^{\frac{1}{2}} \qquad (12.4)$$

where k_c is the thermal conductivity of the superconducting material, ρ_{st} is the resistivity of the stabilizer material, and α is the stabilizer to superconductor ratio. For NbTi in a highly conductive Cu-matrix this requires d less than 80 µm. The fact that the two latter criteria agree so well with respect to the requirement on the diameter d is purely coincidental.

Unfortunately, the presence of a highly conductive matrix between the superconducting filaments, leads to new affects and possible new instabilities. In a transverse time varying magnetic field parallel filaments are electrically coupled by the matrix, as shown in Figure 12.3. Below a critical length l_c, the induced supercurrents will partly pass through both the matrix and the filaments. Above l_c, on the contrary, the current loops are closed inside the matrix. The critical length l_c depends on the rate of field change dB/dt, and is given by

$$l_c = 2 \left(\frac{d \rho_{st} J_c}{dB/dt} \right)^{\frac{1}{2}} \qquad (12.5)$$

This coupling phenomenon, and instabilities related to it, can be substantially reduced by twisting of the multi-filamentary wire with a twist pitch $l_p \ll l_c$. Twisting does not, however, prevent coupling of filaments in the self-field of the transport current in the wire. To avoid self-field instabilities the wire diameter d should be limited to about 2–3 mm. Conductors with high current carrying capacity are therefore built up from several wires, called strands, in the form of transposed cables or braids. By designing a transposed conductor in such a way that all strands occupy the same position in the cross section within a 'transposition length' l_t self-field instability can be avoided.

The methods described above are important for intrinsic stabilization of the superconducting state. However, in case of large external disturbances, e.g. by

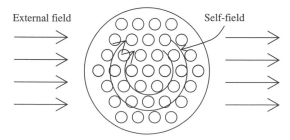

Figure 12.3 Cross-section of a multifilamentary conductor together with field directions of a transverse external field and conductor self field. After Krauth in Fossheim [167].

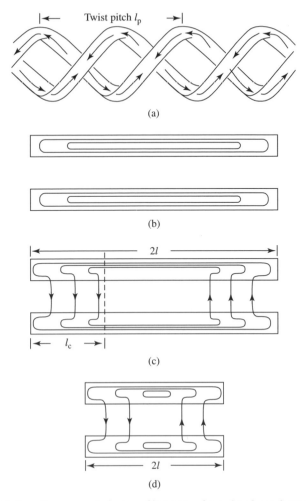

Figure 12.4 (a) Coupling currents in two filaments of a twisted conductor. Partial decoupling occurs when $l_p \ll 2l_c$. (b) Uncoupled filaments ($B = 0$ or $\rho_{ST} \to \infty$). (c) Coupling currents between two parallel filaments of length $2l \gg 2l_c$. (d) Coupling currents for a conductor length $2l < 2l_c$. After Krauth in Fossheim [167].

mechanical movement they may not be sufficient in large systems like magnets. Full stability may then be achieved by adequate cooling so that the superconductor recovers from a localized transition event to the normal state. Achieving such *cryogenic stability* requires that the capability for heat transfer to the cryogenic coolant is larger than the Joule heating in the normal conducting state. This can be expressed as

$$I^2 \frac{\rho_{st}}{A_{st}} < Pq \tag{12.6}$$

where I is the conductor current, A_{st} is the stabilizer area, P is the cooled perimeter, and q is the heat transfer to the coolant per unit conductor surface area. This criterion requires large amounts of stabilizer, $\alpha = 10-30$, and efficient cooling by large surface areas and good heat transfer. As a consequence, overall critical current densities in cryogenically stabilized conductors are limited to rather low values, a few times 10^3 A/cm^2. For this reason, additional criteria have been developed requiring smaller amount of stabilizers. We refer here to concepts like cold end recovery, and the minimum propagating zone.

12.1.6   AC losses

Time-varying magnetic fields lead to different types of losses in filamentary superconductors. The most important ones are hysteretic losses, and coupling losses which we already mentioned. Hysteretic losses occur during cycling of the magnetization curve of the superconducting filaments. The loss per field cycle can be calculated from the area of the hysteresis loop as

$$Q_h = \oint M \, dB \tag{12.7}$$

The resulting expressions depends on whether the field amplitude B_m is smaller or larger than the penetration field B_p as given in Eq. 12.1. In the Bean model J_c is constant within the range of the field profiles and one finds for the hysteresis losses per cycle when $\frac{B_m}{B_p} < 1$

$$Q_h = \frac{B_m^2}{2\mu_0} \frac{1}{3} \frac{B_m}{B_p} \tag{12.8}$$

while for $\frac{B_m}{B_p} > 1$ the result is

$$Q_h = \frac{B_m^2}{2\mu_0} \left(\frac{B_p}{B_m} - \frac{2}{3} \left(\frac{B_p}{B_m} \right)^2 \right) \tag{12.9}$$

For $\frac{B_m}{B_p} \gg 1$ this gives, when combined with the Bean equation (Eq. 12.1)

$$Q_h = B_m J_c a n \tag{12.10}$$

The demagnetization factor n has been included here. For circular filaments in a transverse field configuration $n = 2$. As can be seen from Eq. 12.10 magnetization and hysteresis losses can only be kept small by using fine filaments. For most applications $d = 5-10\,\mu m$ is sufficient, but for 50 Hz applications sub-micrometre filaments are required.

The coupling losses Q_c can be estimated for slow field changes, i.e. for $T_m \gg \tau$, according to

$$Q_c = \frac{B_m^2}{2\mu_0} \frac{8\tau}{T_m} \tag{12.11}$$

where T_m is the field rise time and τ is the conductor time constant given by

$$\tau = \frac{\mu_0}{8\pi^2} \frac{l_p^2}{\rho_{tr}} \qquad (12.12)$$

Here l_p is the twist pitch (or transposition length l_t) as before, and ρ_{tr} is the effective transverse resistivity of the matrix (or of the cable or braid). Since l_p typically is limited to values larger than about 10 times the respective diameter if each stage (strand or cable stage), the only means to reduce the coupling losses further is by introduction of resistive barriers, for instance CuNi, into the composite.

12.1.7   Mechanical characteristics

Important design criteria are related to mechanical loads that a superconductor has to accommodate during building and operation, i.e. during coil winding, cool-down, operation, and warm-up. The stresses during operation are dominated by magnetic forces. Typically the stress level is given by

$$\sigma = JBR \qquad (12.13)$$

where J is the overall current density, B is the magnetic field, and R is the coil radius. The conductor must withstand these stresses without mechanical damage and without degradation of the critical current density. NbTi conductors show little degradation of J_c as a function of stress and strain. Nb$_3$Sn-based conductors are much more sensitive to strain, especially in high magnetic fields. Also, Nb$_3$Sn conductors are very brittle after reaction (discussed below) and allow only strain levels below 0.3% for safe operation.

12.1.8   Fabrication technology

As described above, a technical superconductor consists of a composite of superconducting filaments and a stabilizing matrix, with additional optional components. A suitable fabrication process must be capable of producing long length of composite conductors in a reliable and economic way. The process must allow to optimize the superconducting properties, especially the current carrying capacity. The completed composite wire must be processable into more complex conductor geometries, like cables and magnet coils. In spite of the fact that many superconducting materials are known, the complexity of all these requirements has so far resulted in only two low-T_c materials of technical importance for bulk applications. These are the ductile alloy NbTi and the brittle intermetallic compound Nb$_3$Sn. Due to its difficult handling Nb$_3$Sn is only used at high fields (greater than 10 T) where the current carrying capacity of NbTi decreases very strongly.

The fabrication of a multi-filamentary wire is shown schematically in Figure 12.5. In the case of NbTi/Cu composite, prefabricated NbTi and Cu parts are bundled inside a Cu-tube to form a billet with an outer diameter of typically 100–200 mm. The billet is evacuated and warm extruded to a rod of 20–70 mm diameter. Further area reduction is achieved by cold working on drawing benches, bull blocs, and multiple drawing machines. Twisting of the wire is performed close to final dimension. The number of NbTi filaments and the copper to superconductor ratio is chosen according to the application.

The composite wire can be further processed into rectangular and other profiles and cables required for special applications. Co-processing of NbTi and Cu is possible because both materials are quite ductile and are similar with respect to hardness and resistance to deformation. With other, less favourable materials combinations other solutions had to be found. In some cases stabilization with ultrapure aluminum is desirable due to very low residual resistance. A process to achieve Al-stabilized conductor is by co-extrusion of Al onto a prefabricated NbTi/Cu composite wire or cable as shown in Figure 12.6. In Figure 12.7 some examples of NbTi conductors may be seen.

The brittle Nb_3Sn compound cannot be processed by either coldworking or warmworking. The solution is to use a composite of ductile precursor materials and to develop the superconducting phase after the deformation process. The most reliable method is by the bronze route where Nb rods are embedded in a CuSn bronze, and the Nb_3Sn is formed by solid sate diffusion and reaction

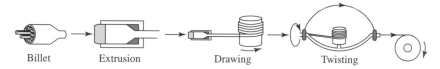

Figure 12.5 Fabrication of a multifilamentary conductor. Conductors with a very large number of filaments ($>10\,000$) are produced by a two stage bundling and extrusion.

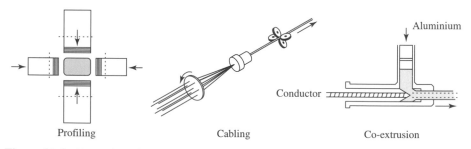

Figure 12.6 Examples of processing of multifilamentary wires for special applications (rectangular profiling, cabling, co-extrusion with aluminium).

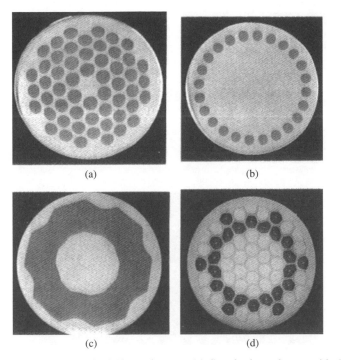

Figure 12.7 Configuration of NbTi conductors. (a) Standard conductor with 54 filaments and a Cu:NbTi ratio $\alpha = 1.35$ for Nuclear Magnetic Resonance (NMR) and laboratory magnets. Typical wire diameter 0.7 mm. (b) Conductor with 24 filaments and $\alpha = 6.5$ for whole body magnetic resonance imaging (MRI) magnets. Typical wire diameter 1 to 2 mm. (c) Conductor with 6000 filaments with 5 µm diameter at a wire diameter of 0.65 mm for accelerators with low magnetization requirements for the filaments. (d) Conductor with 1200 filaments with 10 µm diameter and mixed Cu/CuNi matrix with low hysteresis and coupling losses e.g. for pulsed field applications. After Krauth in Fossheim [167].

process at about 700 °C. This is the reason why the Nb$_3$Sn configuration differs from that of the NbTi conductor. To allow effective transformation of the Nb into Nb$_3$Sn within a reasonable time, in practice a few days, the filament diameter is limited to about 3 to 5 µm. Consequently, a very high number of filaments is required, typically 10 000, such that the bundling and deformation process has to be performed twice. In addition, the CuSn matrix does not stabilize and protect the conductor, so that additional copper has to be added, preferentially in the second bundling process. The pure Cu has to be protected from poisoning by Sn during the reaction heat treatment by means of a diffusion barrier, for instance of Ta. Figure 12.8 shows examples of resulting multi-filamentary wires.

Optimization of critical current density is part of the conductor fabrication process. It is determined by the microstructure, and can therefore be controlled during the fabrication process. Optimization of J_c is achieved by introduction of finely dispersed microstructural features acting as pinning centres. In the case of NbTi normal conducting α-Ti precipitates in the superconducting β-phase are

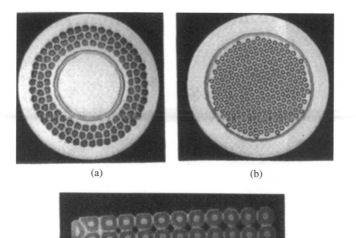

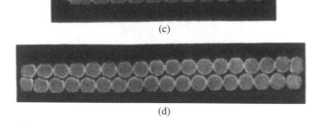

Figure 12.8   Configuration of Nb₃Sn conductors with Ta diffusion barriers. (a) Standard conductor with 6000 filaments in a bronze matrix and internal Cu stabilization. (b) Conductor with 20 000 filaments and external Cu stabilization. Fully transposed flat (keystoned) cables for accelerator magnets (dipoles and quadrupoles). (c) 23 NbTi strands with 0.84 mm diameter. (d) 36 Nb₃Sn strands with 0.92 mm diameter. After Krauth in Fossheim [167].

known as the most effective pinning centres. In the case of Nb₃Sn the grain boundaries act as pinning centres. A microstructure with fine grains is therefore required to get high J_c-values. This is achieved by performing the reaction heat treatment at about 700 °C for three days. The reaction treatment can be performed either before the coil winding (react-and-wind) or after coil winding (wind-and-react). The latter is mostly used for small magnets, and the former for large magnets with low bending strain during coil winding.

12.2   High-T_c wire and cable

12.2.1   High-T_c wire and tape

Successful development of high-T_c wire as a basis for high current transport was so far achieved by use of Bi2223, and to some extent using Bi2212 material. The dominating methodology is to produce wires by the powder-in-tube (PIT) technique. This is done by filling the pre-reacted powder into silver a tube. Its dimension will determine how long length can eventually be produced. The resulting filled tube is extruded and drawn to several times its original length. Continued drawing leads to a single filamentary wire which may be rolled to the dimension of about 3 mm width and 0.3 mm thickness. After processing at temperatures approaching 900 °C in oxygen atmosphere, and subsequent cooling to room temperatures at varying rates, a superconducting wire results. In the next stage multi-filamentary wires of 100-odd superconducting tape wires of kilometre lengths is produced. Production of such wires involves a repetition of the drawing stage in which the first extruded and partially drawn tube is cut into short lengths of metre size, and packed into a second Ag tube. The drawing and rolling process is now repeated, resulting in a multi-filamentary tape shaped wire. Thermal processing is similar to the single filament case. To illustrate the sophistication required to develop second generation of conductors of the quality required on a large scale, we refer to the Topical Contribution, by Thompson and Christen (Section 13.9).

12.2.2 Full-scale high-T_c cable

The technology for high-T_c cable manufacturing is maturing gradually. Several projects are under way. Full scale cables have so far been manufactured from Bi2223 tapes only. The next generation of tapes based on deposition of YBCO film is expected to provide great improvements in the long run. The speed of the manufacturing process is still far too low for this next generation of tapes to be applicable in cables.

A good example of high-T_c cable implementation using Bi2223 tapes is the Danish HTS power cable project [173], the first HTS cable operated in a public network. In 2001 installation of a 30-m, $30\,kV_{rms}$, and 2000 A three-phase HTS cable was carried out in a substation of the Copenhagen utility network. One year later, the operating experiences were deemed satisfactory. The cable supplied electric energy corresponding to the average consumption of about 25 000 households, over 100 000 MWh of energy.

The advantages seen for a HTS power cable may be listed as follows

- High current and power rating

- Compact cable dimensions and low weight (about 1/3 of conventional cable)

- Low losses

- No thermal or magnetic interaction with surroundings.

HTS power cables have a potential for reduced cost per transferred unit of energy compared to conventional technology due to low material consumption and potentially lower installment cost. The cables may be used as short internal

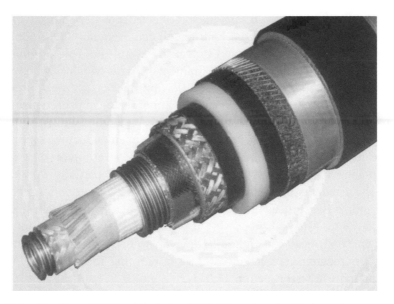

Figure 12.9   The Danish HTS cable design [173]. Reproduced with the authors' permission.

substation connections and retrofits (replacement) of low capacity conventional cables in existing ducts. They may further be applied for urban feeder cables, bulk power transmission, low-voltage high-current cables, DC power, etc.

The design of the Danish HTS cable is shown in Figure 12.9. The superconducting tapes are wound in a helical pattern on the inner corrugated flexible steel tube. The superconductor is cooled by liquid nitrogen flowing through the steel tube at a rate of 700 kg/hour at 4 bar. The HTS tapes are nowhere in direct contact with liquid nitrogen. The main structure from inner to outer layers is briefly characterized by the following sequence: The inner tube, called the 'former', on which the superconducting tapes are wound, has an outer diameter of 30 mm. Then follow first the superconductor tape layers, next a thermal insulation layer, an electrical insulation layer, an electrical Cu-screen, and finally a polyethylene sheath of ID/OD dimensions 95/105 mm. The critical current I_c was measured as 2770 A at 79 K, and 5000 A at 66 K. No degradation was detected during one year of operation.

12.2.3   HTS induction heater

Electromagnetic induction heating is commonly used for industrial heating of metals. Figure 12.10 illustrates the principle involved in induction heating [174]. A metal workpiece is placed inside a solenoidal coil, and when an AC current is passed through the windings, the time-varying magnetic field generated by the coil induces electric currents in the metal, causing resistive dissipation and

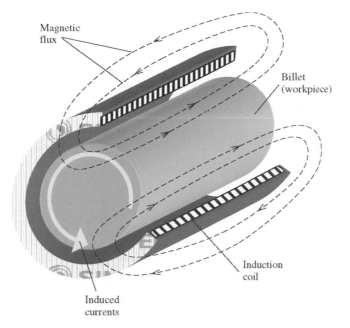

Figure 12.10 Principles of billet heating using a superconducting coil [174]. Reproduced with the authors' permission.

heating of the workpiece. For aluminum extrusion the typical billets used for extrusion are of about 20 cm diameter and 1–2 m length. The billet is heated to temperatures of around 500 °C in about 2 min to soften the metal just before extrusion takes place.

Runde and coworkers [174] have introduced a new concept in induction heating by use of high-temperature superconductor (HTS) windings in the induction coil. The great advantage is that the losses in the coil can be reduced to a small fraction of those in conventional coils. A simple analysis shows that the overall electrical efficiency η may be expressed as

$$\eta = \frac{1}{1 + \sqrt{\dfrac{\rho_c}{\rho_w \mu_w}}} \tag{12.14}$$

where ρ_c and ρ_w are the resistivities of the coil and the workpiece, respectively, and μ_w is the permeability of the workpiece. From Eq. 12.14 we observe that if the resistivities of the workpiece and the coil are similar, the efficiency approaches 1/2. However, if the resistivity in the coil is eliminated, the efficiency approaches 1, or 100%. These are of course ideal values, but the effect is in any case very significant. A demonstrator HTS induction heater has been shown to work as expected. The coils were of so-called stacked pancake design made of stainless steel reinforced multi-filamentary Bi-2223/Ag tape with a self-field

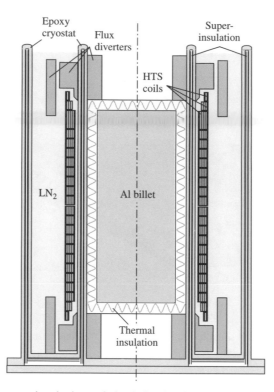

Figure 12.11 Cross-sectional view of the induction heater. The heater is concentrically built up around the dash-dotted symmetry line. The current leads and various structural parts made in fibreglass reinforced epoxy are not included in the drawing [174]. Reproduced with the authors' permission.

critical current of 115 A. Figure 12.11 shows the design used. The reduced scale demonstrator prototype reached 500 °C in about 5 minutes. In the original tests, DC HTS tapes were used. With AC tapes one would expect to be able to apply higher currents, and reduce the heating period substantially. An important feature of the design is the insertion of iron rings at each end of the coil. These are placed there to prevent solenoidal stray field from taking a short route through the coil near its ends. This would have allowed some field to cross the pancake coil windings normal to the tape, a field configuration which has a strongly detrimental effect on the critical current in Bi2223 tapes.

12.3 Magnet technology

Superconducting magnet technology is a mature field today, and well described in textbooks. In the present context we have found it interesting to present two outstanding unconventional examples of recent development in superconductor magnet applications. The frist one is a description of a fullscale complex magnet system, the worlds strongest, 45 T, stationary field hybrid magnet, built and operated by the National High Magnetic Field Laboratory at Florida Sate University in Tallahassee, Florida. For this outstanding example of magnet technology we refer to the Topical Contribution by Schneider-Muntau in Section 13.5. The second example is a pure materials development which has been carried out at the Superconductivity Research Laboratory at ISTEC in Tokyo. Murakami has led the efforts to tailor the pinning properties of YBCO-type cuprate superconductors for storage of truly high magnetic field. We refer to the topical contribution in by Muralidhar and Murakami (Section 13.4) on this subject.

PART IV

Topical Contributions

13

Topical Contributions

This Chapter contains nine invited contributions written for this book by particularly outstanding scientists from Europe, Japan and USA in various fields of superconductor research. The authors have been asked to write their contributions in as short form as seemed pedagogically possible. These examples highlight specific subjects at the forefront of research and development at the time of writing this textbook. Scientific references belonging to this chapter are given at the end of the chapter.

13.1 Spin-Triplet superconductivity by Y. Maeno

What is the spin triplet superconductivity?

When two electrons form a pair in the superconducting state, the total spin of the pair is either $S = 0$ (spin singlet) or 1 (spin triplet). Because of the symmetry relation for the permutation of electrons as Fermions, the state vector of a Cooper pair $|\Psi\rangle = \chi(k_1, k_2)|\sigma_1, \sigma_2\rangle$ must be antisymmetric with respect to the exchange of the two electrons involved. Here k_i and σ_i are the momentum and spin of an electron and χ is the orbital wave function. For spin-singlet state with antisymmetric spin part, $|\uparrow\downarrow\rangle - |\downarrow\uparrow\rangle$, the orbital wave function has to be symmetric so that the allowed orbital angular momentum L of a pair contains even-number components only: $L = 0$ is called the s-wave, $L = 2$ the d-wave, etc. It is in principle possible to form spin-triplet pairs. In this case the spin state is symmetric, $|\uparrow\uparrow\rangle$, $|\uparrow\downarrow\rangle - |\downarrow\uparrow\rangle$, or $|\downarrow\downarrow\rangle$, corresponding to $S_z = 1, 0,$ or -1, respectively. Thus the orbital wave function has to be antisymmetric with the orbital angular momentum $L =$ odd number: $L = 1$ is called the p-wave, $L = 3$ the f-wave, etc.

Conventional superconductors are in the spin-singlet s-wave state. However, a number of unconventional non-s-wave superconductors have been found. For

Superconductivity: Physics and Applications   Kristian Fossheim and Asle Sudbø
© 2004 John Wiley & Sons, Ltd   ISBN 0-470-84452-3

example, it has been shown that even the high-transition temperature copper-oxide (high-T_c cuprate) superconductors have spin-singlet electron pairs, but the pairing is in the d-wave symmetry.

Which superconductors are spin triplet?

Investigation of spin-triplet superconductivity started soon after the birth of the BCS theory [1]. It has been well established that superfluidity of ^{3}He is carried by Cooper pairs of ^{3}He atoms with their nuclear spins forming spin-triplet states [2]. However, it has not been easy to prove whether a candidate superconductor is in fact spin triplet, mainly because the Meissner effect masks the magnetic susceptibility of the electron spins.

Nevertheless, there are a number of superconductors for which the spin-triplet pairing is strongly suggested [3]. An earlier example is the heavy Fermion super-conductor UPt_3, for which there are three superconducting phases when the magnetic field and temperature are varied. Recently, a new class of supercon-ductors have been found in which superconductivity occurs below the transition temperature of ferromagnetic ordering and coexists with the ferromagnetism. A good example of this class of superconductors is UGe_2.

A layered ruthenate Sr_2RuO_4 [4] is a unique example because it is probably the only oxide for which spin-triplet nature has been confirmed so far. As shown in Figure 13.1, its crystal structure is common to the first high-T_c cuprate super-conductor discovered by Bednorz and Müller, $La_{2-x}Ba_xCuO_4$, but the T_c of Sr_2RuO_4 is 1.5 K, much lower than the corresponding high-T_c cuprate. Since the electronic properties of its normal state is characterized in detail, it is arguably the most extensively and quantitatively characterized spin-triplet superconductor [3]. From the nuclear magnetic resonance (NMR) Knight shift, the spin suscepti-bility for the magnetic field parallel to the RuO_2 layer was found to be unchanged on entering the superconducting state, marking the definitive evidence for the spin-triplet pair formation with $S_z = 0$ [5]. In addition, from the muon-spin relaxation measurements, spontaneous internal magnetic field was found to emerge below T_c [6]. This suggests that the time-reversal symmetry (TRS) is broken in the superconducting state. The origin of the broken TRS is attributed to the unquenched orbital angular momentum of the pairs; thus $L_z = +1$ or -1. Taking into consideration of consistencies with a number of other experi-mental results, as well as with theoretical requirements, it is believed that the pairing state is described as $S = 1$, $S_z = 0$; $L = 1$, $L_z = +1$ or -1. Such spin and orbital state for this p-wave superconductivity is schematically represented in Figure 13.2. The mechanism of the spin-triplet formation has not been fully understood, but it is believed by many that the strong electron correlations due to Coulomb interactions among the 4d electrons of ruthenium ions play the essen-tial role in the pairing, rather than the conventional electron-phonon interactions.

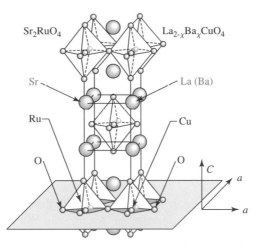

Figure 13.1 The layered crystal structure common to the spin-triplet ruthenate superconductor and high-T_c cuprate superconductor. Image is by K. Deguchi.

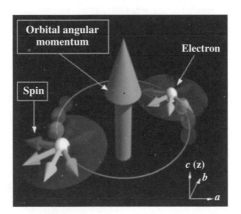

Figure 13.2 The representation of the spin triplet electron pairs in Sr_2RuO_4. Small arrows
depict the spins of an electron pair, while the large arrow represents the orbital angular
momentum. Image is by K. Deguchi.

New physics expected with spin-triplet superconductivity

For spin-triplet superconductors, not only the charge degree of freedom, but also
the spin degree of freedom exhibits "superfluidity". Thus a variety of phenomena not present in spin-singlet superconductors are expected. Because of the
internal degree of freedom of the triplet pairs, multiple superconducting phases
are excepted to emerge and they are in fact observed in Sr_2RuO_4, as well as
in UPt_3 [3]. Although not reported in any superconductor so far, induction of
collective motions of spin or orbital moments without breaking the pairs is also
expected. For Sr_2RuO_4 the *phase* of the wave function should vary with the

azimuthal angle and effects sensitive to this phase are also expected in the tunneling. Furthermore, the TRS broken state is characterized by "chirality", with all the Cooper pairs within a single superconducting domain having a common sign of their orbital moments (plus or minus). If the chirality can be externally controlled, it may be used as a quantum bit, a nanoscale unit for future quantum devices.

13.2 π-SQUIDs – realization and properties
by J. Mannhart

The predominantly $d_{x^2-y^2}$-pairing symmetry of the high-T_c superconductors influences fundamental properties of Josephson junctions and Josephson junction-based devices, such as Superconducting Quantum Interference Devices (SQUIDs). Notably, it offers the possibility to fabricate Josephson junctions which in equilibrium are biased by a phase shift of π, and dc π-SQUIDs which consist of such a π-Josephson junction and a conventional junction connected in a loop [7]. The electrical characteristics of these devices provide detailed insight into the symmetry of the superconducting order parameter, tracing for example possible admixtures of sub-dominant symmetry components. Furthermore, the devices present novel building blocks for superconducting quantum electronics [8].

All-high-T_c dc π-SQUIDs have been fabricated according to the design illustrated in Figure 13.3. A detailed description has been provided in [7]. Exploiting the bicrystal technology [9], the devices are based on two $\sim$10 μm wide, symmetric 45° [001]-tilt grain boundaries formed by $\sim$100 nm thick, c-axis oriented $YBa_2Cu_3O_{7-\delta}$-films grown on $SrTiO_3$-tetracrystals. Due to the orientation of the grain boundaries, across one of the junctions order parameter lobes with opposite signs face each other. This π-phase difference defines the π-junction. In practice the π-junction is the junction with the smaller critical current, because it has the smaller Josephson coupling energy.

The current-voltage, $I(V)$-characteristics of the π-SQUIDs follow the behavior expected according to the resistively shunted junction model, with additional self-induced resonances which disappear for small applied magnetic fields, as shown in Figure 13.4 [10]. The magnetic field dependencies of the π-SQUIDs critical currents $I_c(H)$ (see Figure 13.5) are characterized by a minimum at small fields, exactly opposite to the behavior of the standard SQUIDs' critical currents, which for vanishing applied fields are maximal. This characteristic effect is expected for superconductors with a pure $d_{x^2-y^2}$-symmetry. The $I_c(H)$-dependencies for larger applied magnetic fields are highly symmetric, proving that the measurements are not influenced by spurious trapped magnetic flux [7].

The self-induced resonances shown by the π-SQUIDs' $I(V)$-characteristics reveal that in the voltage state the $d_{x^2-y^2}$-wave symmetry induces circulating

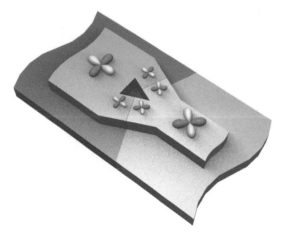

Figure 13.3 Sketch of a π-SQUID fabricated by using the bicrystal technology (after [7]).

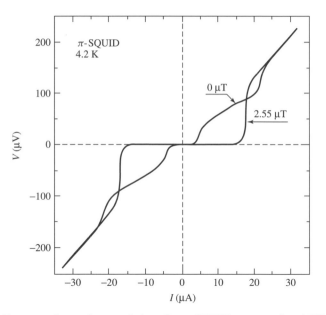

Figure 13.4 Current-voltage characteristics of a π-SQUID measured at 4.2 K at zero field and at the smallest magnetic field that causes a maximum of I_c (from [10]).

ac-currents, which oscillate with the Josephson frequency in the π-ring configuration. This provides evidence that the π-shift as well as the order parameter symmetry are maintained for Josephson frequencies up to several tens of GHz [10].

The technology to fabricate π-Josephson junctions and π-SQUIDs is extendable to the design of circuits with larger numbers of π-type junctions. Well operating π-SQUIDs have now also been fabricated by using the ramp-type Josephson junction technology [11].

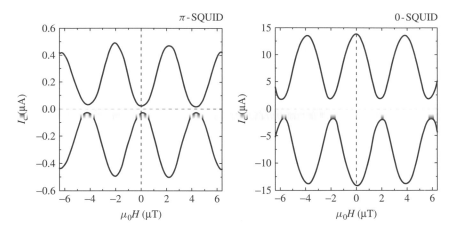

Figure 13.5 Dependencies of the critical current on applied magnetic field of a π-SQUID and of a standard SQUID at 77 K (from [7]).

This work was performed together with H. Bielefeldt, B. Chesca, B. Goetz, H. Hilgenkamp, A. Schmehl, C.W. Schneider, and R.R. Schulz. Support by the BMBF (project EKM 13N6918) and by the ESF (Pishift-Project) is gratefully acknowledged.

13.3 Doppler effect and the thermal Hall conductivity of quasiparticles in d-wave superconductors by N.P. Ong

At temperatures very close to absolute zero, all the electrons in a superconductor are paired. The collection of Cooper pairs constitutes the condensate or super-fluid. At finite temperatures T, however, a few of the Cooper pairs break up to form a gas of 'singles' called Bogolyubov quasiparticles (the entropy of the quasiparticle gas lowers the sample's free energy). With increasing T, the quasi-particle population $n_{qp}(T)$ increases rapidly until all the pairs become singles at the critical transition temperature T_c.

The thermal conductivity κ has proved to be a fruitful way to study the transport properties of quasiparticles. In a temperature gradient $-\nabla T$, the flow of quasiparticles towards the cooler end generates a thermal current $\mathcal{J}_e$ (as the condensate itself has zero entropy, it does not contribute to the thermal conductivity). However, the gradient also produces a parallel phonon current $\mathcal{J}_{ph}$ which greatly complicates the interpretation of measurements of κ.

First, let us consider the normal state of a conventional metal such as Pb. κ is the sum of the electronic term κ_e and the phonon term κ_{ph}, viz. $\kappa = \kappa_e + \kappa_{ph}$ (κ_e is about 100 times larger than κ_{ph} in Pb, but κ_e is 5-8 times smaller than κ_{ph} in the cuprates). As T falls below T_c in Pb, the rapid (exponential) decrease of

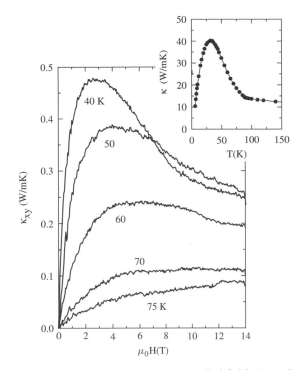

Figure 13.6 (Main Panel) Curves of κ_{xy} versus the applied field H at selected temperatures in high-purity $YBa_2Cu_3O_7$. At high T, κ_{xy} is linear in H, but as T decreases, its field profile displays a prominent peak. The initial slope $\lim_{H \to 0} \kappa_{xy}/H$ increases by a factor of 1000 between 85 and 40 K. The inset shows the zero-field thermal conductivity κ versus T. A prominent peak appears below $T_c = 93$ K [adapted from 13].

n_{qp} leads to a sharp decrease in κ_e. However, this is partially compensated by an increase in κ_{ph} because the decreasing n_{qp} results in a marked decrease in the scattering of phonons. Hence the total κ in a typical s-wave superconductor decreases rather gradually below T_c. If the sample is of exceptional purity, this gradual decrease is interrupted by a resurgent $\kappa \sim \kappa_{ph}$ which rises to a prominent peak below $\sim \frac{1}{2}T_c$. The phonons, now largely free of any scattering by quasiparticles, develop exceedingly long mean-free-paths (mfp) limited only by the size of the crystal. Eventually, at very low T, κ_{ph} vanishes as T^3, reflecting the specific heat of the phonon gas.

A major puzzle in the high-T_c cuprates became apparent shortly after the 1987 discovery of the '90-K' superconductor $YBa_2Cu_3O_7$. Instead of decreasing below T_c, κ is observed to rise sharply, reaching a peak value 2–3 times larger than the value just above $T_c = 93$ K (inset of Figure 13.6). The origin of the giant anomaly in κ has been a source of debate. Recalling κ in Pb, many investigators initially identified the giant anomaly with a strong enhancement of the mfp of phonons. However, this identification was unconvincing because the quasiparticle population in the cuprates falls quite slowly below T_c (in contrast with Pb). How can we distinguish the phonon and quasiparticle currents in the cuprates?

Let us consider applying a magnetic field **B** normal to the plane of the crystal. The field pierces the condensate as an array of vortices. Quasiparticles do not 'see' the applied magnetic field, but they scatter from the vortices (which act like impurities in the otherwise uniform condensate). A quasiparticle incident on a vortex line strongly scatters from the steep decrease in the pair potential at the core. More germane to our discussion, it interacts with the intense azimuthal supercurrent surrounding the core. Generally, the energy of a quasiparticle is shifted in the presence of a superflow. If the quasiparticle moves in a direction parallel to that of the local superfluid velocity (co-moving), its velocity relative to the lattice is increased (this so-called Doppler effect is analogous to viewing a pedestrian walking on a moving ramp). Conversely, a quasiparticle moving against the superfluid has its velocity lowered. The difference in kinetic energies implies that quasiparticles prefer to go around a vortex in the direction *against* the azimuthal superfluid velocity. If the superflow is clockwise (viewed from above), the incident quasiparticle is preferentially scattered to the left, whereas it is scattered to the right when the flow is counterclockwise. The resulting 'skew' scattering produces a net 'Hall entropy current' (if $-\nabla T$ is applied along **x** with **H** $\parallel$ **z**, the Hall current is along **y** and given by $\mathcal{J}_y = \kappa_{yx}(-\nabla T)$, where κ_{xy} is called the thermal Hall conductivity). Phonons, which are charge-neutral, do not experience this asymmetric scattering. Hence the thermal Hall effect acts very much like a selective filter that ignores the phonon current.

The thermal Hall effect was detected in high-purity crystals of YBa$_2$Cu$_3$O$_7$ and investigated in detail from $\sim$12 K to temperatures slightly above T_c [12, 13]. In weak fields, κ_{xy} increases linearly with H (main panel of Figure 13.6). In high-purity crystals, this linear field dependence evolves into a profile with a prominent peak. The profile is consistent with a very long quasiparticle mean-free-path ℓ when the field is absent. If we plot the initial slope κ_{xy}/H ($H \rightarrow 0$) against T, we find that it displays a remarkable thousand-fold increase [13] between T_c and 30 K that dwarfs the corresponding increase in κ. This is consistent with an increase in ℓ of over 100 within this temperature interval. The steep increase is sufficient to produce a giant peak in κ_e that matches the observed giant anomaly. The thermal Hall effect presents the strongest evidence to date that the giant anomaly derives entirely from the quasiparticles.

13.4   Nanometer-sized defects responsible for strong flux pinning in NEG123 superconductor at 77 K by M. Muralidhar and M. Murakami

The bulk superconductor composite of $(Nd_{0.33}Eu_{0.38}Gd_{0.28})Ba_2Cu_3O_y$ with 5 mol% $(Nd_{0.33}Eu_{0.33}Gd_{0.33})_2BaCuO_5$, 0.5 mol% Pt, and 10 wt% Ag$_2$O exhibits

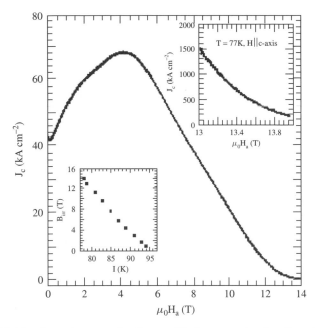

Figure 13.7 Field dependence of the critical current density (J_c) for the $(Nd_{0.33}Eu_{0.38}Gd_{0.28})Ba_2Cu_3O_y$ sample with 5 mol% NEG-211, 0.5 mol% Pt, and 10 wt% Ag_2O at 77 K and $H\|c$-axis. The bottom inset shows the extrapolation of B_{irr} (T).

the irreversibility field (B_{irr}) over 14 T for $B//c$ axis at 77 K [14]. The critical current density reached 70, 49, and 22 kA/cm^2 at 4.5, 7, and 10 T, respectively (see Figure 13.7). The extrapolation of temperature dependence of irreversibility field between 79 K and 94 K measured with a vibrating sample magnetometer indicated the actual B_{irr} value at 77 K about 15 T (bottom inset of Figure 13.7).

Structural analyses made by transmission electron microscopy (TEM) revealed a nanometer-scale modulation structure, which had a lamellar structure with spacing of a few nanometers. Similar nano-structures were observed in the NEG123/NEG211 composite samples with 3-7 mol% NEG-211. All the samples exhibited very high irreversibility fields over 12T at 77K [15]. Chemical analysis of the matrix showed that there is a chemical fluctuation in the (Nd+Eu+Gd)/Ba ratio on a nanometer scale. Scanning tunneling microscopy (STM) also confirmed the presence of nano-scale structure as shown in Figure 13.8a. Here its average spacing is around 3.5 nm, which is close to the coherence length of YBCO at 77K ($\xi_{ab}(77\,K) \approx 4.5$ nm) [16]. The observation with higher magnification showed that the structure consisted of rows of aligned clusters of slightly off stoichiometric composition of 3–4 nm in size (see Figure 13.8b). These clusters were identified by EDX analysis as $(NEG)_{1.015}Ba_{1.985}Cu_3O_y$.

Tunneling current spectra taken on the RE-rich clusters, and the regular matrix showed similar conductivity of both regions. This shows that the composition of the clusters is not much different from that of the 123 matrix.

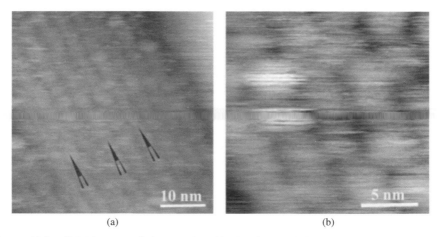

Figure 13.8   STM images of the sample with 5 mol% NEG-211 [(a)-(b)]. The image was recorded with bias voltage $V_s = 1$ V and tunneling current $I_t = 0.3$ nA. The black arrows in (a) mark some nanolamellas. Note the presence of arrays of clusters around 3 to 4 nm in diameter dispersed in an organized manner in the NEG-123 matrix.

Thus superior flux pinning of NEG123/NEG211 composites at higher fields and temperatures is ascribed to the formation of new type of nanometer-scale lamellar array of RE-rich clusters. These new pinning medium led to an increase of B_{irr} up to 15 T at 77 K and thus will make it possible to construct a superconducting magnet generating >10 T at liquid nitrogen temperature.

13.5   Hybrid Magnets
by H. Schneider-Muntau

A hybrid magnet is a combination of a powered resistive magnet (insert) with a surrounding superconducting magnet (outsert). Hybrid magnets are built with the goal to produce the highest continuous magnetic fields possible [17–21].

Design and construction of hybrid magnets are a considerable technological and logistical challenge. Besides the sheer size of the magnet, the interaction of the subsystems, the magnitude of the forces within and between the resistive and the superconducting magnet, the cryogenic requirements, and safe and reliable operation of such a complex system require detailed attention and very specific capabilities in several disciplines.

The essential components of a hybrid magnet are: superconducting magnet, the cryostat, the resistive insert, and the quench detection and protection system (see Figure 13.9).

The *superconducting magnet* (Figure 13.10, 13.11) surrounds the resistive magnet. Because of its bore size of 360 mm (Sendai, Nijmegen), 610 mm (Tallahassee) and 800 mm (Grenoble), the magnet becomes impressive in size,

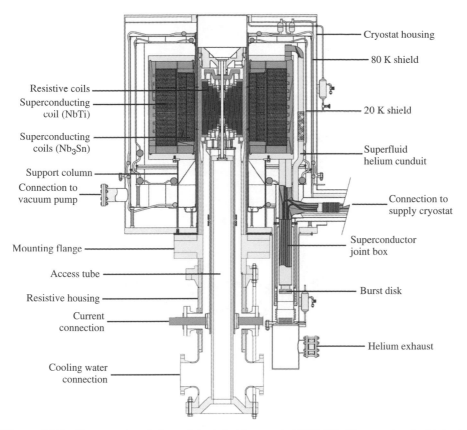

Figure 13.9 Cross-sectional view of the hybrid magnet at NHMFL. It generates the highest continuous field of 45 T in a bore of 32 mm since June 2000 and has since then provided more than 2000 h per year of measuring time to the international science community.

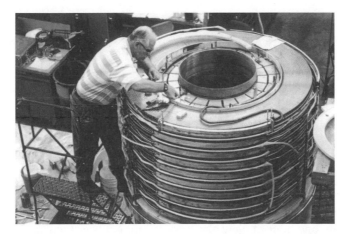

Figure 13.10 The outsert in construction. It generates 15 T in 615 mm.

Figure 13.11   The outsert being put on the central bore tube.

especially if the field contribution is above 8 T. Even more dramatic is the increasing stored energy (in round numbers): from 10 MJ (Nijmegen) to 60 MJ (Tsukuba and future Grenoble hybrid) to 110 MJ (Tallahassee). The energy content and, even more important, the coupling between the two magnet systems make a hybrid outsert much more challenging than a stand-alone, large-bore superconducting magnet. The most important design issues for the outsert are: (a) Changes of the field of the resistive insert result in induced currents. Field changes also induce eddy current heating and coupling losses between the filaments in the superconductor. They are proportional to $(dB/dt)^2$ and define specific requirements for the coupling time constant between filaments. The worst case is an insert trip from full field which can result into an energy deposition of over 20 kJ. (b) The ramp time of the outsert should not be more than approximately one hour in view of its operation in a user facility. The hysteresis losses are proportional to the maximum field and are independent of the ramp rate and define a requirement for the maximum admissible energy deposition as a function of its evacuation time. These losses can reach values of up to 100 kJ for a high field magnet. Wires with a small effective diameter have lower hysteresis losses. (c) To limit the magnet size, the conductor should operate as closely as possible to its critical current. A high index number will reduce the heating from current sharing. (d) The stability of a magnet against quench depends, among other things, on the amount of stabilizer and the available enthalpy. The presence of helium, especially superfluid helium, within or beside the conductor, increases the margin considerably. (e) In case of a quench, the energy must be extracted from the magnet and discharged into an outer resistor. An alternative solution is to dissipate the energy equally over the volume by firing heaters to quench the magnet uniformly.

It is obvious from above that the worst case scenario, for which the outsert magnet must be designed, is an insert trip after a ramp to full field, plus heating through current sharing. The heat must be absorbed and transported out of the conductor with only modest excursion in the local temperature. Therefore, important design criteria are, (a) the operating temperature, and (b) the temperature gradients between current carrying filaments, conductor matrix, surrounding helium, main helium buffer and refrigerator. Three different approaches have been developed for the design of the outsert in response to the requirements listed above. They can be distinguished by the superconductor they employ: the solid, cable-in-conduit and Rutherford cable conductor.

The *cryostat*, together with the cryogenic system, must provide the environment for safe and reliable operation of the superconducting magnet. The operating temperature of the magnet has an important impact on the design of the cryostat and magnet performance. A bath-cooled, ventilated magnet wound from solid conductor and operated at about 4.2 K represents the standard technique and has the great advantage of simplicity. However, operation of the magnet at superfluid helium with its extremely high thermal conductivity and low viscosity, increases magnet performance considerably. The heat transfer between conductor and liquid and, therefore, conductor stability and margin to external disturbances is essentially improved. Also, the critical current density in the superconductor is much higher at lower temperatures. A subcooled 1.8 K superfluid helium bath, including magnet, at atmospheric pressure can be achieved if a 1.8 K heat exchanger is introduced into the main bath. Such a cryostat system consists of a 1.8 K vessel housing the magnet, a 1.8 K heat exchanger, and two radiation shields which are at intermediate temperatures between 4.2 and 100 K, optimized for minimum losses. The magnet is kept at 1.8 K by a thermal exchange loop through expansion of 4.2 K helium in a Joule-Thomson valve using a vacuum pump. The forces between and within the two magnet systems, the superconducting magnet and the resistive insert, must be considered in the case of axial and radial misalignment of any of the magnetic coil centers. These forces are generated between the current carrying conductors of all coils, and are transmitted through the coil winding and support structure and the cryostat. In case of a major failure, they can reach alarming magnitudes of several MN, which is many times the weight of the superconducting magnet system.

The *resistive insert magnets* have lower efficiency than do stand-alone magnets without background fields (Figure 13.12). The major difficulties are: (a) The reduced volume in which the electric power must be dissipated. For economic

Figure 13.12 The hybrid magnet with view on the support and cooling water connections of the resistive insert. It consumes 30 MW, 300l/s of cooling water are needed. The superconducting magnet is located in the top vessel.

reasons, the magnet engineer has to choose the bore of the superconducting magnet as small as possible and, therefore, the resistive magnet must be very compact and efficiently cooled. (b) The additional field from the outer magnet. The background (booster) field from the superconducting magnet increases the Lorentz forces as the insert experiences the sum of its own and the outer field. Therefore, special high-strength conductors, a multi-coil design and an adequate support structure are required. (c) The high alignment forces between the coils. Impressive forces are generated within the insert and, to a lesser extent, between the insert and the outsert. (d) The forces resulting from the termination of the winding. These endforces are extremely large, resulting from the high fields and the high operating currents. They impact strongly on the design because of their magnitude and the limited space available.

An early *detection of a quench* is important to avoid uncontrolled growth of the resistive zone and destruction through excessive heating. Detecting a quench is difficult because the small resistive voltages must be separated from the large inductive voltages induced when either magnet is swept. The inductive voltages are up to a factor of 1000 higher. Inductive voltages can be cancelled if the magnet is divided into zones of equal inductance and the voltages are compared in bridge fashion, for instance, between taps connected to the winding.

In the event of a quench, the *protection system* must be capable of extracting enough energy from the system that high stresses from differential thermal expansion or even overheating/melting are precluded. This is done by discharging the magnet with a defined time constant into an external resistor. Limiting the maximum or hot-spot temperature to about 150 K and the discharge voltage for fully impregnated coils to 5 kV and for ventilated coils to 3 kV results in time constants of between 3 and 20 seconds, depending on the coil energy. The quench detection and control system of a superconducting outsert is best done with several industrial quality computers. As part of the total control system and its reliability, the safety devices such as the power supply, discharge resistor, breakers, and computers, must be checked continuously for their operational conditions.

13.6   Magneto-Optical Imaging of Vortex Matter
by T.H. Johansen

Space-resolved magnetic measurements have during the last decades proved increasingly useful in revealing micro-structural features and physical phenomena that control the overall electromagnetic behavior of superconductors. Among the various methods, scanning probe magnetometry, Bitter decoration etc., the technique of magneto-optical (MO) imaging has two major advantages, namely it allows easy zoom from cm-scale field-of-view down to micron resolution, and secondly, the time response is very quick (<100 ps), so that even very fast processes can be followed in real-time.

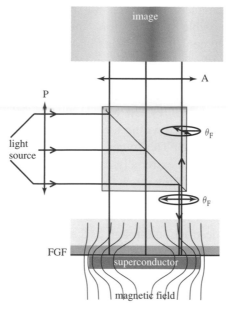

Figure 13.13   Schematic of the MO imaging principle where a Faraday-active ferrite garnet film (FGF) is the basic sensor. At the top is seen an image of flux penetration in $Ba_2Cu_3O_y$, as discussed in the text.

As illustrated in Figure 13.13, the imaging is based on having a Faraday active film placed in close contact with the sample. Polarized light shines down through the MO film, where the light undergoes a Faraday rotation that increases with the magnetic field. After reflection from a mirror the effect is doubled, and when the light hits the analyzer (A) it contains a distribution of rotation angles θ_F corresponding to the perpendicular field on the face of the superconductor. The analyzer is set at $90°$ relative to the polarizer (P) and filters the light producing an optical image where the brightness is a direct map of the surface field distribution.

A typical MO image of flux penetration into a superconducting film made of $YBa_2Cu_3O_y$ is seen in Figure 13.13. The image shows how the field distributes over a middle section of a long strip, where the two parallel edges are easily identified as the bright lines of field enhancement (demagnetization effect). The dark central region is the flux-free Meissner-state part of the strip. This image, revealing very smooth and regular flux patterns, proves immediately that the uniformity of the film is excellent. Moreover, an image like this can also be used to measure very reliably the critical current density, J_c. For a long thin strip the Bean model implies that the fraction of unpenetrated area at given applied field B_a equals $1/\cosh(\pi B_a/\mu_0 J_c t)$, where t is the strip thickness. Hence, J_c is obtained by simply taking a ruler and measure the ratio of two distinct areas in the MO image. Alternatively, the image can be transformed via Biot-Savart's law into a 2D current distribution. By this procedure, which requires careful calibration, we could on a *model-independent basis* determine the entire current flow over the superconductor area [22].

In the majority of cases MO imaging reveals that flux penetration in superconductors is not as regular as in the previous example. When built-in microstructure and extended defects perturb the flow of the shielding current, their effect shows up directly in the image. An example of such diagnostic use of the technique is given in Figure 13.14(a), where one sees how flux penetrates into a 3 mm wide Ag-sheathed monofilament Bi-2223 tape (the arrows point at the edges of the core). In addition to the gross behavior, which is thin-strip-like, the image contains considerable fine structure consisting in a nearly periodic set of bright lines with a characteristic curvature. These lines of enhanced flux penetration are due to connectivity damage originating from the rolling step in the production of such tapes.

In some cases, complex flux patterns can form independent of the defect structure in the material. Dramatic examples are the flux avalanches occurring in films of MgB_2, see Figure 13.14(b). We discovered that below 10 K the flux penetration develops in the form of dendritic structures, which one at a time burst into the Meissner state region as the applied field is slowly increased. When such an experiment is repeated in all details we see that the exact dendrite pattern varies widely, implying that the behavior is not controlled by the quenched disorder. Instead, recent work suggests that the behavior is the result of a thermo-magnetic instability (flux jump) [23].

Very recently, the MO imaging technique made a leap forward when it proved possible to resolve individual vortices, and thereby directly visualize their motion [24]. Our efforts towards this goal consisted in optimizing both the garnet MO film, its mounting on the sample as well as improving the optical system with respect to polarization contrast. Shown in Figure 13.15(a) is an MO image of a vortex lattice in a 0.3 mm thick single crystal of $NbSe_2$. This material was selected for its excellent reflectivity (the light was reflected directly from the crystal surface) and also because the London penetration length is fairly

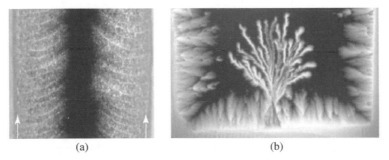

(a)   (b)

Figure 13.14 (a) MO image of flux penetration in a Bi-2223 monofilament tape revealing its granular magnetic microstructure. (b) Flux dendrites formed in a $4 \times 4 \, mm^2$ MgB_2 film during field increase near 10 K

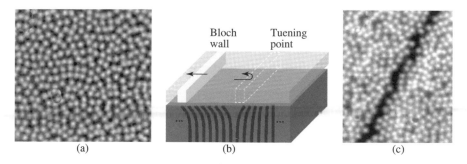

Figure 13.15 (a) MO image ($25 \times 25\,\mu m^2$) of vortices in NbSe$_2$ at 4 K and 0.7 mT frozen-in field. (b) Schematic of a "vortex brush" effect, and (c) MO image showing that a Bloch wall in the FGF allows such manipulation of the vortices.

small making the vortices well focussed. Since the flux density in this image is small the vortex-vortex interaction is too weak to let them form a regular lattice.

A most interesting extension of this work is to explore the possibility of doing vortex manipulation i.e., generate controlled vortex displacement while simultaneously monitoring their motion. Even this was successfully accomplished recently using the fact that our FGFs contain mobile Block walls separating areas of opposite in-plane magnetization. These walls act essentially like mesoscopic bar magnets with one pole adjacent to the superconductor surface, hence interacting very directly with the vortices. Figure 13.15(b) illustrates our idea to use such a wall, which we displace by applying a tiny in-plane field, to force the vortices to move (or bend) while the same FGF is used for imaging of their motion. The result of such a "vortex brush" experiment is seen in the MO image in (c), recorded after the wall first came to a turning point and then retreated, leaving a gap among the vortices along the retreat line. Future work will show if more sophisticated manipulation schemes can be found. VIDEO clips of real-time motion of vortices can be viewed at *http://www.fys.uio.no/super*

13.7 Vortices seen by scanning tunneling spectroscopy, by Oystein Fischer

The scanning tunnelling microscope is well known as a tool to observe the topography of a surface with atomic resolution. As illustrated in Figure 13.16, using a vertical piezoelectric drive, a tip is brought very close to the surface to be observed so that a tunnel current can circulate between the tip and the surface. When moving the tip over the surface with the help of two other, horizontal piezoelectric drives the tip height is constantly adjusted so that the tunnel current is kept constant. The tip height then follows the contours of the surface and the

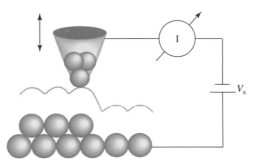

Figure 13.16 Schematics of the scanning tunnelling microscope used in constant current mode.

voltage on the vertical piezoelectric drive then can be used to make a map of the surface.

If instead of varying the height one keeps the height constant and one monitors the current, it can be shown that the derivative of the current is related to the local density of states as defined for a superconductor by

$$N(\mathbf{r}, E) = \Sigma(u_i^2(\mathbf{r})\delta(E - E_i) + v_i^2(\mathbf{r})\delta(E + E_i)) \tag{1}$$

Where u_i and v_i are the quasiparticle amplitudes and E_i is the quasiparticle energies. The STM measures to a good approximation

$$\mathrm{d}I/\mathrm{d}V(\mathbf{r}, V) \propto \Sigma(u_i^2(\mathbf{r})f'(E - E_i) + v_i^2(\mathbf{r})f'(E + E_i)) \tag{2}$$

where f' is the derivative of the fermi function. At $T = 0$, $\mathrm{d}I/\mathrm{d}V(\mathbf{r}, V) \propto N(\mathbf{r}, eV)$. In addition to the magnetic field attached to it, a vortex is also characterized by the fact that the order parameter goes to zero in its centre. As a result, the quasiparticles resulting from breaking the Cooper pairs can exist at an energy lower than the gap in the core of the vortex. These localized states can be directly seen by the STM as given by Equation 2. Using this difference between the vortex core and the surrounding superconductor it is possible to map the vortex lattice also by this method (Figure 13.17).

Compared to other techniques using the magnetic field to map the vortex lattice, this technique has the advantage that it can be used at high magnetic fields. Figure 13.18 show examples of: (a) the hexagonal vortex lattice in MgB_2 [25] and (b) the anisotropic vortex lattice in the basal plane of $YBa_2Cu_3O_7$ [26].

Another advantage of this technique is that it also allows to study the electronic states in the vortex core. These are characteristic of the specific superconducting state of the material and can be used as a signature in our search to understand different superconducting materials. In a BCS s-wave superconductor the reduction of the superconducting order parameter act as a potential well in which the quasiparticle states are quantized as follows (the energy is measured

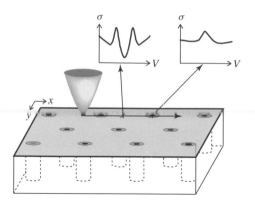

Figure 13.17   Spectroscopic imaging of the vortices. The imaging contrast relies on the fact that the tunnelling spectra are different outside and inside the cores.

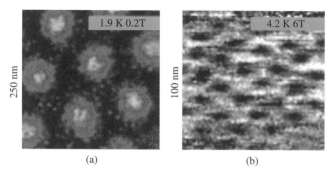

(a)   (b)

Figure 13.18   Spectroscopic images of vortices in a) MgB_2 and b) YBCO.

with respect to the Fermi level) : $E_n \propto (\Delta^2/E_F)(n + \frac{1}{2})$ where $n = \pm 0$, 1, 2, 3,...

In ordinary low temperature superconductors this spacing is only in the range of μeV and the individual states have so far not been seen individually. However, if the tip is situated at the centre of the vortex only states close to $E = 0$ will be observed thus resulting in a zero bias peak [27] as shown in Figure 13.19 for $NbSe_2$ [28].

The cuprate superconductors have a remarkably different behaviour. Since this material is found to be a d-wave superconductor, we would expect, based on the BCS theory, to find a behaviour similar to the one seen in $NbSe_2$. What is actually seen is very different. One finds two peaks, one below and one above the Fermi level, as if the vortex core would only contain two localized states (Figure 13.20) [26]. These states appear at an energy which is found to be proportional to the superconducting gap and this is one of the signatures of the cuprate superconductors which still needs to be explained by a future theory of high temperature superconductivity [29].

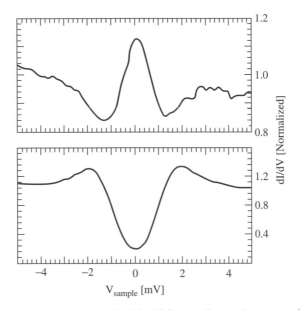

Figure 13.19 Tunneling spectra acquired in NbSe$_2$ at. Top:at the center of the vortex core, a zero bias peak is observed. Bottom: outside the vortex.

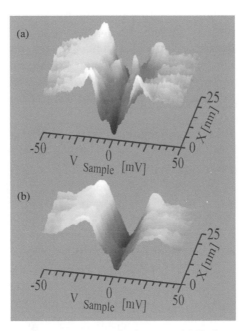

Figure 13.20 Set of tunnelling spectra acquired at $T = 4.2\,$K along a 25 nm path (a) across a vortex core, and (b) between the vortices.

13.8   Resistivity in Vortex State in High-T_c Superconductors by K. Kadowaki

One of the most significant and useful properties of superconductors is zero resistance. DC electric currents, therefore, flow persistently without energy dissipation. This well-known superior property, however, seems to be lost in high temperature superconductors, especially in the presence of magnetic fields at temperatures close to the *zero-field* critical temperature, hereafter denoted T_c. To study this unusual feature resistivity measurements [30] give important clue information, as described below.

In conventional type II superconductors in magnetic fields above the lower critical field B_{c1} magnetic flux can coexist with the superconducting state by allowing flux penetration into the superconductor as quantized vortices, each with $\phi_0 = 2.0678x10^{-15}$ Wb. However, motion of vortices causes resistivity and energy dissipation due to irreversible quasiparticle excitations. When currents flow in superconductors, the Lorentz force $F = J \times B$ acts on the vortices perpendicularly to them. They will start to move when the currents reach the critical current density J_c, because at this point the Lorenz force overcomes the pinning force. In the presence of strong pinning, the pinning force may be strong enough to carry current density higher than 10^6 A/cm^2. Pinning therefore plays a key role in maintaining zero DC resistance in magnetic fields.

In ideal type II superconductors, i.e. without pinning, on the other hand, the critical current density is zero. In such cases the material is resistive at any finite current. Thus, the resistive behavior in the vortex state of high-T_c superconductors has often been thought to arise from weak pinning. Although this seems logical, it is based to a large extent on an oversimplification of the physics of these superconductors.

The typical resistivity data for single crystal $Bi_2Sr_2CaCu_2O_{8+\delta}$ are shown in Figure 13.21 [30]. We notice first of all, that a finite resistivity is present in a wide temperature region below T_c even in a small field such as a few mT, in sharp contrast to conventional superconductors, where a parallel shift of the resistive transition to lower temperatures is commonly observed in a magnetic field. This broadening phenomenon is not due to motion of vortices caused by depinning by the Lorentz force, but is largely due to strong superconducting critical fluctuation effects. These fluctuations can be described as thermally generated vortex loops [31]. Flux motion is due to the dynamics of these loops, and the movements of the externally generated flux which occurs when superimposed by the thermally generated vortex loops.

We notice in Figure 13.21 the extraordinary behavior that resistivity in low magnetic fields shows a sharp drop at some temperature which is strongly dependent on the applied field. The higher the applied field, the farther away from T_c does the drop occur. The corresponding field where the sharp drop occurs is called B_m. In Figure 13.22 we show the locus of points in the (B, T)-plane

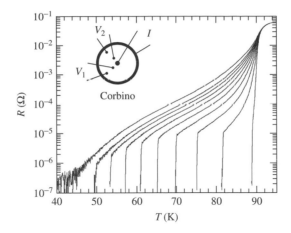

Figure 13.21   (a): Temperature dependence of the in-plane resistivity of single crystal $Bi_2Sr_2CaCu_2O_{8+\delta}$ with $T_c = 89.1$ K in various fixed magnetic fields $B_{ext} = 0$, 4, 8, 12, 16, 20, 24, 28, 32, and 50 mT (from right to left) applied to the c-axis. The inset schematically shows the electrode arrangement of the Corbino geometry with the dimensions of the diameter $d = 2.7$ mm and the thickness $t = 20\,\mu$m.

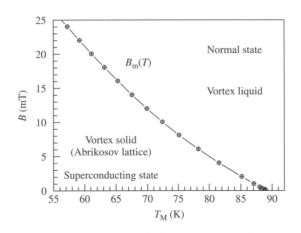

Figure 13.22   The vortex phase diagram in single crystalline $Bi_2Sr_2CaCu_2O_{8+\delta}$ in magnetic fields parallel to the crystallographic c-axis. The vortex line lattice melting transition (VLMT), $B_m(T)$, is determined by the resistivity measurement using Corbino geometry. Here, the lower critical field, B_{c1}, is not shown. Adapted from [32].

where the jump occurs. It describes a line, descending towards T_c. When interpreted as a possible thermodynamic line, the jump in resistivity gives the impression of a first order phase transition, in contrast to a continuous phase transition line. There is now broad consensus that the jumps seen in Figure 13.21 are due to freezing of the vortex system to a vortex lattice, or, on heating, melting of the vortex lattice, hence the name *vortex lattice melting transition*, VLMT given to this line. Since this represents a sudden change of the state of the vortex line aggregate, this is a true phase boundary between two different states of the vortex system, an Abrikosov type *vortex solid* below the melting line, and a *vortex liquid* above that line. It turns out that this feature in the case of $Bi_2Sr_2CaCu_2O_{8+\delta}$ can only be observed by the Corbino geometry, depicted schematically in the inset of Figure 13.21, and cannot clearly be obtained in the conventional strip geometry due to edge current effects. The observed jump is approximately 10^{-4} times smaller than the normal state resistivity, compared with more than 10% in the case of $YBa_2Cu_3O_{7-\delta}$.

When the VLMT temperature, T_m, shifts down below 70 K with increasing magnetic field, the resistivity jump at B_m becomes weaker and weaker, and then completely disappears below about 45 K. This means that the first order VLMT changes its character, possibly to the second order-like phase transition, perhaps due to the pinning effect, which resides even in the purest single crystal sample available at present. When the pinning is introduced in addition, the phase diagram becomes surprisingly rich and much more complicated, providing opportunities for further research, especially for practical applications.

From the results of resistivity measurements [30], the vortex phase diagram in $Bi_2Sr_2CaCu_2O_{8+\delta}$ for $H//c$ is constructed as shown in Figure 13.22. The first order VLMT-line, B_m, discriminates between the two aggregate states of vortex matter: Liquid above the melting line, characterized by a zero shearstiffness of the vortex system, and solid or frozen below, characterized by a nonzero shearstiffness. Stated in terms of phase coherence of the superconducting wavefunction, the vortex solid phase is characterized by long range phase coherence along the magnetic field but not transverse to it, while the liquid has lost long range phase coherence in all directions, also parallel to the magnetic field. Such anisotropy is quite rare, and it is as if vortices constitute a new kind of matter, and in fact we may say it does! With increasing magnetic field B_m shifts down as a function of temperature according to $B_m(T) \propto (1 - T/T_c)^\alpha$ with $\alpha = 1.55$. This temperature dependence can be obtained by a model which takes into account both the weak Josephson coupling and electromagnetic coupling between pancake vortices. In a region very close to T_c ($\Delta T/T_c < 0.04$), however, the temperature dependence obeys another relation, $B_m(T) \propto (T_c/T - 1)$, i.e. $\alpha = 1.0$. Since the Josephson coupling has much stronger temperature dependence and vanishes sharply at a region just below T_c, this temperature dependence is attributed to the dominant electromagnetic coupling effect. It is surprising that the jump (or drop) in the resistivity corresponding to the onset of B_m is sharp and can clearly be observed even in a small magnetic field of 0.1 mT as seen in Figure 13.21. This means that the first order VLMT occurs even in such a low density of pancakes with a vortex separation of $a_0 \sim 5\,\mu m$ in the Abrikosov vortex lattice, where the triangular order of the pancake vortices has been observed with strong thermal fluctuations.

13.9   Coated conductors: a developing application of high temperature superconductivity, by James R. Thompson, and David K. Christen

The discovery of high temperature superconducting materials (HTS) in 1985-86 brought great promise and optimism for their rapid usage in technological applications. One of the most obvious areas for use of these new superconductors was for conducting large, high density electric currents with zero-to-minimal dissipation in magnets, power cables and transmission lines, transformers, motors, and other power-handling equipment. Unfortunately, early investigations of HTS materials revealed a vexing feature: the *intra*-grain critical current density J_c within single crystals (especially those containing many defects that "pin" vortices strongly) can be very high, $\geq 10^6$-10^7 A/cm^2, while in contrast, the transport of current across grain boundaries (the *inter*-grain J_c) tends to be very much smaller, often orders of magnitude lower. This phenomenon of "weak-linkage" at grain boundaries is common in HTS materials and has been studied intensively in order to understand its origins. Details of the present understanding of grain boundaries in HTS, together with references to the original literature, are given in a review article by Hilgenkamp and Mannhart. [33] For technological applications, the central problem is that the inter-grain $J_c \sim \exp(-\theta/\theta_0)$ generally decreases exponentially with the angle θ between adjacent grain orientations, where the decay constant θ_0 is only a few degrees ($3° - 6°$). This means that *adjoining HTS grains in a practical wire or tape must be closely aligned crystallographically to achieve high levels of current conduction with low dissipation.* Furthermore, the inherent anisotropy of the superconductor (almost always YBCO) requires that supercurrents flow in the *a-b* planes containing the Cu-O sheets. Hence the physical properties of weak links and anisotropy mean that HTS grains in a tape must have their *c*-axes normal to the tape and adjoining grains should have only small angle grain boundaries: the HTS material must be *bi-axially textured.*

The developing technology of second generation, [34] coated conductors addresses these and other problems using multiple layers or coatings, each having some specific role(s). A superconducting HTS layer is deposited onto "buffer layers," which coat a base metallic substrate that provides strength and flexibility; Figure 13.23 shows a typical architecture used for a coated conductor. Grain alignment of the HTS is achieved by growing it *epitaxially* on a polycrystalline template of some suitable oxide, which itself must be bi-axially textured. For epitaxy, its structure and in-plane crystal lattice constants should closely match the nearly square *a-b* plane dimensions of the HTS, $\sim$0.39 nm. Ideally, the template of buffer layer(s) should be chemically stable and non-reactive with the HTS, nonporous, strongly bonding, and electrically conductive. A second essential

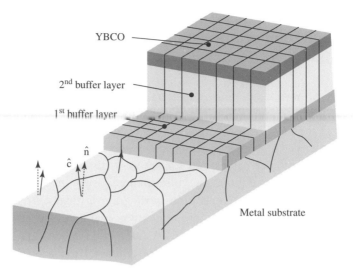

YBCO

2nd buffer layer

1st buffer layer

$\hat{n}$

$\hat{c}$

Metal substrate

Figure 13.23   A cross-sectional sketch of a coated conductor, based on the "RABiTS" technology. A biaxially textured metal substrate is coated with buffer layers of various oxides to serve as a diffusion barrier to substrate atoms and to replicate the template, on which YBCO is deposited with biaxial texture. Thus high density electric currents can flow in the Cu-O planes of the superconductor.

role, of course, is to "buffer" or separate the HTS from the metal substrate; in direct contact, metal atoms diffuse into the HTS, poisoning it and destroying the superconductive properties. To date, no single buffer layer material has been found that meets all of these needs, and great effort is being devoted to finding simple multi-layer architectures that function well.

Today, there are three principal approaches for achieving long lengths of biaxially-textured coated conductor tape. In the technique known as RABiTS (Rolling Assisted Biaxially Textured Substrates), [35] the substrate metal tape (alloys of nickel or copper) is textured by special thermo-mechanical processing procedures, providing an oriented template for the subsequent epitaxial deposition of buffer layer and YBCO coatings. Another approach uses Ion Beam Assisted Deposition (IBAD), [36] where an energetic ion beam irradiates the substrate at a particular angle during the deposition of an oxide buffer layer onto a polycrystalline metal tape (stainless steel or similar alloys). While the mechanism of the IBAD process is not completely understood, the buffer layer grows preferentially oriented with respect the ion beam and excellent biaxial texture can be achieved. The third technique is surprisingly simple. By appropriately inclining the substrate with respect to the incident plume of deposited buffer layer, biaxial texture is achieved by a preferential shadowing of slowly growing crystallites in favor of those that grow rapidly along a particular crystal axis. This Inclined Substrate Deposition (ISD) [37] approach for biaxial texture is still under development at the laboratory level, while both the RABiTS and IBAD approaches are currently being developed for commercialization by several industries worldwide.

Along with the great progress to date, a number of challenges remain. One is to increase the fractional cross-section of HTS in the tape, to increase its "engineering" current density. Increasing the HTS layer thickness is an "obvious" approach, but in practice the critical current often increases more slowly than the thickness, for reasons that are not well understood. A second challenge is to stabilize the superconductor against damage at weak "hot-spots" by providing parallel conduction paths of high conductivity normal metal. Third, time-varying currents or magnetic fields cause vortices to move in and out of the superconductor, leading to dissipation. Controlling ac losses in the superconductor and substrates is a significant concern. A major technological issue is the "scaleup" the existing and developing fabrication methods to produce long tapes with lengths of 100–1000 m. Accompanying this is the formidable economic challenge of cost, with a target price of 10–30 \$US per kiloampere-meter of coated conductor. In conclusion, the goal of high current, low loss superconducting wires and tapes has proven to be far more elusive than first imagined; nonetheless, the progress has been very considerable and it is reasonable to expect wider and wider application of HTS-based coated conductors in the coming years.

References, Chapter 13

[1] P.W. Anderson and P. Morrel, *Phys. Rev.* **123**, 1911 (1961).

[2] D. Vollhardt and P. Wölfle, *The Supefluid Phases of* ^3He (Taylor & Francis, London, 1990).

[3] A.P. Mackenzie and Y. Maeno, *Rev. Mod. Phys.* **75**, 657 (2003) and the references therein for spin-triplet superconductors.

[4] Y. Maeno, H. Hashimoto, K. Yoshida, S. Nishizaki, T. Fujita, J.G. Bednorz, and F. Lichtenberg, Nature **372**, 532 (1994).

[5] K. Ishida, H. Mukuda, Y. Kitaoka, K. Asayama, Z.Q. Mao, Y. Mori, and Y. Maeno, *Nature* **396**, 658 (1998).

[6] G. Luke *et al.*, *Nature* **396**, 242 (1998).

[7] R.R. Schulz, B. Chesca, B. Goetz, C.W. Schneider, A. Schmehl, H. Bielefeldt, H. Hilgenkamp, J. Mannhart, and C.C. Tsuei, *Appl. Phys. Lett.* **76**, 912 (2000).

[8] E. Terzioglu and M.R. Beasley, *IEEE Trans. Appl. Supercond.* **7**, 3670 (1998).

[9] J. Mannhart and P. Chaudhari, Physics Today, Nov. 2001, page 48; H. Hilgenkamp and J. Mannhart, *Rev. Mod. Phys.*, **74**, 485 (2002).

[10] B. Chesca, R.R. Schulz, B. Goetz, C.W. Schneider, H. Hilgenkamp, and J. Mannhart, *Phys. Rev. Lett.*, in press.

[11] Hilgenkamp H, Araindo, Smilde HJH, Blank DHA, Rijnders G, Rogalla H, Kirtley JR, Tsuei CC NATURE 422 (6927): 50–53 MAR 6 2003.

[12] Krishana K. Krishana, J. M. Harris, and N. P. Ong, *Phys. Rev. Lett.* **75**, 3529

[13] Zhang Y. Zhang *et al.*, *Phys. Rev. Lett.*, **86**, 890 (2001).

[14] M. Muralidhar, N. Sakai, N. Chikumoto, M. Jirsa, T. Machi, M. Nishiyama, Y. Wu and M. Murakami: *Phys. Rev. Lett.*, **89** (2002) 237001-1-4.

[15] M. Muralidhar, N. Sakai, M. Nishiyama, M. Jirsa, T. Machi and M. Murakami: *Appl. Phys. Lett.*, **82** (2003) 943–945.

[16] Handbook of Superconducting Materials (Institute of Physics, London, 2002).

[17] M. Morita *et al.* in H.J. Schneider-Muntau, ed. High Magnetic Fields, Applications, Generation, Materials, 333 (World Scientific, Singapore, 1997).

[18] K. van Hulst *et al.*, Physica B 164,13, 1990.

[19] Y. Nakagawa *et al.* in Proceedings 9th Int. on Conf. Magnet Technology, 424, (Swiss Institute for Nuclear Research, Villigen, 1985).

[20] J.R. Miller *et al.*, IEEE Transactions on Magnetics 30, 1563, 1994.

[21] A. Bonito Oliva *et al.*, IEEE Transactions on Applied Superconductivity 10, 432, 2000.

[22] T.H. Johansen, M. Baziljevich, H. Bratsberg, Y. Galperin, P.E. Lindelof, Y. Shen and P. Vase. Phys. Rev. B **54**, 16264 (1996).

[23] T.H. Johansen, M. Baziljevich, D.V. Shantsev, P.E. Goa, Y.M. Galperin, W.N. Kang, H.J. Kim, E.M. Choi, M.-S. Kim and S.-I. Lee. Europhys. Lett. **59**, 599 (2002).

[24] P.E. Goa, H. Hauglin, M. Baziljevich, E.I. Il'yashenko, P.L. Gammel and T.H. Johansen. Supercond. Sci. Technol. **14**, 729 (2001).

[25] M. Eskildsen *et al.* Phys. Rev. Lett. **89**, 187003 (2002)

[26] I. Maggio-Aprile *et al.*, Phys. Rev. Lett. **75**, 2754 (1995)

[27] H. Hess *et al.*, Phys. Rev. Lett. **62**, 214 (1989)

[28] Ch. Renner *et al.*, Phys. Rev. Lett. **67**, 1650 (1991)

[29] B.W. Hoogenboom *et al.*, Phys. Rev. Lett. **87**, 267001 (2001).

[30] J. Mirković, S.E. Savel'ev, E. Sugahara and K. Kadowaki, Phys. Rev. Lett. **86**, 886 (2001); J. Mirković, S.E. Savel'ev, E. Sugahara and K. Kadowaki, Phys. Rev. **B66** 132505 (2002).

[31] A.K. Nguyen and A. Sudho, Phys. Rev. B **60**: 15307, 1999.

[32] K. Kimura, S. Kamisawa and K. Kadowaki, Physica **C357-360** (2001) 442, K. Kadowaki, and K. Kimura, Phys. Rev. **B57** (1998) 11674.

[33] H. Hilgenkamp and J. Mannhart, Rev. Mod. Phys. **74**, 485–549 (2002).

[34] The "first generation" of HTS wires and tapes are based on the bismuth cuprates Bi-2212 and Bi-2223, where the mica-like structure and compatibility with metallic Ag makes it easier to obtain good intergrain transport of current. Unfortunately, the accompanying superconductive anisotropy is very high, which leads to weak vortex pinning and poor performance in high magnetic fields and temperatures.

[35] A. Goyal *et al.*, Journal of Metals **51**, 19 (1999).

[36] J.R. Groves *et al.*, Physica C, **382**, 43–47 (2002).

[37] K. Hasegawa *et al.*, Appl. Superconductivity **4**, 487 (1996).

PART V

Historical Notes

14

Historical Notes on Superconductivity: The Nobel Laureates

This chapter contains some brief historical and biographical facts about Nobel laureates in the field of superconductivity, including two who played an important role in the development of the field, but received the Nobel prize for other contributions to science. All portraits in this chapter were provided by the Nobel foundation (©).

Heike Kamerlingh Onnes

Heike Kamerlingh Onnes was born in Groningen, Holland in 1853, and received his bachelor's degree from University of Groningen in 1871. Already at the age of 18 he received a gold medal in a science competition sponsored by the University of Utrecht, in which he investigated '...methods for determining vapor density...'. The subject was to have importance in his later research. In 1908 he became the first to liquify the inert gas helium with a boiling point of 4.2 K, thereby opening the door to investigations of all kinds of matter at low temperatures, and in particular laying the foundation for the discovery of superconductivity. The interest in low-temperature physics was by no means a new. Michael Faraday had taken deep interest in condensation of gases. James Dewar in Edinburgh, and later at the Royal Institution, was a prominent expert and had, among other things, invented his famous thermos flask, commonly called a 'dewar' even today. Onnes was influenced early by the theoretical work of van der Waal, which pointed in the direction of low temperature physics in gas-liquid systems. He was appointed professor at Leiden in 1882,

Superconductivity: Physics and Applications   Kristian Fossheim and Asle Sudbø
© 2004 John Wiley & Sons, Ltd   ISBN 0-470-84452-3

a post he would hold for 42 years. His slogan *Door meten tot weten* (through measurement to knowledge) defined the style and spirit of his laboratory. Having gained access to temperatures down to just above 1 K by reducing the vapour pressure above the helium bath, a natural task was to continue the investigations of the low temperature electrical resistance in metals. A whole new domain of temperatures was at hand. A series of investigations by Dewar and coworkers had given resistance curves for a number of metals down to about −200 °C. Their properties in the new temperature domain needed to be measured. Mercury was chosen by Onnes for its high purity which could be obtained through evaporation. Contained in long capillary glass tubes it would freeze to a solid filament. Its electrical resistance could be measured with standard experimental methods. Superconductivity was discovered in 1911. Reports given at the prestigious Solvay conference the same year did not cause the stir we might have expected. The fact that one was dealing with a new state of metals was not quite appreciated. The available theoretical apparatus had not been forged. But this did not distract Onnes and coworkers from pursuing the subject which became known as superconductivity. Gradually it was realized that the transition to the superconducting state was a fairly normal occurrence among metals. Onnes had high hopes of using superconductors to build high-field magnets, up to 100 000 G. Unfortunately, since he worked with type I superconductors this was impossible due to limitations in critical current. Heike Kamerlingh Onnes received the Nobel prize in physics in 1913 for 'his investigations on the properties of matter at low temperatures which led, *inter alia*, to the production of liquid helium'. The whole history of the discovery of superconductivity by Kamerlingh Onnes and coworkers, as well as many other related historically interesting events, have been vividly described by Dahl [2].

John Bardeen

John Bardeen was born in Madison, Wisconsin in 1908. He studied electrical engineering at the University of Wisconsin, receiving his BS degree in 1928, and his MS degree in 1929. After a few years in geophysics he did his graduate studies in mathematical physics at Princeton University, and received his PhD in 1936. After research periods at Harvard University, at University of Minnesota, and at the Naval Ordonnance Lab in Washington DC, he came to Bell Labs in New Jersey in 1945. Here he joined the solid state physics group, and became interested in semiconductor research. In a collaboration with Brattain and Shockley he discovered the transistor effect in semiconductors in 1947, and together they laid the foundation for the modern age of electronics and computers. In 1951 he left Bell Labs to become Professor of electrical engineering and physics at University of Illinois, Urbana. Here he set up the team with Cooper and Schrieffer which was to develop the first successful microscopic theory of superconductivity, later referred to as the BCS theory. John Bardeen's influence on solid state physics, electrical engineering and technology was monumental. He received a number of prestigious awards and prizes. In 1956 he was awarded the Nobel prize in physics with Brattain and Shockley for research leading to the invention of the transistor, and in 1972 he shared the Nobel prize in physics with Cooper and Schrieffer for the theory of superconductivity. John Bardeen is the only person to have received the Nobel prize twice in the same prize domain. The transistor is often called the most important invention of the 20th century. John Bardeen was named by Life Magazine among the 100 most influential people of the 20th century. [Sources: The web pages of the Nobel e-museum is a rich source for further information about the scientific career of John Bardeen, and the impact of his work.]

Leon N. Cooper

Leon N. Cooper was born in New York in 1930. He attended Columbia University where he received his AB in 1951, AM in 1953, and PhD in 1954. During 1954–55 he was a member of the Institute of Advanced Study at Princeton. He held a post doctoral position as a Research Associate with John Bardeen at Urbana, University of Illinois during 1955–57, and served as an Assistant Professor at Ohio State University 1957–58. Since 1958 he has been Professor at Brown University. In his own account, his interest in superconductivity began with meeting John Bardeen at Princeton in 1955. Until then he had no previous knowledge of the field. His background was in field theory, exactly what Bardeen was looking for. His first task upon arriving at Urbana at the age of 26, was to learn the basics of superconductivity. He became convinced, as was Bardeen, that the essence of the problem was an energy gap in the single particle spectrum as evidenced by the exponentially decreasing heat capacity towards $T = 0\,\text{K}$. From a lecture by Pippard he learned that the facts of superconductivity appeared to be simple. In this respect it was an advantage that the isotope effect had been established, while all the exceptions found later were not yet known. Therefore, a phonon mechanism, as had been discussed by Bardeen and Fröhlich, would seem like the right idea. But first he made the important proof, later referred to as the 'Cooper problem': The zero degree instability of the fermion system against formation of a bound electron pair in the presence of the slightest attractive interaction between two electrons placed outside an already full Fermi distribution. An intense collaboration with Bardeen and his young student Robert Schrieffer started, with the aim to develop a theory for the electron-phonon interaction, and for superconductivity. In 1957 their famous 'BCS'-paper was published. The pairing due to the previously envisaged hypothetical attraction between two electrons was identified as an electron–phonon scattering event by which electrons with opposite momenta and spin in a thin shell near the Fermi-surface formed a shortlived binding. The effect of this, happening all over the Fermi-surface, was to create a new ground state, the superconducting state. The new theory had all the right properties, the energy gap, the Meissner effect, the penetration depth, the coherence length, the isotope effect, the prediction for ultrasonic attenuation, the coherence factors in NMR, etc. Cooper was appointed Professor at Brown University in Providence, Rhode Island in 1958 and has remained there since. He has later changed field entirely, becoming the Director of Brown University's Center for Neural Science, founded in 1973 to study animal nervous systems. The center created an interdisciplinary environment with students and faculty interested in neural and cognitive sciences towards an understanding of memory and other brain functions. Professor Cooper holds a number of honorary doctorates. He shared the Nobel prize in physics with Bardeen and Schrieffer in 1972. [Sources: A personal interview for this book conducted by one of the authors (K.F.) in 2001. In addition: The Nobel e-museum, and the scientific literature]

J. Robert Schrieffer

Robert J. Schrieffer was born in Illinois in 1931. In 1940 the family moved to New York, and in 1947 to Eustis, Florida. Schrieffer's original plan was to make a career in electrical engineering, and started on an engineering education at MIT in 1949. His interest in this field came from personal experience as a radio amateur on a homemade 'ham' radio in his young teenage years. In those years he had also made somewhat daring experiments in rocket science. But at MIT he discovered, through his own reading, the challenges and fascinations of physics, and made the switch to physics after two years. Under John C. Slater he did his bachelor's thesis on the structure of heavy atoms. He became interested in solid state physics, and began graduate studies with John Bardeen at University of Illinois. He did research, both theoretical and experimental, on semiconductors the first two years. In his third year on the advice of Bardeen he started collaboration with Leon Cooper, and the three together were committed to solving the superconductivity problem. While the three men struggled with the superconductivity problem, the young PhD student Robert Schrieffer felt uneasy about progress, and without telling his adviser, Professor Bardeen, he conducted a separate research project in ferromagnetism as a safeguard against a possible total failure to solve the superconductivity problem. When Bardeen was about to go away for a meeting in December 1956 he suggested Schrieffer should go on working on the superconductivity problem for yet another month before changing subject, because he felt they might be able to solve the problem. While Bardeen was away, Schrieffer happened to be on a visit to New York. Sitting on the subway he realized that the Tomonaga approach for the integration between pions and nucleons might be the way to go in a consistent way. He wrote down the wavefunction, now known as the BCS wavefunction, and calculated the energy of the system. It had the same form as the Cooper solution, but was exponentially stronger. He felt this was an interesting development. On his return he told Cooper, and then Bardeen. Bardeen said: 'I think that's the answer. That solved it!' During the next 11 days they worked out the thermodynamics and other properties. First they calculated the condensation energy in terms of the gap. Using the results of Tinkham and Glover who had recently measured the energy gap by infrared absorption, numerical values could be determined. Their analysis agreed with experiment. The paper was published in *Physical Review* in 1957, the famous 'BCS' paper. Schrieffer emphasizes that the BCS theory has a much wider validity than just the phonon mechanism, referring to the applicability of the BCS theory in totally different systems like in nuclear matter and in neutron stars. Schrieffer has had a distinguished career. He spent the first couple of years after his thesis work on the BCS theory at the University of Birmingham and at the Niels Bohr Institute in Copenhagen, and then at the University of Chicago and the University of Illinois. In 1962 he joined the faculty of the University of Pennsylvania and became a professor there. In 1980 he was appointed Professor at University of California in Santa Barbara where he served as Director of the Institute of Theoretical Physics from 1984 to 1989. He was later called on to become University Professor at Florida State University in Tallahassee, Florida, and is Chief Scientist at the National High Magnetic Field Laboratory (NHMFL) since 1992. Professor Schrieffer holds several honorary doctorates, and a number of prestigious awards. He received the Nobel prize in physics for 1972, shared with Bardeen and Cooper for the theory of superconductivity. [Sources: A personal interview for this book conducted by one of the authors (K.F.) in his office at NHMFL in 2001. In addition: The Nobel e-museum, and the scientific literature.]

Ivar Giaever

Ivar Giaever's career in physics is a highly unusual one. He was born in Bergen, Norway in 1929, but the family moved to Toten north of Oslo within a year. After high school, his priority was to study electrical engineering at Norway's leading engineering school, the Norwegian Institute of Technology, in Trondheim (since 1995 incorporated into the Norwegian University of Science and Technology). However, the competition to get in was very strong, and due to the equivalent of a C in a Norwegian language course, he was not admitted to study in the Department of Electrical Engineering, but had to accept Mechanical Engineering, a subject which he simply was not interested in. Consequently, by his own account, he spent his student years in Trondheim from 1948 to 1952 mostly doing other things than study, like playing bridge, billiard, chess, and poker. He became a local champion in all but the last one. Still, when he showed up for exams, he passed, and made it to a degree in mechanical engineering, after which he married. Now, the postwar housing situation in the cities of Norway after the Nazi occupation was extremely difficult. Having found a job in Oslo turned out to be of no help. The young family decided to emigrate to Canada in 1954, where Giaever, after a brief period in an architect's office, joined the Candian General Electric's Advanced Engineering Program. Soon Giaever discovered that salaries were better south of the border, and moved to General Electric Company in Schenectady where he as an employee, and now also as a serious and hard working student, took the company's demanding engineering courses, level A, B and C. Next, he worked as an applied mathematician on various assignments. He was greatly attracted by the opportunity to do research within the company with its impressive staff of skilled scientists at the GE Research and Development Center. Having joined the center in 1958, and concurrently started to study physics at Rensselaer Polytechnical Institute, he earned his PhD degree in 1964. It was during class in a physics course several years earlier, where superconductivity was being taught, that the idea struck him how to measure the superconducting energy gap which the BCS theory had recently predicted. His mentor at the research centre had told him that electrons could tunnel through thin barriers between semiconductors, a notion he could hardly believe since he was not yet familiar with quantum mechanics. Giaever now saw the possibility to try it out in a superconductor–insulator–metal contact, and in the process measure this important quantity, the gap. To his relief he also calculated that the predicted gap size, being in the millivolt range, was perfectly suited for the experiment he planned. All experimental facilities he needed were around, along with the support of highly skilled scientists. Giaever made his thin film structure as planned, and could soon, in 1960, 'measure the energy gap in a superconductor with a voltmeter', as he put it. Against the background sketched above, this was quite an achievement, and became next to a scientific sensation in the physics community. The mechanical engineer, now physics student, had done an experiment the experts could only have wished to do, but did not conceive. Adding to this the great ability Giaever has to communicate his work orally, has made him an attraction at meetings, at universities and research institutes. No doubt, his story can be taken as yet another example of the 'American dream' come true. The fact that he is always open and candid about his unusual background as a physicist has added a special flavour to his story and his work. His follow-up research on the density of states near the superconducting gap demonstrated even further that his discovery was no accident. But he is the first to insist that some element of luck was involved, and comments that this is needed to succeed. One should not be tempted to think his success came easy, however. Many years of hard work at General Electric was behind it all. Ivar Giaever continued his tunneling experiments for several years and contributed immensely to the progress in superconductivity. Giaever also was the first to published measurement showing a finite current between superconductors in a zero voltage condition, what later became known as the DC Josephson effect. However, due to the fact that the whole idea of a zero voltage supercurrent across a barrier between two superconductors had not yet been formulated, Giaever never laid claims on having discovered the Josephson effect. Ivar Giaever shared the Nobel prize in physics in 1973 with Leo Esaki and with Brian D. Josephson. Giaever had by then already left superconductivity, and had started work in biophysics during a stay at Cambridge, UK in 1969. His special area has been the behavior of protein molecules at solid interfaces. He left General Electric in 1988 to become an Institute Professor at Rennselaer Polytechnique Institute in Troy, and has for a number of years, concurrently, been a Professor at the University of Oslo, Norway. He is the recipient of numerous honorary degrees and prestigious prizes. [Sources: A personal interview for this book conducted by one of the authors (KF) in Schenectady, 2001. In addition: The Nobel e-museum, and the scientific literature]

Brian D. Josephson

Brian D. Josephson was born in Cardiff, UK in 1940. He received his BA degree at the University of Cambridge in 1960, and his PhD, also in Cambridge, in 1964. Josephson had already shown exceptional talent as a teenage scientist, but in an area far removed from where he was to gain international fame. He had the good fortune to do his PhD research with Pippard at the Royal Society Mond Laboratory. Furthermore, already during his second year as a research student the laboratory had Professor Phil Anderson as an inspiring visitor, in 1961–62. Josephson gives much credit to Anderson for his own interest in superconductivity, in particular for introducing him to the concept of 'broken symmetry' in superconductors. This concept was already inherent in Anderson's pseudospin formulation of superconductivity theory from 1958. In particular, Josephson wondered if the broken symmetry could be observed experimentally. He concluded that while the absolute phase angle of pseudospins would be unobservable, the consequences of a phase difference might not. At this point he learned about Giaever's tunnelling experiment from 1960. However, Pippard had considered the tunnelling of Cooper pairs through a thin barrier and found the probability to be very small. When Anderson showed him a paper by Cohen, Falicov and Philips where they had confirmed Giaever's formula for the current in his superconductor–insulator–metal contact, Josephson understood how he could calculate the current through a barrier between two superconductors. The expression he arrived at contained three terms, two of which were already known from previous work, but a third one was new. This term was unexpected: a current which was proportional to the sine of the phase difference across the barrier. The coefficient of this term was an even function of the voltage, and could not be expected to vanish at zero voltage. The obvious interpretation was that this was a supercurrent, and it appeared with the same order of magnitude as the quasiparticle current seen by Giaever. This was surprising, considering earlier suggestions by Pippard. At the age of 22, Josephson made the famous prediction of the supercurrent through an insulation barrier between two superconductors, known today as the Josephson effect. He made the prediction of both a DC effect and an AC effect. While the former would appear at zero applied voltage the latter should be present under the application of a small DC voltage. His predictions were confirmed, and became the basis for whole new fields of superconductivity research and technology. In later years there have been discussions among scientist whether the Nobel prize to Josephson should rightly have been shared with Anderson. To this question Anderson answers a clear 'no'. He explained us that such opinions might stem from the fact that he had rederived some of the results that Josephson had already found. The reason for doing so, he explained us, was that Josephson had not published all his findings, some of which were only reported in his thesis. Brian Josephson has held academic positions at the University of Illinois, the University of Cambridge and various visiting professorships. He is Professor of Physics at the University of Cambridge since 1974. [Source: The Nobel e-museum, and interview by one of the authors (K.F.) with P.W. Anderson in 2001.]

J. George Bednorz

J. George Bednorz was born in Neuenkirchen, Germany in 1950 as the fourth child. His parents had involuntarily been separated during the turbulences after World War II, but were happily reunited in 1949. In his youth he was influenced by his mother's music interest, and came to play both piano, violin and trumpet. His fascination with science was awakened not by physics, but by chemistry. He felt that doing experiments in chemistry stimulated his practical interests, and could have unexpected results. He started to study chemistry at the University of Münster in 1968, but ended up with majoring in crystallography. During two periods as a summer student at the IBM Zurich Research Laboratory in Rüschlikon, and later as a diploma student in 1974, he worked under the guidance of Hans Jörg Sheel in the Physics Department headed by K. Alex Müller, a scientist he deeply respected. His diploma work was on $SrTiO_3$, a great specialty of Müller's who became so pleased with young Bednorz' work that he encouraged him to continue his research on perovskite materials towards a PhD, supported by IBM, at the Swiss Federal Institute of Technology (ETH) under the combined supervision of Professor Heini Gränicher and Müller. His thesis work was on the crystal growth and solid solutions of perovskite type compounds, investigating structural, dielectric and ferroelectric properties. Upon completion, he joined the IBM lab in Rüschlikon in 1982. This would not seem like a good background for superconductivity research. But already while Bednorz was a student at ETH in 1980, Heini Rohrer at the IBM laboratory had asked him if he could prepare crystals of $SrTiO_3$ doped with Nb for the purpose of studying the superconducting properties of this material under varying doping conditions, with Gerd Binnig. Bednorz responded that 'if Nature allows, you will get it'. After a couple of days the material was ready, and the superconducting transition temperature had increased by a factor 4! This also had the interesting implication that the gradient of T_c versus doping was very steep. But when Bednorz joined the IBM laboratory in 1982, this line of research had been stopped, since now Rohrer and Binnig were working on the scanning tunnelling microscope, also a work to be awarded the Nobel prize. However, in 1983 Alex Müller, having spent two sabbatical years at the IBM laboratory in Yorktown Heights, New York where he had done work on granular superconducting Al, approached Bednorz again, and asked if he would join him in an attempt to go new ways in superconductivity. The idea was to exploit a polaronic interaction using Jahn–Teller ions, a field championed by Harry Thomas. Müller thought the mechanism might work in perovskites. Bednorz immediately agreed to collaborate. From then on a systematic effort was being made. This was a low cost project carried out as a side effort along with other ongoing work by both. Naturally, the first attempt was to go for classical Jahn–Teller systems like the lanthanum nickelates. Here, La was replaced by Y. Later also the B-position was modified. The idea was to modify the bandwidth. After one year the project was in danger of being stopped since the results were discouraging: All compounds were insulators. Bednorz now suggested to use copper instead of nickel to achieve the Jahn–Teller effect. Electrical conduction was obtained, but no superconductivity. Bednorz needed a break and went to the library. Here, he discovered the work by a French group, Raveau and coworkers, on the Ba-La-Cu-O compounds, and realized they should modify the A-position of the ABX_3 instead of the B-position. Already in the first measurement, in January 1986, a dip in the resistivity was found at 11 K. Since they did not have a magnetometer at the time, the test for diamagnetism could not be performed until a SQUID magnetometer had been acquired in September. However, the results were stable and reproducible. They felt confident that superconductivity had been discovered. Still, when each of them gave talks in different places in Germany in the fall of 1986 there was almost no response. This changed dramatically after the Japanese group headed by Tanaka at University of Tokyo in the fall of 1986 announced results that confirmed superconductivity in lanthanate. Their own work also showed the Meissner effect. From now on superconductivity was a matter of public interest. A new era had started. George Bednorz has continued as a scientist at the IBM laboratory in Rüschlikon near Zurich. He is the recipient of numerous awards and prizes, and shared the Nobel prize in physics with K. Alex Müller in 1987. [Sources: Interview for this book by one of the authors (K.F.) in 2001, and scientific collaboration. In addition, the Nobel e-museum and the scientific literature.]

K. Alex Müller

K. Alex Müller was born in Basel, Switzerland in 1927, and lived first in Saltzburg where his father studied music, later in Lugano where he became fluent in the Italian language. His mother died when he was eleven, after which he attended Evangelical College in Schiers, in the Swiss mountains. He remained there until the end of the war. He was fascinated by the radio, and wanted to become an electrical engineer, but his chemistry tutor, Dr Saurer, convinced him to study physics. After military service he enrolled in the Physics and Mathematics Department of the Swiss Federal Institute of Technology (ETH). The freshman class was three times too big, and the process of elimination was correspondingly tough. They were called the 'atom bomb semester' for obvious reasons. Müller had excellent teachers, like Scherrer, Kanzig and Pauli, and did his diploma work with Professor G. Busch on the Hall effect in grey tin, followed later by PhD work on paramagnetic resonance (EPR) in Busch's group. Here he identified the impurity present in the perovskite $SrTiO_3$ a fact he took much advantage of later. Upon completion of his PhD and after graduation in 1958 he worked at Battelle Memorial Institute in Geneva, whereafter he came to the IBM laboratory in Rüschlikon in 1963. He remained there until his official retirement from IBM, after which he is a professor at University of Zurich. He was a key person in the research which took place in the late 1960s and in the 1970s and early 1980s on understanding the critical properties of phase transitions. Again $SrTiO_3$ was the vehicle, and it became the best studied of all, specially its properties related to the structural phase transition near 105 K. With Thomas he identified the order parameter and worked out the Landau theory for this system. He was and is a world leading scientist as far as structural transitions in perovskites is concerned. This competence was not wasted, as it turned out, when he undertook the challenge with Bednorz to discover high-T_c compounds. From the time superconductivity was discovered in oxygen deficient $SrTiO_3$ at Bell Labs in 1964, he had an eye on this subject, but did not get directly involved in superconductivity until his 2-year long sabbatical at the IBM lab in

Yorktown Heights at the end of the 1970s, at which time he studied Tinkham's textbook from A to Z, as he said to us: '. . . like a graduate student after the age of 50'. Now he started research on superconductivity for the first time, in granular Al. His interest in the subject did not diminish after this. Some time after his return to the IBM lab in Rüschlikon, having heard a talk by Harry Thomas at a meeting in Erice, he was inspired to invite George Bednorz to collaborate in a search for superconductors among Jahn-Teller perovskites. We refer to his own account in Chapter 2, and to the account given by Bednorz above about the ensuing progress. The work that Binnig and Bednorz had done on his 'old friend' among perovskites, $SrTiO_3$ – a work he had followed closely as a manager – was also on his mind when he suggested the collaboration which would turn out such spectacular results, ending with the sensational developments in late 1986 and in early 1987: the discovery of record breaking high-T_c perovskite superconductivity in $La_{2-x}Ba_xCuO_4$. Alex Müller has achieved the rare position to be a world-leading scientist in two totally different fields of physics. Those who have the privilege to know him, have experienced his profound ability to combine knowledge from different areas of physics into a penetrating understanding of complicated subjects. The award of the Nobel prize in physics to Müller and Bednorz in 1987, attests to the fact that the spectacularly important and unexpected is often to be found in such combination of knowledge. Alex Müller holds on to his original ideas about the (bi)polaronic mechanism for superconductivity in the cuprate superconductors, a view that undeniably led to their great success. In his view, the observed isotope effect as well as the so-called stripe domains attest to the correctness of this basis for superconductivity in cuprate superconductors. Müller is the recipient of numerous awards in addition to the Nobel prize. He holds an honorary doctor degree at 17 universities. [Sources: Interview for this book by one of the authors (K.F.) in 2001, scientific collaboration and personal correspondence; the Nobel e-museum, and the scientific literature.]

Alexei A. Abrikosov

Alexei A Abrikosov grew up in a well known family in Moscow, where he was born in 1928. He lived in Moscow all his life until he emigrated to the US in 1991. Both his parents were medical doctors. His father was quite famous, and received the Golden statue of socialist labor, and the Stalin prize, but was not politically active. Upon Lenin's death, he performed the autopsy. Young Alexei's mother told him that under no circumstance should he become a medical doctor, for reasons he still does not know. Consequently, he excluded a medical career from the start, but already at the age of ten he was convinced he would become a scientist. He was already very interested in the life of great scientists and inventors, like Faraday and Edison, and already at that age he dreamed of winning the Nobel prize, and becoming a member of The Royal Society, completely unrealistic, he thought. With time he was to achieve both. He graduated from high school at the age of 15. He had great talents in mathematics, but entered, at this young age, the Institute of Power Engineering, partly to avoid the looming danger of being drafted in the future when he would reach such age; and then transferred to Moscow University after the war ended, still only 17 years old. Already as a very young man, still looking like a small boy, according to his own description, he was accepted by the great Lev Landau who understood what talents were at hand. At an unusually early age he passed Landau's famous "theoretical minimum", and stayed close to him. However, the KGB did not allow him to work on the hydrogen bomb with the Landau group due to suspicions against an uncle of his, a diplomat, of whose existence young Alexei had himself, at that time, no idea! Eventually, he did his PhD with Landau after all, and later was a postdoc in his group. He received his first degree in 1951 from the Institute for Physical Problems (Moscow, Russia) for the theory of thermal diffusion in plasmas and then the next degree, Doctor of Physical and Mathematical Sciences, in 1955, from the same Institute for a thesis on quantum electrodynamics at high energies. In 1975 he was awarded the Honorable Doctorate from the University of Lausanne, Switzerland.

During his long scientific life he has explored successfully many fields but mainly the theory of solids: superconductors, metals, semimetals and semiconductors. He is very famous for the discovery of the theoretical foundation for Type II superconductors and their magnetic properties (the Abrikosov vortex lattice), for which he was awarded the Noble Prize in Physics 2003. This work was published in 1957, but the results were achieved already in 1953, without Abrikosov being allowed to publish them. His boss, Landau, did not initially believe the results, and was not convinced about them until he learned that Richard Feynman in the US had the idea that quantized vortices in superfluids could be responsible for driving the lambda transition from a superfluid phase in Helium II to a normal fluid. In this case, contrary to normal practice, Landau read the Feynman paper himself, and believed the results. Landau never apologized. In his view Abrikosov had not come up with the simple physical arguments which he required, and which would make it obvious why Abrikosov's solution was correct. Even after Landau's acceptance, there were many more obstacles ahead before recognition was fully achieved. In Abrikosov's own mind it was only reached upon the publication of decoration experiments by Essmann and Träuble in 1967, in which a regular lattice of vortices was clearly demonstrated.

In 1991 Abrikosov moved to the US and joined the Materials Science Division as an Argonne Distinguished Scientist at the Condensed Matter Theory Group of the Materials Science Division where he is still active. In Argonne he has worked on the theory of high-Tc superconductors, properties of colossal magnetoresistance in manganates and, together with experimentalists there, discovered the so called "quantum magnetoresistance" in silver chalcogenides. Abrikosov has been elected a Member of the National Academy of Sciences (USA) and the Russian Academy of Sciences, Foreign Member of the Royal Society of London and the American Academy of Arts and Sciences. He has been awarded numerous Russian and International Awards and the Honorable Citizenship of Saint Emilion (France). [Sources: A personal interview with Abrikosov by one of the authors of this book (KF) in 2003, the Nobel e-Museum, and the scientific literature].

Vitaly L. Ginzburg

Vitaly L. Ginzburg was born in 1916, and grew up in Moscow during revolutionary times, under the establishment of the Soviet Union. His father was an engineer, and his mother a medical doctor. Very unfortunately she died when he was still only four years old. Except for two years of evacuation during the war, he has lived all his life in Moscow. Times were difficult after the revolution. Before the revolution their family had a four room apartment, after it they had to share it with two more families. They did not starve, but the food they had to eat was far below traditional Russian standards. In 1931 the government decided that those who had finished seven years of elementary school should go to a special school to be trained to be workers, instead of receiving higher education. But Ginzburg went to work as a technician in a laboratory instead, and educated himself enough to enter Moscow University in 1933 at the age of 17. He finished there in 1938. He originally doubted his abilities to be a theoretical physicist, but after some encouraging work on quantum electrodynamics he was accepted by the famous physicist I.E. Tamm, head of the P. N. Lebedev Physical Institute, belonging to the Academy of Sciences. From 1938 Ginzburg studied to be a theorist, and defended his Candidate of Science thesis in 1940, and his Doctor of Science in 1942. He became a deputy under Tamm, and remained in the Lebedev institute for the rest of his career and life, still active there at the age of 87. After Tamm died in 1971, Ginzburg became the director of the institute until 1988, when he retired. Andrei Sakharov was at the same institute, but could not be the head since he was a dissident.

In 1943 Ginzburg started work in superconductivity, trying to follow up Landau's work in superfluids which in its turn had been inspired by Kapitza's discovery of superfluidity in helium. First he worked on the thermoelectric effect. Eventually his interest focused on the application of Landau's general theory of second order phase transitions. His first application of this theory was in ferroelectrics where he used polarization as the order parameter and established the famous Ginzburg criterion for the validity of the Landau expansion. Superconductivity was a far less obvious case. He wanted to expand the energy in the superfluid density. But in quantum mechanics the density is the square of the wavefunction. So he had to use the square of the still unknown ψ-function for the density. Hence the energy was expanded in a series in even powers of ψ. Landau agreed with this development. But according to his recollection they disagreed on the matter of the charge to put into the quantum mechanical momentum in the kinetic energy term. Ginzburg thought of the charge as an effective charge which could be different from unity. Landau insisted there was no reason why it would not be unity. Hence that is stated in the paper. Out of modesty Ginzburg prefers to call their theory "ψ-theory" instead of Ginzburg-Landau theory. This theory has become monumentally important in superconductivity. It is usually applied as a mean field theory, but computationally it can be generalized to include fluctuations, and to also treat dynamical problems in superconductivity. Its wide applicability in high-T_c superconductivity has come as both a surprise and a blessing to this field where the coherence length is so short that initially there were serious doubt as to its validity in such cases. Theoretical progress in the field of high-temperature superconductivity, particularly on the microscopic origins of the phenomenon, has been very slow indeed. It has been one of the major outstanding issues in physics for nearly two decades, since its discovery in 1986. However, the Ginzburg-Landau model has been enormously fruitful in uncovering and understandning the plethora of novel vortex phases that can appear in extreme type-II superconductors such as the high-T_c cuprates, where disorder and thermal fluctuation effects are pronounced. This is extremely important for intelligently engineering of superconductors for large-scale applications. Vitaly L. Ginzburg shared the Nobel prize in physics with Abrikosov, and with Anthony Leggett in 2003, for inventing the Ginzburg-Landau model. It is fair to say that the Nobel prize for this work was extraordinarily well deserved, and much overdue. Ginzburg has in additon, received a number of awards and honors. [Sources: Interview of Ginzburg by one of the authors of this book (KF) in 2003, the Nobel e-Museum, and the scientific literature].

Pierre-Gilles de Gennes

Pierre-Gilles de Gennes was born in Paris in 1932. In the 1960s he was one of the leading scientists in the field of superconductivity, culminating his research in that field by publishing his famous textbook, *Superconductivity of Metals and Alloys* in 1966, still a classic in the field. He did not receive the Nobel prize in superconductivity, but rather for his contributions to the understanding of ordering in soft matter, in 1991. However, his impact on the field of superconductivity could well be characterized as being at the Nobel prize level. de Gennes' background had some unusual elements: During the war his family moved from Paris to a small village, Barcelonette in the French mountains, partly because of the German occupation, but more importantly because of a health problem. This had the consequence that the young de Gennes did not go to school until the age of 11 to 12. Instead, his mother taught him literature and history which she was very interested in, but no science. He was admitted to high school at an unusually early age. He liked science, but felt no particular push. However, as he explained us: 'The attraction of science was perhaps that it allows a precision test. In our field, when you say something you may advance bold assumptions. Later you can check it out.' Before studying at the university he attended a school which gave untraditional science schooling with a direct observational approach to nature. He did his PhD in magnetism, and was influenced by several prominent scientists, among them Abragam and Friedel. He mentions also Edmund Bauer as a specially influential figure in his career. During his military service he studied the BCS theory, and was ready to enter the field upon completion of the service. He set up a very powerful group at Orsay where he created an unusually effective collaboration between experimentalists and theorists. Later, in 1968 he undertook research in liquid crystals, followed by studies of polymers. He moved on to fields like the dynamics of wetting, the physical chemistry of adhesion, and granular materials. Much of his research has been in what we now call complex systems. Pierre-Gilles de Gennes has written 10 textbooks on different subjects in physics. Few scientists have mastered such a broad palette. de Gennes is a towering figure in French and international science. He is a Professor at the Collège de France since 1971, and Director of Ecole Superieure et de Chimie Industrielle de la Ville de Paris. [Sources: Interview for this book by one of the authors (K.F.) in 2001. The Nobel e-museum, and the scientific literature.]

Philip W. Anderson

Philip W. Anderson, born in 1923, grew up in an intellectually stimulating and outdoors loving college environment, with college teachers in the near family on both sides. His father was a professor of plant pathology at University of Illinois in Urbana. His mother came from a similar background. Among the family friends were several physicists. After high school he had an intention of majoring in mathematics, but at Harvard it turned out differently. This was during the wartime, 1940-43, and electrical engineering and nuclear physics were important subjects. Anderson chose electronics and went to the Naval Research Laboratory in Washington DC to build antennas during 1940-43. Back at Harvard from 1945 to 1949 he enjoyed both the courses, and the friendship of people like Tom Lehrer, the mathematician turned popular singer with a knack for political humor. He chose van Vleck as his thesis adviser due to greater accessibility than Schwinger, got married and settled down to learn modern quantum field theory which turned out to be useful even in experimental problems. This was at the birth of many-body physics, an area where he was later to be a major participant and leading scientist. Having completed his thesis he went to Bell Labs to work with a number of outstanding scientists like William Shockley, John Bardeen, Charles Kittel, Conyers Herring, Bernd Matthias, and Gregory Wannier. Here he also became acquainted with the work of Neville Mott and Lev Landau. At about the same time both he and his wife became quite active politically in the Democratic party. They worked enthusiastically for the candidacy of Adlai Stephenson towards the presidential election in 1952, and were active in several other connections.

Anderson's initial interest in superconductivity came from association with the experimentalist Bernd Matthias at Bell Labs with whom he first worked on ferroelectricity. After the BCS-paper came out he made a study of gauge invariance which they had not considered, and which was a concern among theorists. Also, he was a key figure in the development of a pseudo-spin formalism for superconductivity towards the end of the 50's. This line of thinking has later been successful in completely different fields of physics. His paper on superexhange from 1959 is a landmark piece of work. He contributed to the development of a theory for d-wave and p-wave superfluid phases of helium-3. Anderson's name is also associated with the Higgs phenomenon. With Kim he did highly original studies of the dynamics of quantized magnetic flux in superconductors in the early 60's. He coined names like "dirty superconductor", "spin glass" and probably also the name "condensed matter", and of course was the inventor of the theory for "Anderson localisation", producing the famous paper on Scaling Theory of Localization together with the "Gang of Four": Abrahams, Anderson, Licciardello and Ramakrishnan. His stay in Cambridge around 1962 was instrumental in inspiring Brian D. Josephson to develop his theory for Cooper pair tunneling between superconductors, the DC and the AC Josephson effects. He has worked extensively on the Kondo problem, solving it by a "poor man's scaling" approach, as well as inventing the co-called Anderson impurity and Anderson lattice model for heavy fermions. From more recent years his efforts to create a theory for high-T_c cuprate superconductivity, the socalled RVB-theory, stands out as a major effort in his career. Anderson's influence on condensed matter physics has been of profound importance. He is often characterized as one of the most influential minds in all of theoretical physics in the second half of the 20th century. In short, there is hardly an area in condensed matter physics worth mentioning which this truly outstanding scientist has not contributed significantly to. Anderson shared the Nobel prize in physics with John van Vleck and Sir Neville Mott in 1977. [Sources: Interview with Anderson by one of the authors of this book (KF) in 2001, the Nobel e-Museum, and the scientific literature].

References

1. H. Kamerlingh Onnes. Leiden Comm. 120b, 122b, 124c (1911).

2. Per Fridtjof Dahl. *Superconductivity. Its Historical Roots and Development from Mercury to the Ceramic Oxides.* American Institute of Physics, New York, 1992.

3. W. Meissner and R. Ochsenfeld. *Naturwissenschaften*, 21:787, 1933.

4. A. A. Abrikosov. *Zh. Eksp. Teor. Fiz.*, 32:1442, 1957.

5. A. A. Abrikosov. *Sov. Phys. JETP*, 5:1174, 1957.

6. L. W. Schubnikow et al. *Physikalische Zeitschrift der Sowjetunion*, 10:165, 1936.

7. J. Bardccn, L. N. Cooper, and J. R. Schrieffer. *Physical Review*, 108:1175, 1957.

8. B. D. Josephson. *Physics Letters*, 1:251, 1962.

9. J. George Bednorz and K. Alex Müller. *Z. Phys. B*, 64:189, 1986.

10. V. L. Ginzburg and L. D. Landau. *Zh. Eksp. Teor. Fiz.*, 20:1064, 1950.

11. J. D. Jackson. *Classical electrodynamics*. John Wiley and Sons, Inc., New York, third edition, 1999.

12. C. P. Pool, Jr., H. A. Farach, and R. J. Creswick. *Superconductivity*. Academic Press Inc., London, 1995.

13. T. Ishiguro, K. Yamaji, and G. Saito. *Organic superconductors*. Springer Verlag, Berlin, Heidelberg, New York, second edition, 1998.

14. W. A. Little. *Phys. Rev.*, 134:A1416, 1964.

15. D. Jérome, A. Mazaud, M. Ribault, and K. Bechgaard. *J. Physique Lett.*, 41:L95–98, 1980.

16. A. F. Hebard, M. J. Rosseinsky, R. C. Haddon, D. W. Murphy, S. H. Glarum, T. T. M. Palstra, A. P. Ramirez, and A. R. Kortam. *Nature*, 350:600, 1991.

17. R. Chevrel, M. Sergent, and J. Prigent. *Solid state chem.*, 3:515, 1971.

18. J. F. Schooley, W. R. Hosler, and M. L. Cohen. *Phys. Rev. Lett.*, 12:474, 1964.

19. Ø. Fischer. In Krusius M. and Vuorio M., editors, *Proceedings of the LT14 Conference*, volume 5, page 172, Amsterdam, 1975. North Holland Publ. Comp.

20. C. W. Chu, P. H. Hor, R. L. Meng, L. Gao, Z. J. Huang, and Y. Q. Wang. *Physical Review Letters*, 58:405, 1987.

21. R. J. Cava, A. W. Hewat, E. A. Hewat, B. Batlogg, M. Marezio, K. M. Rabe, J. J. Krajewski, W. F. Peck, and L. W. Rupp. *Physica C*, 165:419, 1990.

22. J. D. Jorgensen, B. W. Veal, A. P. Paulikas, L. J. Nowicki, G. W. Crabtree, H. Claus, and W. K. Kwok. *Phys. Rev. B*, 41:1863, 1990.

23. T. T. M. Palstra, B. Battlogg, L. F. Schneemeyer, and J. V. Waszczak. *Physical Review Letters*, 61:1662, 1988.

24. G. Rian. *A thermodynamic study of the phase stability regions in the system* $Y_2O_3 - BaO - CuO_x$. PhD thesis, Norwegian Institute of Technology, 1992.

25. P. W. Anderson. *Science*, 235:1196, 1987.

26. P. G. Radelli, D. G. Hinks, A. W. Mitchell, B. A. Hunter, J. L. Wagner, B. Dabrowski, K. G. Vandervoort, H. K. Viswanathan, and J. D. Jorgensen. *Phys. Rev. B*, 49:4163, 1994.

27. C. E. Gough, M. S. Colclough, E. M. Forgan, R. G. Jordan, M. Keene, C. M. Muirhead, A. I. M. Rae, N. Thomas, J. S. Abell, and S. Sutton. *Nature*, 326:855, 1987.

28. K. Alex Müller. In B. Batlogg, C. W. Chu, W. K. Chu, D. Ü. Gubser, and K. A. Müller, editors, *Proc. 10th Anniv. HTS Workshop*, page 3, March 1996.

29. A. K. Nguyen and A. Sudbø. *Physical Review B*, 60:15307, 1999.

30. A. K. Nguyen and A. Sudbø. *Europhys. Lett.*, 46:780, 1999.

31. Z. Tesanovic. *Physical Review B*, 51:16204, 1995.

32. Z. Tesanovic. *Physical Review B*, 59:6449, 1999.

33. U. Thisted, J. Nyhus, T. Suzuki, J. Hori, and K Fossheim. *Phys. Rev. B*, 67:184510, 2003.

34. F. Steglich, J. Aarts, C. D. Bredl, W. Lieke, D. Meschede, W. Framz, and H. Schäfer. *Phys. Rev. Lett.*, 43:1892, 1979.

35. J. Akimitsu. *Symposium on Transition Metal Oxides*. Semdai, Japan, 2001.

36. J. Nagamatsu, N. Nakagawa, T. Muranaka, Y. Zenitani, and J. Akimitsu. *Nature*, 410:63, 2001.

37. C. M. Varma, Z. Nussinov, and W. van Saarloos. *Physics Reports*, 360:353, 2002.

38. I. Giaever. *Physical Review Letters*, 5:147, 1960; 5:164, 1960.

39. I. Giaever, H. R. Hart Jr., and K. Megerle. Tunneling into superconductors at temperatures below 1K. *Phys. Rev.*, 126:941, 1962.

40. R. W. Morse and H. V. Bohm. *Physical Review*, 108:1094, 1957.

41. K. Fossheim. Unpublished data.

42. K. Fossheim, N. T. Opheim, and H. Bratsberg. *Supercond. Sci. Technol.*, 15:1252, 2002.

43. B. I. Halperin, S.-K. Ma, and T. C. Lubensky. *Physical Review Letters*, 32:292, 1974.

44. J. Bartholomew. *Physical Review B*, 28:5378, 1983.

45. J. Hove, S. Mo and A. Sudbø. *Physical Review B*, 65:104501, 2002.

46. L. T. Claiborne and R. W. Morse. *Physical Review*, 136:A893, 1964.

47. K. Fossheim. *Physical Review Letters*, 19:81, 1967.

48. D. C. Mattis and J. Bardeen. *Physical Review*, 111:412, 1958.

49. J. R. Cullen and R. A. Ferrell. *Physical Review*, 146:282, 1966.

50. S. Mo, J. Hove, and A. Sudbø. *Physical Review B*, 65:104501, 2002.

51. K. Fossheim and B. Torvatn. *Journal of Low Temperature Physics*, 1:341, 1969.

52. L. C. Hebel and C. P. Slichter. *Physical Review*, 113:1504, 1959.

53. M. Tinkham. *Introduction to Superconductivity*. McGraw-Hill, New York, second edition, 1996.

54. L. D. Landau and E. M. Lifschitz. *Statistical Physics*, Pergamon, Oxford, third edition, 1980.

55. R. Feynman, R. B. Leighton, and M. Sands. *The Feynman Lectures on Physics*, volume III. Addison-Wesley, Reading, Massachusets, 1965.

56. S. Hasuo and T. Imamura. *Digital logic circuits*, volume 77. IEEE Proc., 1989. 1177-1193.

57. M. Kitamura, A. Irie, and G.-I. Oya. *Phys. Rev. B*, 66:054519, 2002.

58. R. Kleiner and P. Müller. *Phys. Rev. B*, 49:1327, 1994.

59. C. C. Grimes and S. Shapiro. *Phys. Rev.*, 169:397, 1968.

60. R. A. Ferrell and R. E. Prange. *Phys. Rev. Lett.*, 10:479, 1963.

61. N.F. Pedersen. *Solitons*. Elsevier, Amsterdam, 1986.

62. Masashi Tachiki. *Physica C*, 282:383, 1997.

63. Y. Matsuda, M. B. Gaifullin, K. Kumagai, K. Kadowaki, and T. Mochiku. *Phys. Rev. Lett.*, 75:4512, 1995.

64. G. L. deHaas Lorentz. *Physica*, 5:384, 1925.

65. F. London and H. London. *Royal Society of London*, Proceedings A149:71, 1935.

66. D. Shoenberg. *Royal Society of London*, Proceedings A175:66, 1940.

67. C. J. Gorter and H. B. G. Casimir. *Physicalische Zeitschrift*, 35:963, 1934.

68. A. L. Schawlow. *Phys. Rev.*, 109:1856, 1958.

69. A. B. Pippard. *Proc. Roy. Soc.*, A216:547, 1953.

70. J. Waldram. Master's thesis, Cambridge University, 1961.

71. E. A. Lynton. *Superconductivity*. Methuen's Physical Monographs, 1962.

72. I. S. Gradshteyn and I. M. Ryzhik. *Table of integrals, series and products. English translation.* Academic Press Inc., London and San Diego, 1980.

73. A. Sudbø and E. H. Brandt. *Physical Review B*, 43:10482, 1991.

74. A. Sudbø and E. H. Brandt. *Physical Review Letters*, 67:3176, 1991.

75. E. Sardella. *Physical Review B*, 53:14506, 1996.

76. A. K. Nguyen and A. Sudbø. *Physical Review B*, 53:843, 1996.

77. A. Sudbø and E. H. Brandt. *Physical Review Letters*, 67:3176, 1991.

78. E. Sardella. *Physical Review B*, 45:3141, 1992.

79. G. E. H. Reuter and E. H. Sondheimer. *Proc. Roy. Soc.*, A195:336, 1948.

80. J. R. Leibowitz and K. Fossheim. *Physical Review Letters*, 21:1246, 1968.

81. D. E. Farrell, R. P. Huebener, and R. T. Kampwirth. *Solid state commun.*, 11:1647, 1972.

82. T. E. Faber. *Proc. Roy. Soc.*, A248:460, 1958.

83. L. D. Landau. *Nature London*, 141:688, 1938.

84. L. D. Landau. *J. Phys USSR*, 7:99, 1943.

85. W. E. Lawrence and S. Doniach. *12th Int. Conf. Low Temp. Phys.* Kyoto, Japan, 1970.

86. T. Tsusuki. *J. Low Temp. Phys.*, 9:525, 1972.

87. C. P. Bean. *Phys. Rev. Lett.*, 8:250, 1962.

88. P. W. Anderson and Y. B. Kim. *Rev. Mod. Phys.*, 36:39, 1964.

89. J. Bardeen and M. J. Stephen. *Phys. Rev.*, 140:A1197, 1965.

90. P. H. Kes, J. Aarts, J. van der Berg, C. J. van der Beek, and J. A. Mydosh. *Supercond. Sci. Technol.*, 1:242, 1989.

91. A. I. Larkin and Yu. V. Ovchinnikov. *J. Low. Temp. Phys*, 34:409, 1979.

92. M. V. Feigel'man, V. B. Geshkenheim, A. I. Larkin, and V. M. Vinokur. *Phys. Rev. Lett.*, 63:2303, 1989.

93. Svein Gjølmesli. PhD thesis, Norwegian Institute of Technology, 1995.

94. S. Gjølmesli, K. Fossheim, Y. R. Sun, and J. Schwartz. *Phys. Rev. B*, 52:10447, 1995.

95. M. P. Maley, J. O. Willis, H. Lessure, and M. E. McHenry. *Phys. Rev. B*, 42:2639, 1990.

96. E. Zeldov, M. M. Amer, G. Koren, A. Gupta, M. W. McElfresh, and R. J. Gambino. *Appl. Phys. Lett.*, 56:680, 1990.

97. G. Blatter, M. V. Feigel'man, V. B. Geshkenbein, A. I. Larkin, and V. M. Vinokur. *Review of Modern Physics*, 66:1125, 1994.

98. P. L. Gammel, L. F. Schneemeyer, J. V. Wasczak, and D. J. Bishop. *Physical Review Letters*, 61:1666, 1988.

99. D. R. Nelson. *Physical Review Letters*, 60:1973, 1988.

100. D. S. Fisher, M. P. A. Fisher, and D. A. Huse. *Physical Review B*, 43:130, 1991.

101. H. Träuble and U. Essmann. *J. Appl. Phys.*, 39:4052, 1968.

102. R. H. Koch, V. Foglietti, W. J. Gallagher, G. Koren, A. G. G. Kozew, A. Gupta, and M. P. A. Fischer. *Phys. Rev. Lett.*, 63:1511, 1989.

103. M. G. Karkut, M. Slaski, L. K. Heill, L. T. Sagdahl, and K. Fossheim. *Physica C*, 215:19, 1993.

104. G. Eilenberger. *Physical Review*, 153:584, 1967.

105. A. Houghton, R. A. Pelcovits, and A. Sudbø. *Physical Review B*, 40:6763, 1989.

106. E. Sardella. *Physical Review B*, 44:5209, 1991.

107. R. E. Hetzel, A. Sudbø, and D. A. Huse. *Physical Review Letters*, 69:518, 1992.

108. H. Safar, P. L. Gammel, D. A. Huse, D. J. Bishop, J. P. Rice, and D. M. Ginsberg. *Physical Review Letters*, 69:824, 1992.

109. V. M. Vinokur, M. V. Feigel'man and V. B. Geshkenheim. *Phys. Rev. Lett.*, 67:915, 1991.

110. J. Gilchrist and C. van der Beek. *Physica C*, 231:147, 1994.

111. J. Rhyner. *Physica C*, 212:292, 1993.

112. L. Landau and E. Lifshitz. *Electrodynamics of continuous media.* Pergamon, New York, 1981.

113. J. Clem. *Magnetic Susceptibility of Superconductors and other spin systems*, page 177. Plenum Press, New York, 1991.

114. K. Fossheim et al. *Physica Scripta T42*, 20, 1992.

115. J. D. Livingston. *Phys. Rev.*, 129:1943, 1963.

116. E. Zeldov, A. I. Larkin, V. B. Geshkenbein, M. Konczykowski, D. Majer, B Khaykovich, V.M. Vinokur, and H. Shtrikman. *Phys. Rev. Lett.*, 73:1428, 1994.

117. E. H. Brandt, R. Kossowsky, S. Bose, V. Pan, and Z. Durosoy eds. *Physics and material science of vortex state, flux pinning and dynamics.*, volume 356 of *NATO ASI*. Kluwer Academic Publisher, the Netherlands, 1999.

118. H. Hilgenkamp and J. Mannhart. Grain boundaries in high-T_c superconductors. *Rev. Mod. Phys.*, 74:485, 2002.

119. K. A. Müller, M. Takashige, and J. G. Bednorz. *Physical Review Letters*, 58:1143, 1987.

120. Y. Yeshurun and A. P. Malozemoff. *Phys. Rev. Lett.*, 60:2202, 1988.

121. Ellen D. Tuset. PhD thesis, Norwegian University of Science and Technology, 1997.

122. E. D. Tuset, K. Endo, H. Yamasaki, S. Miasawa, S. Yoshida, K. Kajimura, and K. Fossheim. In Tuset E.D., Thesis, Norwegian Institute of Science and Technology, 1997.

123. L. T. Sagdahl. PhD thesis, Norwegian Institute of Technology, 1994.

124. L. Krusin-Elbaum, G. Blatter, J. R. Thompson, D. K. Petrov, R. Wheeler, J. Ullmann, and C. W. Chu. *Phys. Rev. Lett.*, 81:3948, 1998.

125. V. V. Moshchalkov, M. Baert, E. Rosseel, V. V. Metlushko, M. J. Van Bael, and Y. Bruynseraede. *Physica C*, 282:379, 1997.

126. A. M. Troyanovski, M. van Hecke, N. Saha, J. Aarts, and P. H. Kes. *Phys. Rev. Lett.*, 89:147006, 2002.

127. Kristian Fossheim, Ellen D. Tuset, Thomas W. Ebbesen, Michael M. J. Treacy, and Justin Schwartz. *Physica C*, 248:195–202, 1995.

128. P. C. Hohenberg. *Physical Review*, 158:383, 1967.

129. N. D. Mermin and H. Wagner. *Physical Review Letters*, 17:1307, 1966.

130. J. M. Kosterlitz and D. J. Thouless. *J. Phys. C*, 6:1181, 1973.

131. J. M. Kosterlitz. *J. Phys. C*, 7:1084, 1974.

132. A. P. Young. *Physical Review B*, 19:1855, 1979.

133. J. V. José, L. P. Kadanoff, S. Kirkpatrick, and D. R. Nelson. *Physical Review B*, 16:1217, 1977.

134. P. Minnhagen. *Reviews of Modern Physics*, 59:1001, 1987.

135. H. Kleinert, F. S. Nogueira, and A. Sudbø. *Nuclear Physics*, 666:361, 2003.

136. J. M. Kosterlitz. *J. Phys. C*, 10:3753, 1977.

137. L. Onsager. *Physical Review*, 65:117, 1944.

138. P. W. Anderson. *Basic Notions of Condensed Matter Physics*. Benjamin Cummings, London, 1984.

139. K. Huang. *Quantum Field Theory. From operators to Path Integrals*. Wiley, New York, 1998.

140. D. R. Nelson and J. M. Kosterlitz. *Physical Review Letters*, 39:1201, 1977.

141. D. J. Bishop and J. D. Reppy. *Physical Review Letters*, 40:1727, 1978.

142. M. Tinkham. *Introduction to Superconductivity*. Mc Graw-Hill, New York, second edition, 1996.

143. V. J. Emery and S. A. Kivelson. *Nature*, 374:434, 1995.

144. H. Kleinert. *Lett. Nuovo Cimento*, 35:409, 1982.

145. H. Kleinert. *Gauge fields in Condensed Matter, Vol. 1*. World Scientific, Singapore, 1989.

146. L. Onsager. *Nuovo Cimento Suppl.*, 6:249, 1949.

147. Z. Xu, N. P. Ong, Y. Wang, T. Kakeshita, and S. Uchida. *Nature*, 406:486, 2000.

148. J. Hove, S. Mo, and A. Sudbø. *Physical Review Letters*, 85:2368, 2000.

149. J. Hove, S. Mo, and A. Sudbø. *Physical Review B*, 65:104501, 2002.

150. M. A. Anisimov, P. E. Cladis, E. E. Gorodetskii, D. A. Huse, V. E. Podneks, V. G. Taratuta, W. van Saarloos, and V. P. Voronov. *Physical Review A*, 41:6749, 1990.

151. B. I. Halperin, T. C. Lubensky, and S. K. Ma. *Physical Review Letters*, 32:292, 1974.

152. T. C. Lubensky and J.-H. Chen. *Physical Review B*, 17:366, 1978.

153. C. W. Garland and G. Nouneis. *Physical Review E*, 49:2964, 1994.

154. H. Kleinert and V. Schulte-Frohlinde. *Critical properties of ϕ^4 theories*. World Scientific, Singapore, 2001.

155. C. Dasgupta and B. I. Halperin. *Physical Review Letters*, 47:1556, 1981.

156. J. Hove. PhD thesis, Norwegian University of Science and Technology, 2002.

157. W. C. Stewart. *Appl. Phys. Lett.*, 12:277, 1968.

158. D. E. McCumber. *J. Appl. Phys.*, 39:3113, 1968.

159. J. Clarke. in H. Weinstock and R. W. Ralston eds. *The new superconducting electronics*, pages 123–180. Kluwer publishers, The Netherlands, 1993.

160. Heikki Säppä, Tapani Ryhänen, Risto Ilmoniemi, and Jukka Knuutila. in K. Fossheim ed. *Superconducting Technology. 10 case studies*. World Scientific Publishing, Singapore, 1991.

161. A. I. Ahonen, M. S. Hämäläinen, M.J. Kajola, J. E. T. Knuutila, O.V. Lounasmaa, J. T. Simola, V. A. Vilkman, and C. D. Tesche. in K. Fossheim ed. *Superconducting Technology. 10 case studies*. World Scientific Publishing, Singapore, 1991.

162. W. G. Lyons and R. S. Withers. in K. Fossheim ed. *Superconducting Technology. 10 case studies*. World Scientific Publishing, Singapore, 1991.

163. Z.-Y. Shen. *High-temperature superconducting microwave circuits*. Artech House Microwave Library, Boston, London, 1994.

164. W. L. Holstein, L. A. Parisi, Z.-Y. Shen, C. Wilker, M. S. Brenner, and J. S. Martens. *J. Supercond.*, 6:191, 1993.

165. R. H. Dicke. *Rev. Sci. Instr.*, 17:268, 1946.

166. D. Winkler, Z. Ivanov, and T. Claeson. in K. Fossheim ed. *Superconducting Technology. 10 case studies.* World Scientific Publishing, Singapore, 1991.

167. K. Fossheim, editor. *Superconducting Technology. 10 case studies.* World Scientific Publishing, Singapore, 1991.

168. W. Buckel. *Supraleitung.* Physik Verlag, Weinheim, 1972.

169. M. N. Wilson. *Superconducting magnets.* Clarendon Press, Oxford, 1983.

170. R. P. Reed and A. F. Clark, editors. *Materials at Low Temperature.* American Society for Metals, Metals Park, Ohio, 1983.

171. E. W. Collings. *Applied Superconductivity, Metallurgy and Physics of Titanium Alloys*, volume 1 and 2. Plenum Press, New York and London, 1986.

172. S. Foner and B. Schwartz, editors. *Superconductor Material Science.* Plenum Press, New York and London, 1983.

173. J. Østergaard and O. Tønnesen. *Cryogenic engineering*, 36:N09, 2002.

174. M. Runde and N. Magnusson. *IEEE Trans. Appl. Superconductivity*, 13:1612, 2003.

Author Index

Superconductivity: Physics and Applications Kristian Fossheim and Asle Sudbø
© 2004 John Wiley & Sons, Ltd ISBN 0-470-84452-3

Subject Index

Superconductivity: Physics and Applications   Kristian Fossheim and Asle Sudbø
© 2004 John Wiley & Sons, Ltd   ISBN 0-470-84452-3